諸凡過往，皆為序曲
WHERE OF WHAT'S PAST IS PROLOGUE.

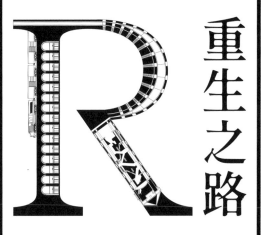

重生之路

The Road to Rebirth

基礎設施的死與生
全球經典案例圖解

賴伯威◎著
WillipodiA都市研究團隊

聯經

目錄 Contents

推薦序◎侯君昊·················4
Foreword

0 | 序·················6
Introduction

1 | 基礎設施史觀·················10
A History of Infrastructure

2 | 基礎設施屬性·················14
The Attributes of the
Infrastructure Projects

3 | 重生案例紀錄·················30
The Collection of
Reborn Infrastructure

4 | 重生演化·················154
Rebirth and Evolution

5 | 重生之路·················164
The Road to Rebirth

A
交通·················32
Transportation

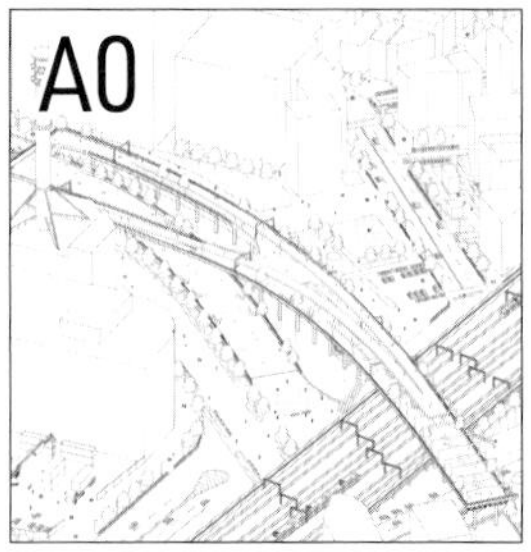

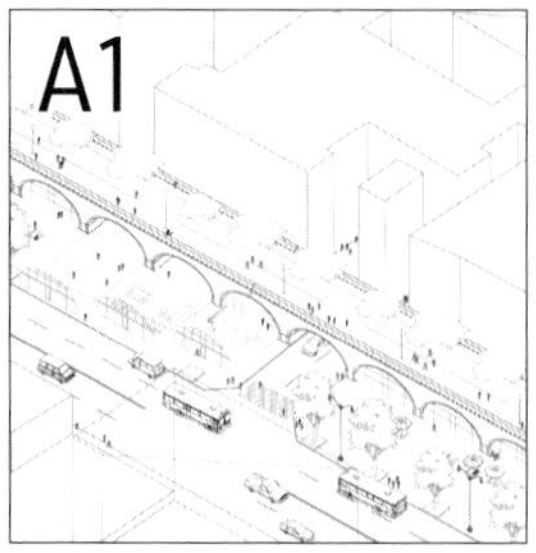

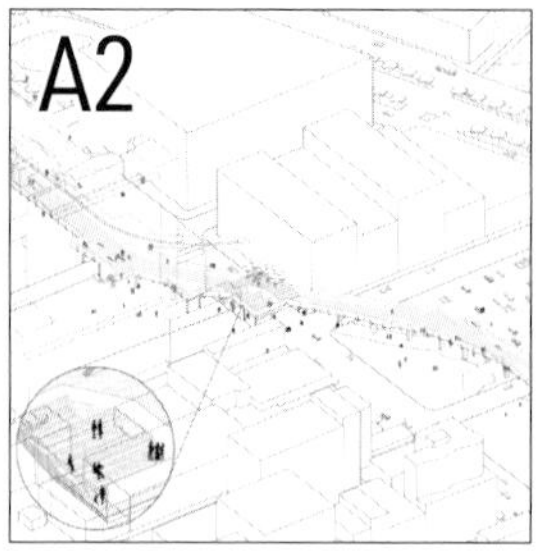

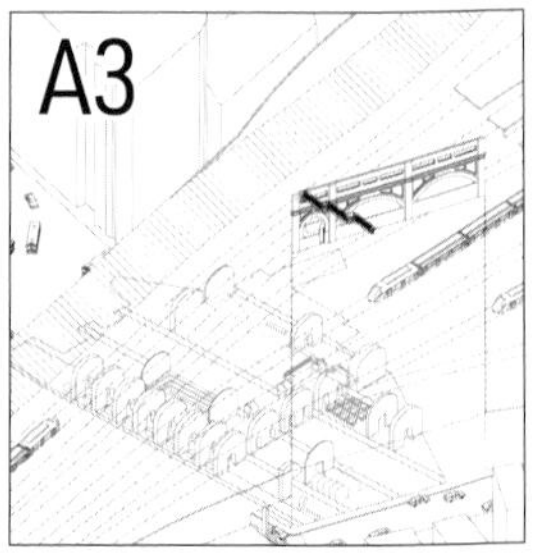

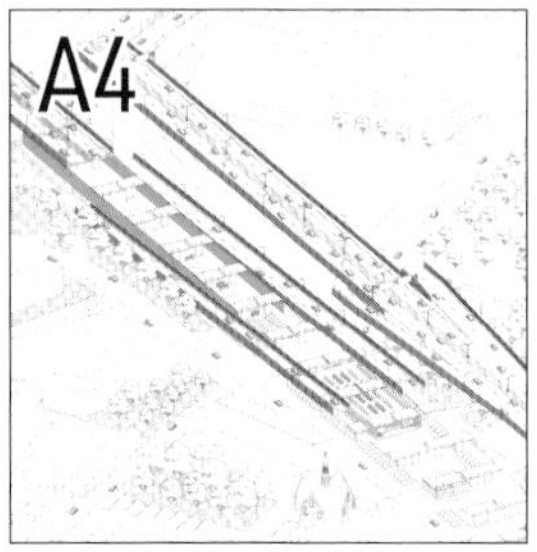

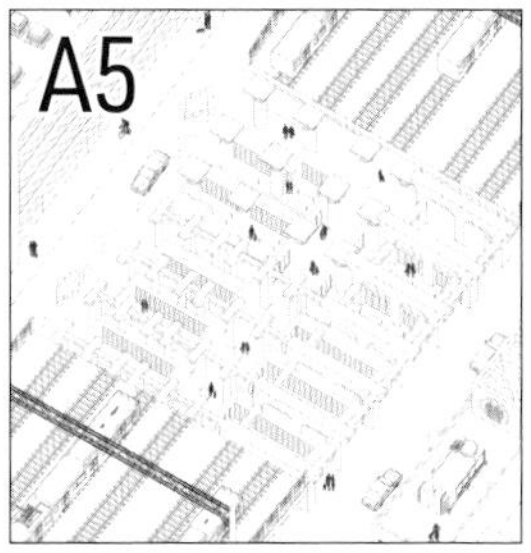

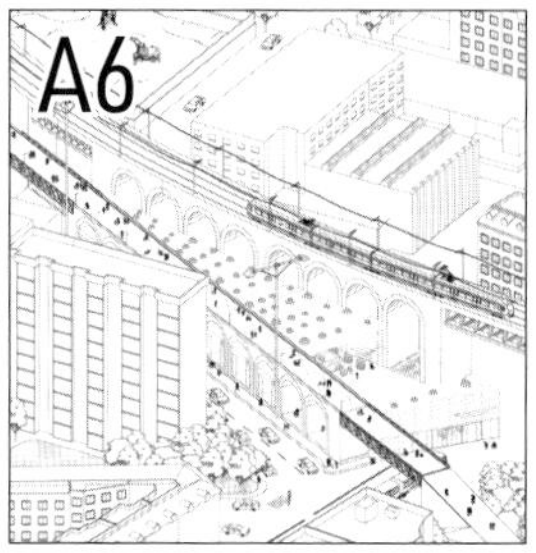

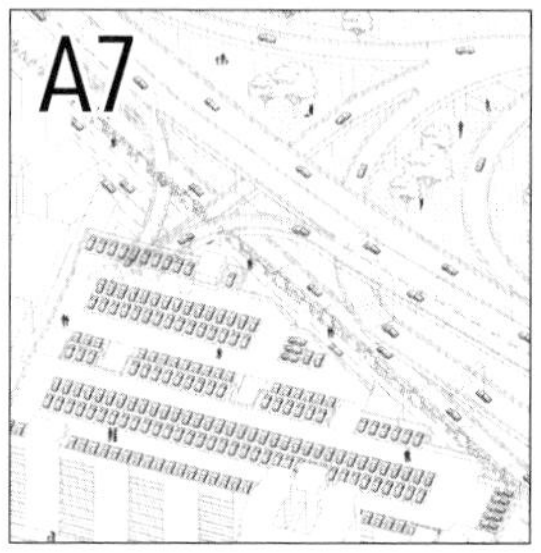

A0 A1 A2 A3
A4 A5 A6 A7 A8

B
港口·················52
Port

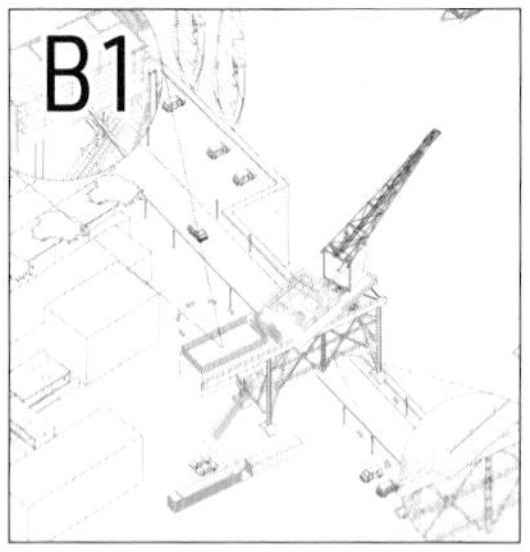

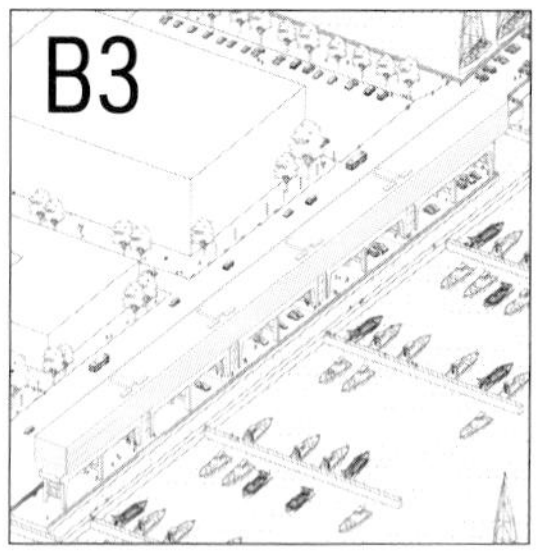

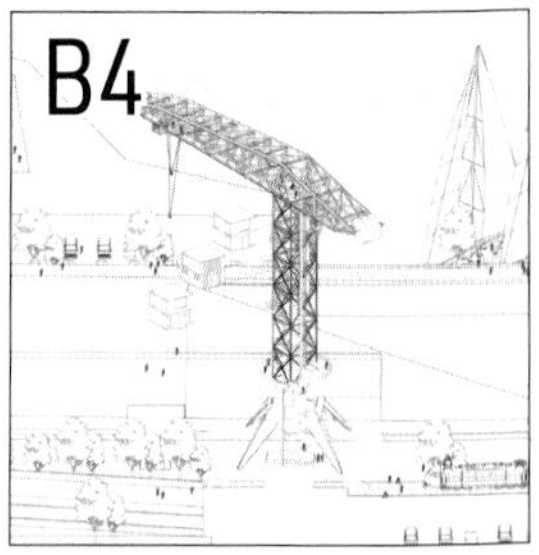

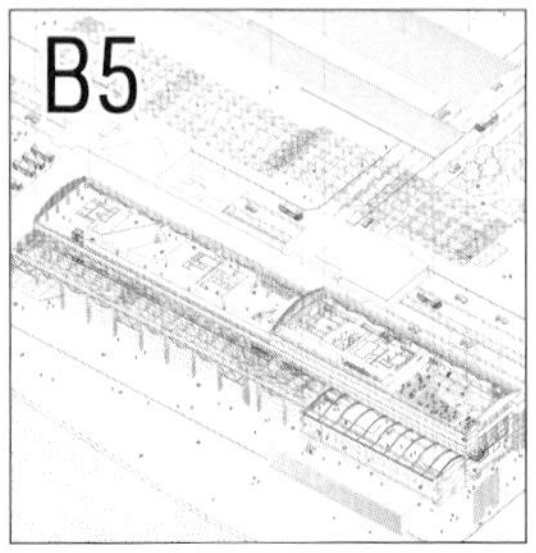

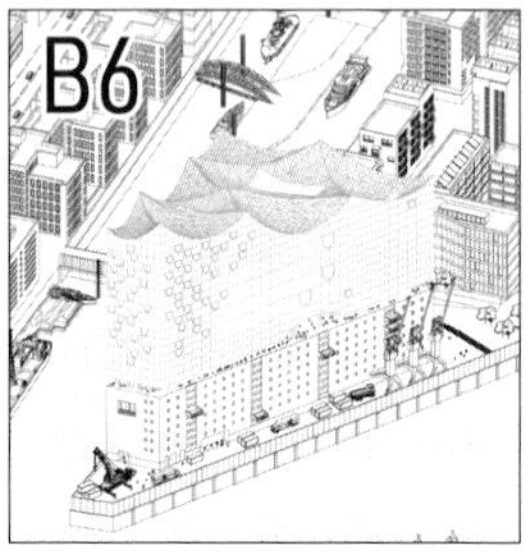

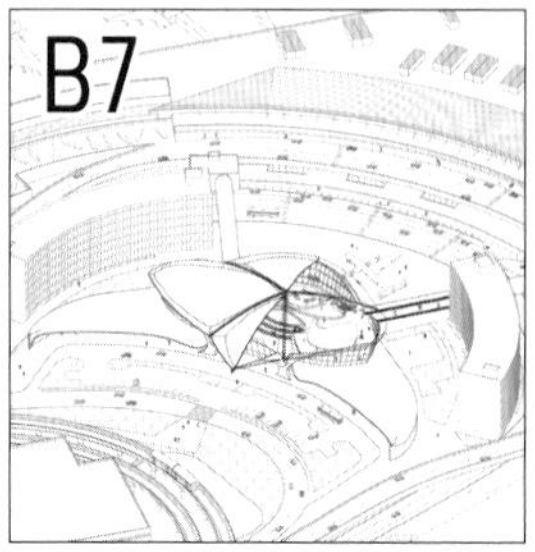

B1 B2 B3 B4
B5 B6 B7

C
資源開採·················68
Mines

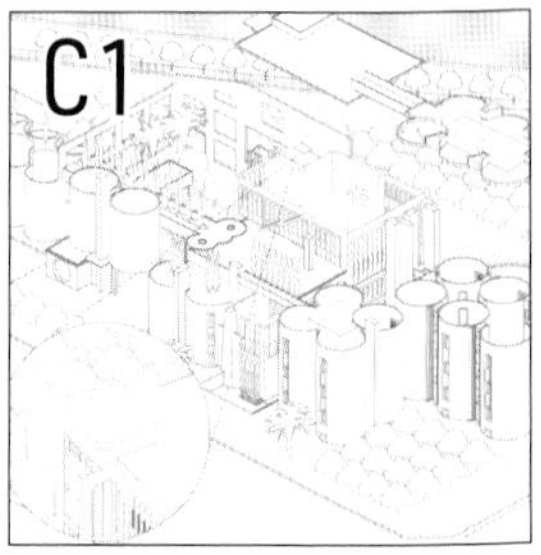

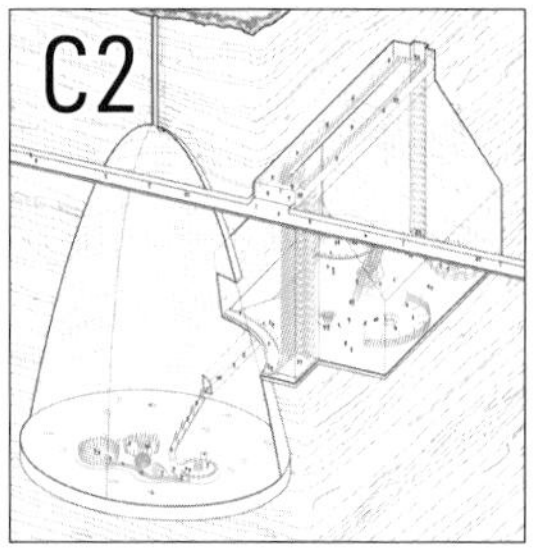

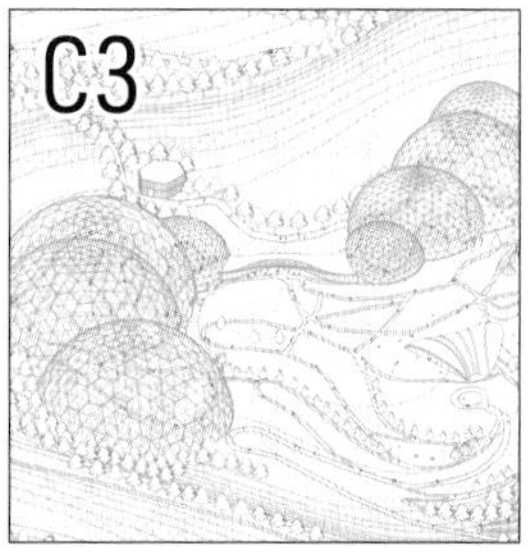

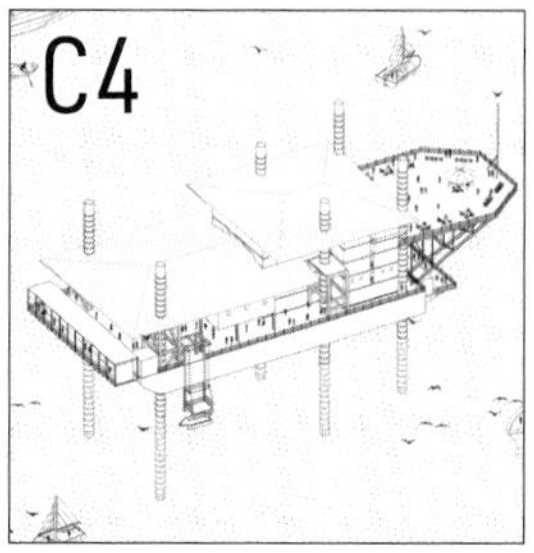

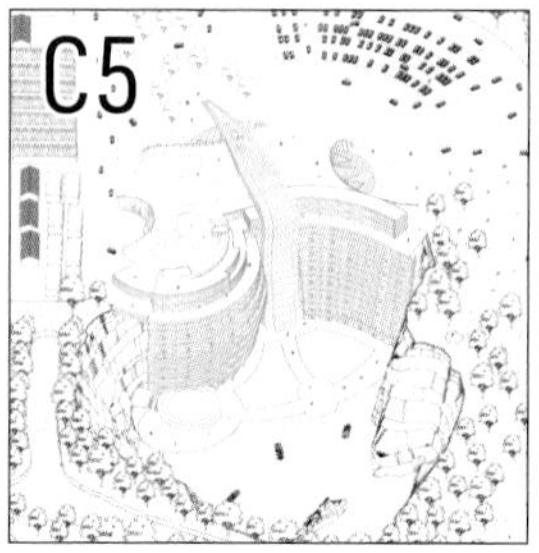

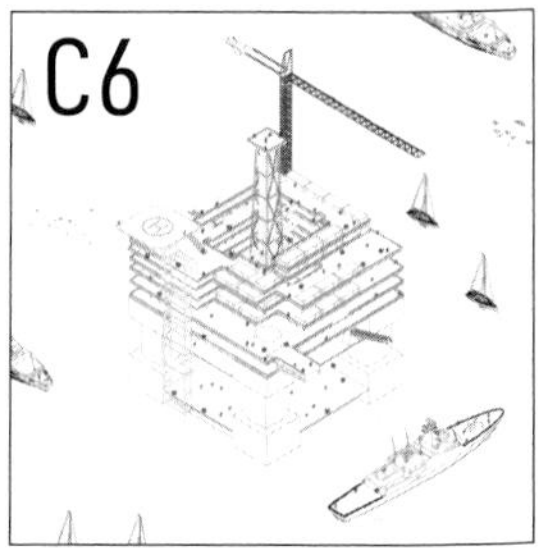

C1 C2 C3 C4 C5 C6

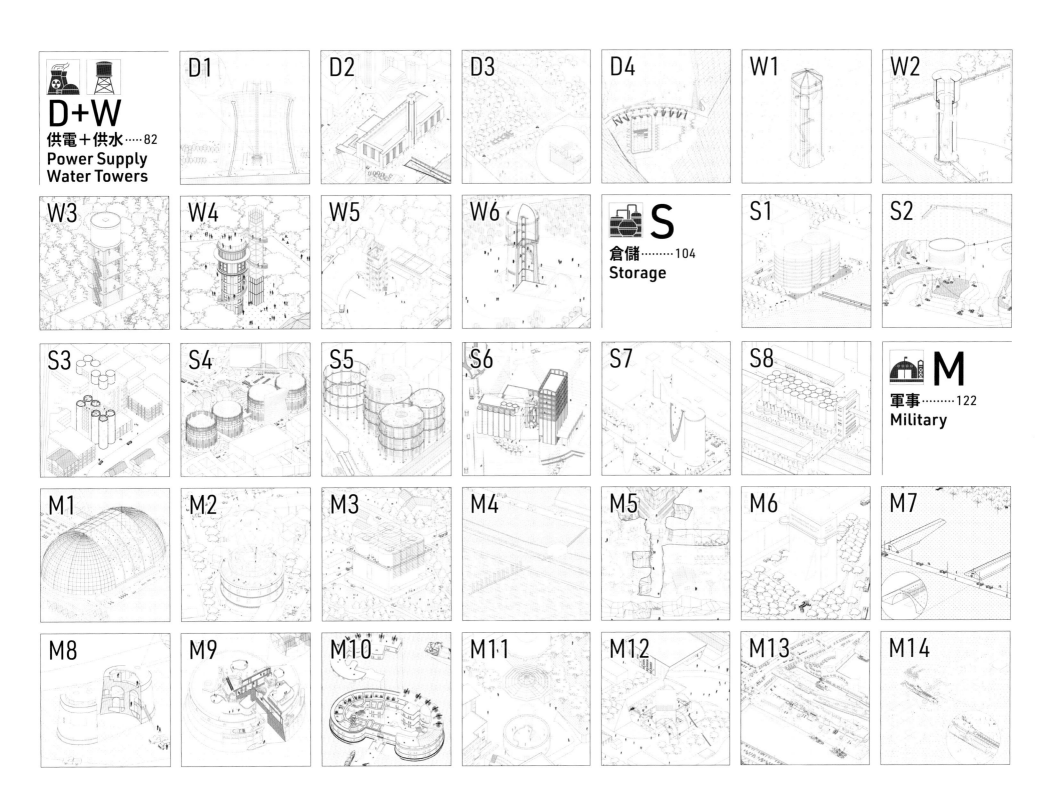

D+W
供電＋供水 ·····82
Power Supply
Water Towers

D1 D2 D3 D4 W1 W2

W3 W4 W5 W6

S
倉儲 ········104
Storage

S1 S2

S3 S4 S5 S6 S7 S8

M
軍事 ········122
Military

M1 M2 M3 M4 M5 M6 M7

M8 M9 M10 M11 M12 M13 M14

你是否察覺到世界的人造地景正在改變，重生成為常態？可能的原因包括：
- 責任開發（Responsible Development）觀念興起，全球一致檢討各種無節制的開發，間接減少、減緩非急迫性大型開發與建設；
- 環保意識興起，政府資源流向綠色產業與無形、微型建設，產業也逐漸接納與生態共生的事實；
- 公民意識的抬頭，促進參與式治理，公共資源的管理效益和合理共享也引發了公民莫大興趣；
- 物件與設施的時代性與文化關聯日益受到重視，引發大規模的文資保存與再加值運動。

在此波變革中，資訊透明化、知識普及化、教育重視思辨與創新、庶民力量透過科技賦能，都扮演了關鍵的推波助瀾之力，藉此助力，人人都可以是吹哨者與行動者。這可以說是公民權的擴張，或是世界公民意識的覺醒。

人們從關注歷史建物與老街區，進一步注意到城市中極其顯眼、卻閒置或荒廢的基礎設施。廣義來說，所有維繫城市、社會、經濟、教育、軍事等機制運作的有形無形基礎建設，都在討論範圍。

基礎設施的重生與再利用是近兩個世紀來各國不得不重視的課題。無論因為何種原因造成設施閒置，只要在人口密集或有能見度的區域，荒煙蔓草的景象對政府與區域發展都是負面印象，甚至有安全疑慮。無論因為何種動機進行整建再利用，民眾都希望重生的設施為城市帶回些許活力與生命力。有趣的是，這與它原本的使命是一致的。然而，這個過程中，願意正視閒置基礎設施的價值並紀錄其活化過程的人卻很少。

基礎設施宛如沉默的巨人，在我們腳下穩定運作，要刻意感受才會察覺它的存在，或者當災難發生才警覺它作為城市維生系統的重要性，才會浮出檯面成為不得不處理的問題。就像生物體不會時時刻刻感知自己血液脈動，只有在無響室隔絕外界雜音時才聽見自己脈搏與血液流動聲音、在生病診療時從各種生理數據察覺自己心肺問題、在嚴重受傷時看到汩汩鮮血間歇湧出才知道生命的脆弱與掙扎。就基礎設施而言，除了政府與監督者之外，只有對建築與都市生態體系有濃厚興趣的人，才會持續關注眼下沉默的眾多基礎設施，以及它們的第二人生。

賴伯威對人造環境的獨特觀察與觀點，展現在他獨特的素材搜集與分析過程上，往往令人佩服「原來這些被遺忘

Have you noticed a change in the man-made landscape? Specifically, are you aware that "rebirth" has become a thing in the field of architecture? There are a number of drivers behind this trend:
- Responsible development: Governments around the world have agreed upon the negative impact of limitless development and therefore reached a consensus on the minimization of unnecessary construction projects.
- Environmentalism: Resources from the government are going to the green industry and intangible micro-development projects. Many other industries are also coming around to the fact of a symbiosis between industry and environment.
- Citizenship: People nowadays put more and more emphasis on the importance of citizenship and participatory governance at a time when they become more and more interested in the management of and how to reasonably distribute public resources.
- Cultural heritage preservation: Increasingly, attention is paid to the historical value and cultural connotation of public facilities, thus triggering large-scale preservation movements.

The transparency of information, the availability of knowledge, the emphasis on making intellectual enquiries and innovating in education, and the technological empowerment given to the common people are also key pillars of bringing about the change. Everyone can take action. We have entered an era of citizenship expansion, or, one might say, reached the awakening moment of global citizens.

Before, people only took notice of buildings, structures and regions with a direct historical legacy; now, they gradually shift part of their focus onto the ubiquitous yet abandoned infrastructure facilities found in cities. Generally speaking, structures that are vital to the functioning of a city and its social, economic, educational and military systems are within the scope of discussion.

For the last and current centuries, the rebirth and reuse of infrastructure have been a major topic no countries can truly ignore. The reasons for its desolation put aside, as long as the derelict structure is in an area with a dense population or high visibility, it is bad for the city's image at the best while disruptive for public security at worst. Therefore, no matter why an infrastructure building is scheduled for refurbishment, people want the new version to reinvigorate their city, which actually matches what the original version is designed for – to keep the city alive. Nevertheless, few have observed and recorded the rebirth process of these valuable constructions.

Just like we don't usually hear our blood flowing or our pulse beating unless we are in an anechoic room, or we are not aware of our cardiovascular problems until we go to the doctor, or we don't become convinced that life is this fragile unless we are suffering fatal injuries, our infrastructure facilities work on their own in silence without anyone ever noticing – unless you put your mind to it or until disasters remind us of their role in keeping the city alive. Aside from the government, only those who are deeply in love with architecture and the urban ecosystem are likely to direct their attention to these mute structures and their second lives.

The way Po Wei Lai gathers and analyses the material gives us unique

的事物還可以這麼看、這麼想啊！」三年前他在《寄生之廟》中為文化界帶來的意外與驚喜，在此書中延續到另一個場域。《寄生之廟》探索的對象，是理所當然存在的信仰匯集點，卻以一種自然又迥異的樣貌融合在日常生活環境中。本書《重生之路》的對象則是全然低調又龐然的基礎設施，它們曾經是理所當然的存在，也曾走過沉寂的黯淡時光，卻在改建再利用後宛如新生，以一種既懷舊又新潮的姿態呈現在世人面前。生物的蛻變過程是生態學家眼中的絕美詩篇，基礎設施的蛻變過程必然是賴伯威眼中的一種建築詩集，因此應該稱賴伯威為建築生態學家吧。

《重生之路》一書提出了幾個有趣的議題：
• 重新檢視形與機能的關係
基礎設施是人造物件中體積最為龐大、影響最為深遠，而且是機能優先的建設（形隨機能）。然而當設施除役，再利用的契機就從僅剩的空殼（形）開始。尤其當代人們對於空間的想像逐漸從刻板的機能與形式必然性中解脫，從形式設想機能的可能性，或適應形式而發展新機能，反而成為迷人的議題。

• 時代的助產士
文化與潮流的累積和推進，形成時代的意識與內涵，建築專業者不過是協助讓它具體化，亦即時代的助產士。時代助產士雖無法阻擋必然發生的趨勢，卻是順利生產的關鍵角色，隨勢而行、解決阻礙時代前進的問題，讓世人目睹時代的瓜熟蒂落。

• 基礎建設史觀
基礎建設在各類產業發展之前就已經就緒，它是城市發展的維生系統，這套系統的硬體架構隨著時間推移而迭代增長，軟體也不斷進步。然而這種基礎建設演化史只有在正史上零星出現，就好像戲劇演出，一般人只看到幕前，卻鮮少關注幕後，但如果沒有人投入幕後，也無法成就幕前。所以若要產業健全、文化完整，勢必要有人願意參與並提供此類「幕後」訊息的接觸管道，揭露、並加以研究。

傳統基礎設施皆為鋼鐵與水泥的巨大量體，隨時代變遷，當代基礎建設中數據與通訊逐漸成為必要核心。從巨型基礎建設邁向微型基礎建設，不禁令人好奇未來都市樣貌與生活型態會有什麼轉變。此外，基礎建設將不再只是單純的基礎建設，未來將面臨老舊基礎建設的「重生」或「轉生」，甚至像自然界一樣的「融合型基礎建設」，所有的人為建設「遺跡」都會逐漸成為整個大環境的基礎與養分。

基礎建設巨人在我們腳下，它們習慣被遺忘，但當人類想看得更高更遠，便不能忘記人人都站在這些巨人的肩膀上、也將共同打造出新的巨人。在巨人的重生與未來之路上，人類既是促成者也是受益者，這種共生、共存、共榮的相依關係，是我們在閱讀此書時，值得謹記與省思的。

國立交通大學建築研究所副教授兼所長
侯君昊
2020.06.05

---

insights into the man-made landscape. I cannot help but think admiringly, "What a wonderful approach to read and interpret these long-forsaken buildings!" Three years ago, his Parasitic Temples made a pleasantly surprising appearance to the cultural circle; this year, we have The Path to Rebirth, the same line of thought for a different subject. While Parasitic Temples talks about places of worship that blend with their surroundings in a harmonious yet unusual way, The Path to Rebirth tells the stories of the low-key infrastructure. These buildings used to be part of our daily life. They have undergone a period of decline and then transformed into a new form, futuristic as well as nostalgia-inducing. To ecologists, the metamorphosis of organisms is the poetry of life; to Po Wei Lai, that of infrastructure must be like the poetry of architecture. Ergo, we can probably call him an "architectural-ecologist."

The Road to Rebirth points out some interesting issues:
• The relationship between form and function:
Infrastructure is a large, impactful and purposeful (form follows function) type of architecture when it is first made. However, after it retires from its original position, its function is gone. At the same time, people nowadays tend to think outside the box when it comes to form and function – there are no absolute rules. You can design a space based on its shape, but you can also make its form accommodate to a never-thought-of function. It is a very intriguing topic.

• Architects as midwives:
The accumulation and advancement of different trends help form the ideology of a certain era, and architecture professionals are the midwives who materialize it. As midwives, they can never conceive or abort the inevitable, but they surely play a key role in giving a smooth birth to it. They dutifully follow the trends and solve any problems that come to be roadblocks to the advancement of the time, so that people can witness and share its fruit.

• A history of infrastructure:
Infrastructure is ready before anything else happens in the development of a city and its industries. It is the life support of a city and, with time, the hard- and software of this system improves and evolves. Yet, there is never a history dedicated to the evolution of infrastructure, until now. In the past, one could only glean some fragments of the topic from other, more formal histories, just like most people who watch the performance of a drama seldom care to know about its production. Meanwhile, how can the actors perform on stage if not for those behind the scenes? Therefore, to make the discipline whole, someone must be willing to provide and study the relevant information.

Traditionally, infrastructure facilities are huge buildings made of steel and concrete; nowadays, data and communication have become the focal point of their modern counterparts. With a shift from macro-infrastructure to micro-infrastructure, one cannot help but wonder what a city and the lifestyle in the future will look like. One thing is certain: Old infrastructure will most definitely be reborn and even become the basis and nutrition of the wider environment.

We usually take these giants for granted, but when we want to see farther places from a higher position, we mustn't forget that we are already standing on their shoulders and are on the way of making new giants. On the path to their rebirth and future, we human beings are both the trigger and the beneficiary, and I hope you can remember this when reading this book.

June-Hao Hou
Associate Professor &
Director at the Graduate Institute of Architecture, NCTU
June 5th, 2020

0

序

「**重生**」這個詞的意思，根據某宗教經典：是指一個人與神建立了一種新的關係。

本書所指的「重生」，不單只是基礎設施從廢墟狀態中復活，更正確的描述：是指一座基礎設施與人建立了一種新的關係。

本書起源於我2007在哈佛研究所時的最後一個作品，榮獲波士頓建築學會頒發「設計的未來The Future of Design」獎項，從此之後的歲月，我不斷關注世界上類似這種把廢棄基礎設施再利用、賦予新功能的過程，將之視為重生的案例。此外，《重生之路》也在之後成為了我教學生涯的重要指定題目。本書除了精選一些已存在的「重生」案例，記錄這些世界各地的廢棄基礎設施重新建立與人的關係，也收錄了幾件我在台灣科技大學與交大建築研究所指導學生們的作品。

書中所有案例的範圍仍限於實體空間的再利用，因為針對的是在舊時代被淘汰的基礎設施，無論是把A變成B或把A變成甲，都是透過功能與空間的再定義與人建立了一種新的關係。然而人類目前只有對基礎設施實體空間再利用這一種模式，對於非物質基礎設施與非實體的再利用，就我們有限的所知所學尚未有案例可循。

Rebirth, according to a religious canon, is the establishment of a new relationship between God and the people.

In a similar way, the rebirth mentioned in this book is not just about the restoration of infrastructure, but, to be more precise, also implies the establishment of a new relationship between people and the infrastructure that is reborn.

For my last project during my study at Harvard in 2007, I was very honored to be given "The Future of Design" award by the Boston Society of Architects. Since then, I've dedicated a lot of time to the research of infrastructure facilities worldwide that were once neglected but were then reused and reborn. This is how *The Road to Rebirth* came about. Later on, this book became an important work on the required reading list for my students. In addition to a fine selection of reborn infrastructure around the world that had established a new relationship with people, some of the collaborative efforts with architecture students at National Taiwan University of Science and Technology and National Chiao Tung University are also included.

The case studies in the book – the infrastructure facilities that had been left behind, undergone transformation, and then established a new relationship with people by redefining architectural functions and space – are limited to physical structures because, so far, people are only capable of redesigning tangible objects. As for the reuse of intangible infrastructure, there are no precedents that I know of.

Introduction

當代城市賴以運作的根本

城市是有系統有組織提供公共服務的集合體。而由基礎設施所提供的公共服務是所有城市內商品與服務生產、工業化或商業發展不可或缺的「基礎」。需要有基礎設施的硬體建設後才能有所謂的軟體服務，是當代城市賴以運作的根本。有針對性的投資與建基礎設施本身就在現代經濟上扮演了重要角色，早已成為了創造就業和經濟增長的引擎，對國家與城市的繁榮影響甚大。

形塑地貌與時代的力量

基礎設施的出現改變了人類對時間的感覺與對空間的認知。縮短了國與國間、城市與城市間、人與人之間的距離。讓人類能填海造陸，化滄海為桑田，改變了這個行星的地景地貌，過去所認知的城鄉風貌也就此改變，而且目前看來只要人類的城市還存在，就仍是不可逆的改變。

然而這樣的改變不禁讓我想起一個古老的悖論，關於我們所屬專業的悖論：

很多人都嚮往過去的建築，那些沒有建築師的建築；很多人都嚮往過去的城鎮，那些沒有城市規劃師的城市。

相比過去，人們將許多對現代建築與現代城市的缺陷與不滿，歸咎於工程師、建築師與城市規劃師的專業上。這些專業似乎製造的問題比解決的問題多？否則為何許多人「心嚮往之」的總是過去？確實比起現在，感官上與感性上我也更喜歡過去。但問題是在於我們總認為：

「現代」是「過去」的延續？

然而現代城市在很多方面上而言，與過去城市（城鎮）相比可說是兩個不同物種。舉一個最底層最局部的例子：原本沒水沒電、沒有衛生設備的建築，從無到有，變成具備這些設備的建築，要仰賴一個城市從過去無法供水供電、沒有衛生下水道或汙水處理廠，從無到有地開始建造而具備這些基礎設施。

「現代」絕不只是「過去」的「升級」這麼單純。

城市做了水電工程改造後幾乎就變成了另外一個物種，以科幻場景做比喻：人類被改造成生化人之後，在大部分的科幻片中，那個生化人已經不再被視為人類，是另外一個基於人類基礎的新物種。改變之大與涉及之廣，已無法用過去的標準看待它們，這種改變最後影響到人類思維方式的改變，但人類並沒有改變看待它們的方式，至少在與過去的連結上，還是基於一個時間上先後次序而來，如同用薄弱的前後關係，把過去與現在視為「同一個」或「同一種」。

它們所有與過去的不同，是「時代」去設計建造城市成為現在這個樣子。時代之下有更多更大更複雜的力量，讓城市一步步變成目前我們看到的「現在」，並非建築師與城市規劃師將它「設計」成我們看到的現在，這些人不過是時代的助產士。

時代所造的城市早已成形，工程師、建築師、規劃師只是讓時代具體呈現在你我眼前。

所有國家所有城市都會遇到的——城市病

在不斷持續追求增長與進步的前提下，當一個經濟體，無論是從一個國家或城市來看，當它由這個經濟階段往下一個經濟階段演進時，就會過渡而留下過時的基礎設施。從「必需」變成「廢物」！如何處理、升級，或再利用這些被淘汰的基礎設施，是每一個國家與城市都會面臨的問題。

這種城市病是因時代改變，經濟增長過程中所需付出進步的代價，差別只在於不同國家不同城市會因發展進程

The Basis of All Modern Cities

A city is an entity that provides systematic public services. These public services come from the infrastructure and are indispensable for the production of commodities and paid services, the industrialization, and the economic development of a city. The hardware (the infrastructure itself) has to happen first before we can talk about the software (the services provided). Hence, infrastructure is the basis of all modern cities. The investment in and the building of infrastructure plays an important role in modern economy, helping create job opportunities and boosting the economy. In other words, it exerts a huge impact on the prosperity of a city and even the whole country.

The Force Shaping Our Landscape and History

The emergence of infrastructure has changed how we perceive time and space. Infrastructure works bridge the gap between two countries, two cities, and even two people; they also enable human beings to fill the sea and turn it into productive land. As a result, the landscape of the Earth is changed, and so is our cityscape. Furthermore, it seems that the change is irreversible as long as cities are still standing.

I cannot help but think of an age-old debate, a controversy about my field of expertise: When it comes to the bygone days, many tend to wear rose-colored glasses. To these people, buildings and cities from the past when there were no architects or urban planners are better than their modern counterparts. For any defects or flaws found in modern buildings and cities, they blame engineers, architects and urban planners. How is it that with these professionals, more problems are created than solutions? If this is not the case, why do people almost always yearn for the past? To be honest, I myself prefer the past for sentimental and emotional reasons. However, the key point is, people often mistakenly assume that the present is the continuation of the past.

Is that so?

No. From numerous aspects, a modern city and a city/town before modernity are two different "species". For instance, to turn a building with no water and power supply nor sanitary facilities into a building equipped with all of the above, the city where the building is located must start constructing infrastructure to provide water, electricity, drainage, and sewage treatment.

Therefore, the present must not be put simply as evolving from the past.

Once a city builds its own infrastructure, it becomes a new species. If you can't get your head around it, imagine this: In most sci-fi movies, once a human being is transformed into a cyborg, it is not seen as a human being anymore but a new species "based on" human beings. This is because, there are too big a change and too huge a difference for the two to be treated with the same standard. However, even though the transformation of cities eventually changes our thinking too, the way we see them doesn't change. People still link the present to the past and, since one is indeed followed by the other in a temporal sense, they use this weak argument as the reason that the two should be viewed as one and the same, or at least the same species.

But our modern cities are different from pre-modern ones, and the force behind the change is Time. Time and the many powerful and complex factors during a certain period of time are the designer and builder that gradually shapes cities into what they are now; those architects and urban planners are just the midwives, not the one who gives birth.

Modern cities are already being shaped before engineers, architects and urban planners present the finished work in front of you and me.

An Illness All Countries and Cities Suffer from

When a country or a city keeps promoting its economic growth and continues to evolve from one stage to the next, it is inevitable some of its infrastructure will become out of date one day. From a necessity to rubbish, what a cruel twist of fate! How to upgrade or reuse these obsolete structures is a problem every country and city has to address.

This is a cost we all have to bear during economic growth. The only difference between individual cases is, depending on the rate of development, each country or city shows signs or "symptoms" at different times. Some of them catch the disease earlier, while some later; some of them have more to handle, while some have fewer. Some of the infrastructure become a burden and a hassle, while some have a chance to be reborn. Time after time, history shows that it is an illness every city will suffer from sooner or later, no matter what. Fortunately, there is at least one treatment for it – adaptive reuse.

不同，所以發病的時間不同。有些城市因而遇到得早，有些遇到得晚；有些城市處理得多，有些城市處理得少。有些基礎設施從此成為累贅與負擔，有些重生成為機遇與發展。歷史上基礎設施一次次反覆迭代的文明演進已經告訴我們，這種城市病或早或晚終究會遇到，而對基礎設施如重生般的「可適性再利用」至少是現階段可行的一種處方。

微觀基礎設施的演化來看都市演化

許多類型的基礎設施與一般建築物相比之下，都是巨大尺度的系統與構造物。建築物無論多大，都只是城市的一小部分，而許多基礎設施則是跨城市規模的。不同於一棟棟建築物組成城市硬體好比是形式上的分子，基礎設施是支撐這個城市運轉一部分的系統，即便這個系統有時候確實還是以單棟建築物的形式呈現。

目前世界上許多城市都正在對這些退役淘汰的基礎設施本身及其周邊，進行微小的、局部的都市更新。但為何要再利用？如何能再利用？這本身不是建築議題，而是牽涉市政、經濟、開發等層面更複雜的議題。只是表面上讓我們看到的解決手段是屬於建築議題，像是一場又一場工程

師與建築師的世紀接力。但只要有真正商業利益存在的地方，並不需要政府去大力推動，整個市場自然就會跟進，自然就會讓公私部門前仆後繼地再利用再開發、推動老舊區域的再發展。我們不否認，許多老建物或基礎工程的可適性再利用，從表面上看來是某一種懷舊情懷，但大多數造成這個重生動機的真正理由都是發現該地點存在投資價值，才會評估物件是否拆除亦或再利用。

以長遠的時間軸來看，城市內「不成功的開發」至少都比「不開發」形成死域要好，重點在於保持城市活力。對城市的經營者與土地的持有者而言，這些基礎設施再利用計畫都是他們所樂見的。所以，重生再利用的開發案就算最後失敗，只要城市土地有在使用與運作，就意味都市演化仍持續著，而最終，城市本身都是贏家。

基礎設施與建築不管有幾次的重生最後皆有一死，
最終只有城市本身是不死的，
它只會不停演化。

建築的形隨機能，是否不存？

19世紀由芝加哥學派建築師路易士‧蘇利文提出的 Form follows Function 理論，強調建築的型態必須隨著機能需要而設計或改變，宣告合理的外型源自合理的機能。理想上，建築師被要求遵守此原則設計。

西方建築類型學（Typology）的研究也說明了各種不同功能的建築會演化出特化類型：學校會有學校的樣子，博物館會有博物館的樣子，與旅館就不會是一個樣。這些不同類型的建築彼此是異中求異，或異中求同。

中式和日式等東方建築則近乎相反，中式建築存在一種基本原型，它們鮮有所謂特化類型。有的是一個基本原型（單元）去擴充增長，以不同排列不同組織方式去滿足不同功能，但可能都是源自同一個基本原型。所以大部分不同功能建築的樣子基本上都還是大同小異，同中求異。

從西方建築類型學角度來看，由基礎設施重生的建築物都是混生種或突變種，機能前後天差地遠。重生前的基礎設施多由工程師主導，形式與機能之間有著必然的關係，並無對美學的過分執著。但多數由建築師接手重生後變成的建築，機能雖與原本天差地遠卻依然能運作良好。絕大

From the Evolution of Infrastructure to the Evolution of Cities

Compared to regular buildings, a lot of infrastructure projects are gigantic systems or structures. For regular buildings, however big they might be, they consist only a small part of the city, while many infrastructure works cross multiple cities. While regular buildings sometimes seem to be more like an accessory component of a city, infrastructure, on the other hand, is functional, allowing a city to operate properly, although some of it can be as small-scale as a single building.

Nowadays, cities all over the world are making an attempt at urban renewal programs to reuse obsolete infrastructure buildings that is out of service and regenerate its surrounding area. But why should we reuse it? And how do we do it? The issue is not only architectural, but also civic, economic, developmental and much more, albeit the fact that on a superficial level, it does look like engineers from one generation are handing off the baton to the architects from the next generation in a relay race. As complex and intimidating as the issue is, where the potential of making a profit lies, nobody misses the opportunity even if the government does not particularly encourage it. Both the public sector and the private sector naturally surge forward to promote the reuse and redevelopment of obsolete infrastructure and blighted areas in cities. There is no denying that nostalgia plays a certain role in the adaptive reuse of infrastructure or old buildings, but at the end of the day, the real motivation behind most reconstruction projects is the lucrativeness of the location itself. In short, location determines whether an

object is to be demolished or reused.

In the long term, "failed development" is always better than "no development." The important thing is to keep the city's vigor. Therefore, city officials and land owners welcome these proposals to reuse infrastructure facilities. Even if the development fails, as long as there is activity on the land of the city, the city is still evolving. In the end, the city will triumph.

Some infrastructure facilities can undergo the process of rebirth several time in their lifetime, but ultimately they fade into oblivion. However, cities are immortal. They will only continue their evolution.

Is There Not Such a Thing as Form Follows Function?

In the 19th century, Louis Sullivan, an architect belonging to the Chicago School, put forth the "form follows function" theory. He emphasized that the form of a building must be designed based on its intended function, proclaiming that a form only makes sense when it suits with its function. All ideal designs should adhere to this rule.

The study of typology in the field of western architecture also emphasizes that buildings and structures always evolve into certain types according to their functions, so that schools will look like a school and museums will look like a museum; they won't bear resemblance to, say, a hotel. To sum up, buildings that are different from each other may focus on the differences or on the similarities among themselves, but either way, they are surely of different types.

On the contrary, in eastern architecture, particularly Chinese architecture, you will find a basic archetype rather than several specialized types. Buildings may fulfill different purposes with different arrangement and organization of each element while the basic archetype remains the same. In short, buildings with different functions are largely similar in form, with minor differences only.

From the perspective of typology in western architecture, all the buildings and structures reborn from infrastructure are hybrids or mutated species whose intended functions are poles apart before and after the rebirth. Before the rebirth, the engineer is the dominant figure, there is a clear association between form and function, and you don't take aesthetics into consideration; on the other hand, during and after the rebirth, the architect is the leading character and the intended function is totally different from before, although that doesn't stop it from working properly. The majority of reborn buildings do not cater to what the original infrastructure is suitable for, but simply try answer the question of what the already dead infrastructure needs to be transformed into, so there is no necessary link with the properties of the former structure. Be it a water tower turned house, a power station turned gallery or a crane turned hotel, numerous examples of infrastructure facilities reborn into hybrids or mutated species have proved that the way typology categorized buildings into specific types is far from absolute.

All the cases of reborn infrastructure point to the conclusion that the "form follows function" theory in the field of architecture could be no more than an ideology, a visionary principle in the architect's own head.

多數的重生機能都不是基於這些基礎設施「適合」被變成什麼，而是這些已死的基礎設施「需要」被變成什麼。與重生前的基礎設施之空間或特質並沒有必然關係。一次又一次基礎設施的重生，如水塔住宅、發電廠美術館、起重機旅館等等每一個案例，每一個混生種或突變種都在說明西方建築類型學所謂的特化建築類型並沒有絕對。

從一個個基礎設施如重生般的案例結果來看，「形隨機能」在建築領域可能只是一種意識形態，一種建築師自己想像中的原則。或著說，當空間或時間大到一定數量級以上便不存在了。

這是問題的提出，而不是結論的宣告，答案還在路上。

基礎設施的進化，
實體之外「再利用」本身的進化

智慧城市正在為我們帶來未來的基礎設施，但撇開新型態的基礎設施不談，光是既有的基礎設施本身技術的升級與進化就足以再次改變地貌，例如高效低價無汙染的電力就能讓人類突破以前受限能源成本而做不到的事：淡水很可能不再是地球上的稀缺資源，在低成本的能源下，人類可以更大規模電解海水取得淡水，環境工程上因成本的大大降低便能將沙漠變綠洲，再次重塑地貌。然而本書談的是基礎設施的過去與當下再利用的成果，不是下一代的基礎設施與未來趨勢。基礎設施本身的未來，例如從新的融資模式與管理模式，到新型態的工程基礎設施、綠藻供電網、自動充電道路、VR 體感站、太空機場與太空站，乃至外星殖民地的再利用等等，那是另一個領域及另一本書了。至於《重生之路》這本書，則像是一個戀物癖般的紀錄集合體。

當科技文明逐漸進步，人類的基礎設施早不再局限於鋼筋混凝土的物質實體，而基礎設施一詞其所定義與涵蓋的範圍也不斷更新，所謂的可適性再利用也不會局限於對新型態基礎設施的實體再利用，因為再利用本身也會跟著進化。真正新型態的基礎設施會帶來新形態的再利用。在智慧城市浪潮持續下，新型態的非物質基礎設施與它們非物質超越實體的再利用，是我們目前無法想像的。

下一代的基礎設施，未來的基礎設施再利用型態的趨勢，存在更未來的未來。

Or we should say, once the scale becomes too big in terms of time and space, it might no longer apply.

Nevertheless, I am just posing a question instead of jumping to conclusions here. We are still searching for an answer.

Future Infrastructure and the Reuse Thereof That Transcend Physical Boundaries

Smart cities start to bring us futuristic infrastructure. Leaving the new forms of infrastructure aside, with just the upgrade and improvement of the technology involved, we are able to transform the landscape once again. Electric power with high efficiency, low cost and nearly zero pollution will enable human beings to make a breakthrough and accomplish what they cannot do when the cost is too high: fresh water might not be a rare resource anymore since it can be obtained through desalination on a large scale; deserts can be turned into oases when the environmental engineering cost goes down considerably. Both scenarios will reshape our landscape once more. However, in this book, instead of talking about infrastructure of the next generation or future trends in infrastructure, I am going to discuss the infrastructure projects from the past generations and our current re-usage of them. The future of infrastructure, from new modes of financing and management to new forms of buildings such as algae power supply, EV-charging roads, VR kiosks, spaceports, space stations, or even space colonies, is the theme of a whole 'nother book; The Road to Rebirth, on the other hand, should be treated more like an obsession with obsolete infrastructure works.

With the advancement of technology, infrastructure will no longer be limited to physical structures made of reinforced concrete, while the definition and range of infrastructure is going to keep changing. Along the same lines, adaptive reuse will no longer be limited to the physical reuse of new forms of buildings since the concept itself is going to evolve with infrastructure. In other words, new forms of reuse will follow the emergence of new forms of infrastructure. Right now, with the rise of the smart city movement, it is difficult to predict what these new forms of infrastructure will look like and how the reuse of them will transcend physical boundaries.

The infrastructure of the next generation and the new trends in reusing them will grant us a future more futuristic than we imagine it to be.

1

基礎設施史觀

A History of
Infrastructure

基礎設施的史觀

「基礎設施（Infrastructure）」一詞在 1875 年開始出現在法語中，由拉丁語字首「infra（下面）」與法語單詞「structure」組合而成。最初的意思是「設置能以任何形式運作或系統的基礎」。在法語中指的是用來建立路基的最底層基礎材料，鐵軌或人工鋪面皆要架設在上。

基礎設施是為城市生產（也包括城市以外的其他區域）和居民生活提供公共服務的工程，是一個城市賴以生存發展的基本條件。這些基本條件由眾多系統組成，如道路、橋梁、隧道、供水、電力、電信與網際網路等。

由基礎設施所提供的公共服務是所有商品與服務不可或缺的生產要素，若沒有這些公共服務，則大部分促進、維持，或改善生活條件所需的商品和服務便無從提供，這種根本性與先行性就是「基礎」一詞的由來。

西元前 1 世紀的古羅馬水道橋 Roman Aqueduct，證明當時的古羅馬人已有公共基礎設施的觀念。供水系統與城牆系統一樣，被古羅馬人視為城市中必備的基礎設施（說不定對古羅馬人而言，競技場與大浴場也算必備的基礎設施）。大體而言，18 世紀以前的基礎設施主要包括城牆、道路、運河，和灌溉系統。有時還有港口和燈塔，為海上航行提供服務，而一些較為先進的城市則有公共噴泉甚至極為罕見的下水道系統。18 世紀第一次工業革命後，基礎設施的種類大量增加，各種新型態的基礎設施多起源於英國，例如鐵路與地下鐵系統。

A History of Infrastructure

The term "infrastructure" first appeared in the French language in 1875. It combined the Latin prefix "infra" with the French word "structure" to mean "the basis for operation or system of any forms." In French, it means the underlying material on which railroad tracks or pavements are laid.

Infrastructure is a construction that provides public services to aid the production of commodities in a city (or other non-urban regions) and the daily lives of its inhabitants. It is a prerequisite to a city's survival and it consists of various systems including roads, bridges, tunnels, water supply, electrical grids, and telecommunications

The public services provided by infrastructure are essential to the production of commodities and paid services. Without these public services, the majority of the commodities and paid services that help maintain or improve the quality life cannot be produced. This nature of being a prerequisite or a foundation is exactly why it is called the infrastructure.

The existence of Roman aqueducts built during the first century BC proves that the Romans were already familiar with the concept of public infrastructure. Along with the defensive walls, the water supply system was seen as a necessary infrastructure facility for a city (perhaps, those amphitheaters and baths are necessary infrastructure facilities for the Romans too!). Generally speaking, before the 18th century, the main types of infrastructure were city walls, roads, canals, irrigation system, and sometimes ports and lighthouses to provide services for marine navigation. In some more advanced cities, there were public fountains and even the extremely rare sewage system. Since the First Industrial Revolution in the 18th century, new types of infrastructure have sprung up, many of which originated from England, such as railroads and subways.

In addition to the fact that infrastructure can provide public services, nowadays it is also viewed as and has become an asset due to its relatively minor fluctuation in value. Invest-

除了提供公共服務外，基礎設施如今也因其價值波動的穩定性，而被視為一種資產類別，基礎設施逐漸演變成了一種資產形式。投資於國家的基礎設施興建更帶來了直接與長期的經濟效益，能夠增加就業，有助於社會經濟結構的轉變與人民生活水準的提高。例如：交通運輸的發展直接導致產業和人口往城市集中，這種經濟活動在空間上的集中，實現了規模經濟，不僅降低了企業的生產成本與交易成本，相對地也降低了基礎設施的單位服務成本，提高利用效率。此一連鎖反應形成閉合的良性循環。當然，所造成的後果就是──基礎設施越建越多。

這些先後出現的基礎設施構成了人類文明的一部分，人類透過興建基礎設施，一步步克服物種與環境上的限制。從屈服大自然、利用大自然，到駕馭大自然而形成今日地球上人類集中生活的都市環境。不論是非，不論代價，此過程中一次又一次的技術突破，透過基礎設施的興建，人類不斷大規模改變這個行星的地景地貌，在人定勝天的野蠻中，宣告了基礎設施的演進史就等於人類文明的演進史。

基礎設施的類型

基礎設施是城市賴以生存發展的配套，由一個個系統工程所組成，這些系統可分為幾大系統類型，包括交通運輸系統、能源供應系統、供水排水系統、安全防災系統、郵電通訊系統、環保和公衛系統、以及教育／醫療／文化與休閒。

交通運輸系統

例如：鐵路、道路、港口、隧道、高架橋、飛機場、捷運系統等。

能源供應系統

例如：能供應水、電、瓦斯、天然氣等日常生活所需能源的礦場、水壩、水塔、配水池、淨水廠、抽水站、天然氣儲存槽、各類型發電廠、變電所、供電網絡，自來水水管、油管及天然氣管等輸送的管線系統等。

安全防災系統

例如：堤防、城牆、防空洞、軍事設施、網際網路防火牆等。

環保和公衛系統

例如：垃圾收集系統、廢棄物掩埋場、汙水處理廠、垃圾焚化廠等。

郵電通訊系統

例如：郵政系統、電話線路、海底電纜、通訊衛星、網際網路等。

教育／醫療／文化與休閒

例如：學校、醫院、博物館、公園等。雖然這些建築對城市至關重要，但在基礎設施的認定上，我們偏好選擇維持城市基本運作與維生功能，與構成這些系統關鍵部分的建築，意即此項以上的其他項。雖然此項類型不在本書所認定的基礎設施中，但很多基礎設施重生後卻成為這類建築或空間。

可適性再利用的經濟永續

在任何一個現代城市，無論是其產業與提供公共服務的方式都必須不斷適應現代化。各國或各城市為確保其競

---

ment in national infrastructure projects brings direct and long-term economic benefits in the form of creating job opportunities, changing the socioeconomic structure for the better and raising living standards. For instance, the building of transportation infrastructure directly results in industry and population moving toward urban areas, which in turn makes economies of scale possible because of the concentration of economic activities in one place, which reduces the production and transaction costs for enterprises, which lowers the service cost for the infrastructure and enhances its efficiency. This chain reaction produces a closed virtuous circle. However, one of the consequences is that more and more infrastructure facilities are getting built.

Infrastructure is part of human civilization. Through the building of it, human beings have been able to overcome all the biological and environmental limits step by step, first conquering, then exploiting, and finally exerting full control of nature. This is how the urban environment where human beings cluster together today is formed. No matter it is right or wrong and regardless of the price we have paid, it is undeniable that human beings have made a lot of technological breakthroughs building infrastructure and in so doing continued to transform the landscape of the planet in a large scale. Our savage triumph over nature declares that the history of infrastructure is the history of human civilization.

Types of Infrastructure

Infrastructure plays an important role in a city's survival and development and is composed of individual systems and constructions. These can be categorized into several main types, including transportation, power supply, water supply and sewage, defense and disaster prevention, communications, environmental protection and public health, and those related to education / healthcare / culture / leisure activities.

Transportation

Including: railways, roads, ports, tunnels, viaducts, airports, subways and so on.

Power and Water Supply

Including: mines, dams, water towers, distribution reservoirs, water treatment plants, pumping stations, LNG storage tanks, power stations, substations, electrical grids, and all kinds of piping systems that transport water, oil and gas.

Defense and Disaster Prevention

Including: banks and dikes, city walls, air raid shelters, military buildings and structures, as well as firewalls for Internet security.

Environmental Protection and Public Health

Including: waste collection systems, landfills, wastewater treatment plants and waste incineration plants.

Communications

Including: mail and postal systems, telephone cables, submarine communications cables, satellites, and the Internet.

Education / Healthcare / Culture / Leisure Activities

Including: schools, hospitals, museums and parks. Although they are very important to a city, I didn't include them in the book. Rather, I only picked those systems and some of the key parts the systems were made up of that support the basic operation and function of a city, namely those belonging to the previous categories. Having said that, a lot of infrastructure facilities are converted to buildings and structures that belong to this particular category.

Adaptive Reuse and Economic Sustainability

It is true for any modern cities and countries that, to keep up their competitiveness among other cities and countries in the world, they must always adapt the most modern method to develop their industries and provide public services, and hence to continually upgrade their infrastructure in terms of both quantity and quality. Inevitably, when a city or a country is evolving from one economic stage to the next, there is going to be "leftover" infrastructure facilities. Every city and every country seems to catch the same disease that can turn a necessity into rubbish. How to tear down, upgrade or reuse these leftovers is an issue every city and country is facing. At the moment, one of the feasible solutions is adaptive reuse.

The Burra Charter drawn up by the International Council on Monuments and Sites (ICOMOS) dictates that to adaptively reuse an infrastructure site is to look for a new function for it. The new function

爭優勢，將不斷持續提升基礎設施的質量來與世界競爭。在不斷持續提升的前提下，一個經濟體，無論從一個國家或城市來看，當它由一個經濟階段往下一個經濟階段演進時，就會過渡而留下過時的基礎設施。彷彿得了一種從「必需」變成「廢物」的城市病。如何處理、拆除、升級，或再利用這些被淘汰的基礎設施，是每個國家與城市都會面臨的問題，而「可適性再利用（Adaptive Reuse）」至少是目前其中一種可行的解決模式。

由國際古蹟遺址理事會（ICOMOS）所制定的《布拉憲章》為建築尋找「可適性再利用」，目的在於為某一基礎設施遺產找到適合的新用途，這些用途將使該場所的重要性得以最大限度地保存和再現，同時對重要結構做最低限度的改變。從本書中我們所收集的基礎設施重生案例上，可以看到許多基礎設施設計在建造之初，大多精心打造並非常堅固耐用，重生改造後的建築很容易具有引人注目的亮點，但我們相信更決定性的關鍵是透過在資本市場上尋找新的用途，來證明該物件與該場所繼續存在的合理性。

不過，上述基礎設施的類型分類，無法完全對應到現實世界中經歷重生的真實案例，未出現的基礎設施類型不代表它們無法改造再利用，只是尚未發生。書中是以目前已有重生案例可循的結果去分類，例如：核電廠重生成主題樂園、造船廠重生成藝術中心、鑽油平台重生成旅館、水塔重生成住宅、筒倉重生成攀岩場、高射炮塔重生成水族館。針對現有的退役基礎設施改造案例，歸納出六大類：

A	交通系列
B	港口系列
C	資源開採系列
W＋D	供電＋供水系列
S	倉儲系列
M	軍事系列

有意為之的紀念物 vs. 無意為之的紀念物

藝術史學家李格爾（Alois Riegl）在1903年寫下的《紀念物的現代崇拜：其特徵與起源》中，將應受保護的歷史紀念物區分出「有意為之的紀念物」與「非有意為之的紀念物」兩大類，而後者的生成多屬於歷史的偶然，其可遇不可求的難得才是本書中案例選擇記錄的價值。

有意為之的紀念物：

為了紀念特定歷史事件與人物而建造的具體有形物件或空間，多半是為了服務政治目的。例如：猶太大屠殺紀念碑、越戰紀念碑、黛妃紀念噴泉。

無意為之的紀念物：

一開始是為了當時代人或數代人實用需要而建造，並無對歷史文化流傳的思考，但之後逐漸衍生了新的意義與價值，而轉變為紀念物。

本書所收錄的案例多屬於無意為之的紀念物，例如曾經作為生產的水泥廠或造船廠的建築，其原本的生產功能，在城市發展與產業轉型過程中退去，留下如廢墟般的存在，曾經在此空間中的生產行為已隨產業告終而逝去，反而是成為廢墟的工業建築在建造之初，形隨機能的服務工業生產的建造邏輯，成為了印證時代技術的產物，並在其服務時間裡逐漸成為都市涵構、鄰里關係、與地方意識的一部分，例如書中許多過時的水塔。這樣技術上與時間上的歷史結晶不該被囚禁在博物館中，而是盡可能在它們所屬的實體環境中就地保留並透過可適性再利用賦予新生。

---

should be able to achieve maximum preservation and representation of the importance of the site while making minimum changes to the main structure. From the reborn infrastructure examples presented in this book, it is not difficult to notice that a lot of them were originally designed and built to be extremely solid and, after rebirth, they often became attractions themselves. Nevertheless, I am more inclined to believe that the decisive factor is how a new function was found for a site through the lens of the capital market and thus validated the rationale behind its continual existence.

However, the above types of infrastructure do not entirely correspond with the types of infrastructure that has been reborn. Yet, it is simply that the types that are not included in this book as case studies have no precedents; it doesn't mean that they cannot be reconstructed and reused. In this book, I only include buildings and structures that have truly undergone transformation, such as a nuclear power plant turned amusement park, a shipyard turned art center, a rig turned hotel, a water tower turned house, a silo turned rock climbing gym, and a flak tower turned aquarium, and then I categorize them accordingly. The six types of infrastructure that has been renovated are as follows:

A	transportation
B	port facilities
C	mines
W+D	water or power supply
S	warehouses
M	military structures

Intentional and Unintentional Monuments

In 1903, the art historian Alois Riegl stated in his The Modern Cult of Monuments: Its Character and Its Origin that objects under protection due to their historical value can be distinguished into two categories – intentional monuments and unintentional monuments. For the latter, he explicitly pointed out the great value of it and asserted that this was where we should direct our attention and care to.

Intentional Monuments

These are tangible objects or spaces erected to commemorate specific historical figures or events. Most of them serve a political purpose, such as the Holocaust Memorial, the Vietnam Veterans Memorial and the Diana Memorial Fountain.

Unintentional Monuments

These are initially erected to meet the practical needs of the contemporary society and the future generations – the inheritance of history and culture is not taken into consideration. But over time, new values are added to the objects or spaces and they become monuments.

The majority of what is included in this book are unintentional monuments. Take cement plants or shipyards as an example. The productive purpose of these structures was lost during urban development and industrial transformation while the productive activities that had once taken place there were terminated in the wake of the end of an industry, so as a result, they became some sort of ruins. However, the logic and pragmatism of the "form follows function" philosophy also prove that, they were the products of the technology of an era. Moreover, while on active service, they became a part of the urban structure, the neighborhood and the local identity. Many of the old-time water towers in this book are good examples of this. Therefore, we shouldn't lock these technically and historically valuable monuments up in the museum and remove them from the physical environment where they have belonged. Instead, we should protect them in situ and give them new lives through adaptive reuse.

Iteration and evolution

New forms of infrastructure in the future are already happening with the emergence of smart cities. There will be more intangible infrastructure projects not unlike the Internet. But so far, the infrastructure facilities that have undergone rebirth are mostly tangible from earlier generations, and from the very beginning, they were doomed to be destroyed. Oftentimes, they met their end even before they became useless. Most of these facilities didn't go obsolete because they had been in use for an overextended time or because they fell into disrepair. Neither did those reborn and given new functions for more than once; they usually didn't have a chance to complete their life cycle before plunging right into their next life. This is all because the rate of development and change is far faster than the speed in which a building or structure goes through its life cycle. When contemplating on these infrastructure works that have experienced life and death for multiple times, I cannot

迭代與演化

　　未來新型態的基礎設施正伴隨智慧城市發生，新時代更多的會是類似網際網路這種「無形的基礎設施」，但目前有再利用重生案例的基礎設施多半源於舊時代「有形的基礎設施」，而有形的基礎設施的命運，從一出生開始就注定要毀滅，值得注意的是它們的結束大多早於其使用年限，大部分基礎設施都不是因為使用時間太長或年久失修而被淘汰。那些不只一次重生成為新功能建築的基礎設施，它們大多也都在生命週期結束前就進入了下一個迭代。進步與改變的速度，通常比建築與基礎設施本身的生命週期還要快。這些歷經多次死亡而再次重生的基礎設施，對照其建築本身，不禁讓我們延伸了這個問題：

　　基礎設施與建築的重生形式本身也會繼續迭代，
　　建築價值的永恆，與進步和改變的迭代速度相比下，
　　是否仍持續存在？是否仍值得追求？

　　過去的教堂動輒興建百年，今日的地標性文化建築動輒興建數十年，但在信仰之外，人類之所以願意忍受這麼長的工期，排除了技術與資本因素，是因為過去用百年千年時間軸來看，建築帶來的影響是長遠的，建築的價值是永恆的。然而在今日現代社會事事快速迭代，遠遠脫離過去千百年不變的時間維度下，我們還該繼續信仰漫長工期所換來的價值嗎？

　　如果在遙遠的未來，伊斯蘭教、基督教、天主教或任何宗教信仰有一天從地球上徹底消失，那麼未來的人類雖不至於馬上摧毀宗教設施，卻也不可能把每座教堂或廟宇都當成古蹟供奉起來，而不動腦筋去拆除、改造，或再利用。建築的未來也會如同這些基礎設施一樣會不斷改變，不斷重生，而且時間越來越快，週期越來越短。

　　「變」與「不變」之間，
　　從無「不變」，
　　而「變」的頻率只會越來越高，
　　既然它遲早會改變，何苦追求永恆？
　　既然能不追求永恆，何苦忍受這麼長的興建時間？
　　是否快速完成迭代才是價值所在？

　　反正它們終將改變，終將演化。

help but ask these questions:
　　When even the form of rebirth itself is going to evolve from one generation to the next, will it remain true that a building can be eternal? Is it still worth pursuing, the eternity of architecture?
　　In the past, the construction of a cathedral can easily be a period of a hundred years. Today, the construction of a cultural landmark can often be a period of a few decades. The reason why people are willing to endure such a long period of construction is (aside from their religious faith) because when we put things into perspective, the influence of a building is long-lasting. The value of architecture lies in its eternity. However, in modern society where everything changes fast, should we continue to believe that the value of architecture is worth our waiting for such a long time?
　　If, in a distant future, Islam, Christianity and the other religions all disappear from the world, even if people won't try to eliminate those houses of worship right away, it is not possible that they will preserve every single church and temple as a historical site rather than think of a way to remove, renovate or reuse each of them. To sum up, the future of architecture is going to be as ever-changing as those infrastructure facilities, its rate of changing as fast and its cycle of rebirth as short.

　　The only thing that never changes is change itself, and the rate of changing is going faster than ever.
　　Should we pursue eternity if everything is going to change?
　　Should we tolerate long periods of construction if we are not going to pursue eternity in the first place?

　　Who's to say that the high speed in which a building completes its life cycle is not the new value of architecture nowadays?

　　After all, everything is going to change and keep on evolving.

2

基礎設施屬性

The Attributes of
the Infrastructure
Projects

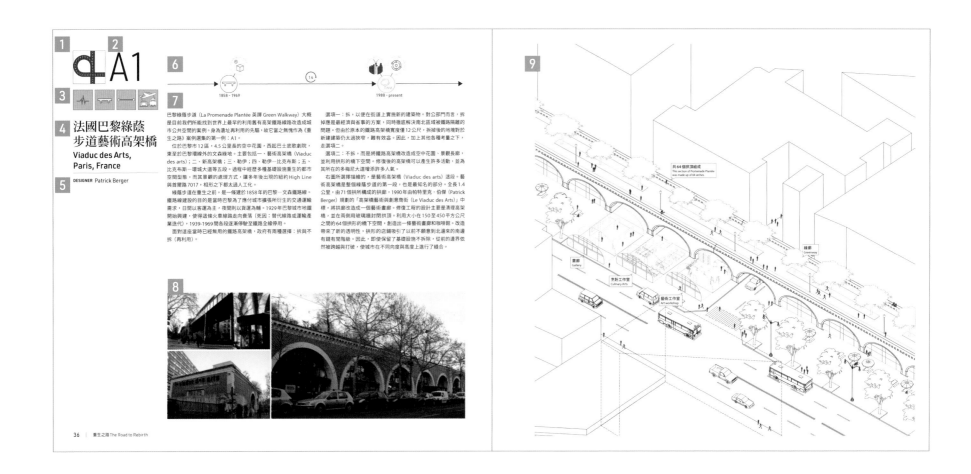

1 基礎設施分類符號
Type of infrastructure

2 案例編號 Number

4 案例名稱 Name of project

5 設計團隊 Designer

3 基本屬性 Basic information
ⓐ 狀態 Current status
ⓑ 基地 Site
ⓒ 殘骸形式 Form & shape
ⓓ 服務標的 Target of service

6 時間屬性 Rebirth information
ⓐ 死亡原因 Cause of death
ⓑ 重生利基 Decisive factor
ⓒ 重生週期 Years passed
ⓓ 原始機能 Before rebirth
ⓔ 改造手法 Purpose of rebirth
ⓕ 重生物種 After rebirth

7 案例說明 Description

8 現場照片 / 效果圖 Pictures

9 藍線圖 The blue-line drawing
ⓐ 黑線 - 原有設施
Black lines: pre-existing infrastructure
ⓑ 藍線 - 重生物件
Blue lines: reborn objects

4個
基本屬性

4 Pieces of
Basic Information

3　基本屬性
　　Basic information

ⓐ 狀態
　 Current status

ⓑ 基地
　 Site

ⓒ 殘骸形式
　 Form & shape

ⓓ 服務標的
　 Target of service

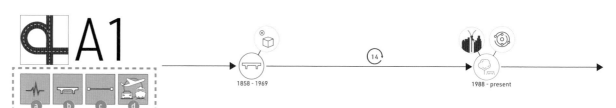

ⓐ ⓑ ⓒ ⓓ

1858 - 1969　　　　　14　　　　　1988 - present

法國巴黎綠蔭
步道藝術高架橋

Viaduc des Arts,
Paris, France

DESIGNER Patrick Berger

巴黎綠蔭步道（La Promenade Plantée 英譯 Green Walkway）大概是目前我們所能找到世界上最早的利用舊有高架鐵路線路改造成城市公共空間的案例。身為遺址再利用的先驅，故它當之無愧作為《重生之路》案例選集的第一例：A1。

位於巴黎市12區，4.5公里長的空中花園，西起巴士底歌劇院，東至於巴黎環線外的文森綠地。主要包括一、藝術高架橋（Viaduc des arts）；二、新高架橋；三、勒伊；四、勒伊—比克布斯；五、比克布斯—環城大道等五段。過程中經歷多種基礎設施重生的都市空間型態，而其景觀的處理方式，讓多年後出現的紐約 High Line 與首爾路7017，相形之下都太過人工化。

綠蔭步道在重生之前，是一條建於1858年的巴黎—文森鐵路線。鐵路線建設的目的是當時巴黎為了應付城市擴張所衍生的交通運輸需求，日間以客運為主，夜間則以貨運為輔。1929年巴黎城市地鐵開始興建，使得這條火車線路走向衰落（死因：替代線路或運輸產業迭代），1939-1969間各段逐漸停駛至鐵路全線停用。

面對這座當時已經無用的鐵路高架橋，政府有兩種選擇：拆與不拆（再利用）。

選項一：拆，以便在街道上實施新的建築物。對公部門而言，拆掉應是最經濟與省事的方案，同時徹底解決南北區域被鐵路隔離的問題。但由於原本的鐵路高架橋寬度僅12公尺，拆掉後的地塊對於新建建築仍太過狹窄，難有效益。因此，加上其他各種考量之下，走選項二。

選項二：不拆，而是將鐵路高架橋改造成空中花園、景觀長廊，並利用拱形的橋下空間。修復後的高架橋可以產生許多活動，並為其所在的多梅尼大道增添許多人氣。

右圖所選擇描繪的，是藝術高架橋（Viaduc des arts）這段。藝術高架橋是整個綠蔭步道的第一段，也是最知名的部分。全長1.4公里，由71個拱所構成的拱廊，1990年由帕特里克·伯傑（Patrick Berger）規劃的「高架橋藝術與創意商街（Le Viaduc des Arts）」中標，將拱廊改造成一個藝術畫廊。修復工程的設計主要是清理高架橋，並在兩側用玻璃牆封閉拱頂。利用大小在150至450平方公尺之間的64個拱形的橋下空間，創造出一條藝術畫廊和咖啡館。改造帶來了新的透明性，拱形的店舖吸引了以前不願意到北邊來的南邊有錢有閒階級。因此，即使保留了基礎設施不拆除，從前的邊界依然被跨越與打破，使城市在不同向度與高度上進行了縫合。

物相之外

Beyond the Physical Boundaries

圖像化的屬性是一種客製化的語言

Icon is a custom-made language

記錄這些基礎設施的同時，我們也正在記錄當下的都市紋理、產業型態與文明進程，這些重生的基礎設施與其周邊是都市變遷的活化石、都市的縮影與切片。在記錄的過程中，除了肉眼所能看見「物相之內」的實體空間之外，在「物相之外」還同時發展了另一套符號，平行於用來單純記錄物相的照片與等角圖，這套圖像符號為基礎設施的「屬性」濃縮了看得見與看不見的部分。

圖像符號的建立，除了更深一層次地記錄基礎設施外，也幫助我們更能清楚梳理不同案例之間橫向比較時的個體異同。重點效果不僅在於用圖像化幫助了解，而是更有效率地用一種符碼簡化需在不同案例間不斷重複出現的同質性文字描述。意即將文字描述化簡成符號，將無論是抽象或具體的概念直接用符號標注使其更簡便，尤其是在不同案例間做橫向比較時效果尤為明顯。這是一種語言，是一種使用範圍只存於本書中、為此研究的客製化圖像化語言。

基礎設施的狀態描述性的4個「基本屬性」，分別是：a.狀態、b.基地、c. 殘骸形式、d. 服務標的。

While the students and I were cataloging the reborn infrastructure facilities, we also recorded the urban environment, the industries and the human civilization at that time. These facilities and their surrounding space survive all the changes that happened in their cities and are thus epitomes of them. During the process, I gradually developed a set of symbols and icons to describe the infrastructure projects both within and beyond the physical boundaries – both visible and invisible to the eye – and I put them side by side with the pictures and blue-line isometric drawing which only record the appearance of infrastructure.

This symbol and icon system is built to record the infrastructure facilities on a higher level and to help us detect the differences between individual examples when conducting cross-section analysis. Not only can iconization help with understanding the reborn objects, it can also ensure, in a more efficient way, a simplified discourse that otherwise would be full of redundancies. In other words, words are simplified into icons. It is more convenient to use symbols and icons to express ideas, whether they are abstract or concrete – it becomes evident when you try to conduct cross-section analysis. These symbols and icons make up a new language, and it is a custom-made language that only applies to this book.

The four pieces of basic information regarding the infrastructure projects are: a. current status; b. site; c. form & shape; and d. target of service.

死亡｜Unreborn and Dead

從第一次死亡（停用）到目前都沒有再利用計畫

Since its death (that is, since it went obsolete), there haven't been any proposals to reuse it.

正在使用｜In Use

該基礎設施目前仍使用中沒有重生過；或該基礎設施目前仍使用中且未來已有明確退役時程，例如：

A5 日本東京秋葉原 2k540 商業藝廊
A6 瑞士蘇黎世翻新高架拱橋
A7 台灣台北環河南路五金街

It is still in use and has not been reborn before or/and arrangements have been made to stop using it at a later date. Examples:

A5 2k540 Aki-Oka Artisan, Tokyo, Japan
A6 Refurbishment Viaduct Arches, Zurich, Switzerland
A7 Huanhe S. Rd Hardware street, Taipei, Taiwan

已重生｜Reborn

從第一次死亡到目前有一次或兩次或三次再利用計畫，例如：

W4 德國布蘭登堡水塔住宅
M2 德國柏林碉堡屋頂住宅
M6 奧地利維也納高射炮塔水族館

Since its (first) death, there have been one or two or three times when it was converted. Examples:

W4 BIORAMA-Projekt, Brandenburg, Germany
M2 Fichte-Bunker, Berlin, Germany
M6 Haus des Meeres Flak Tower aquarium, Vienna, Austria

想像中｜Proposed

設計構想計畫，尚未實現，目前只是設計建議，例如：

A0 日本東京首都高速公路市場（設計提案）
A8 巴西里約叢林 Rio-Santos 高速公路未成段（設計提案）
C6 英國蘇格蘭克羅馬帝灣鑽油平台（設計提案）
D4 台灣桃園榮華壩（設計提案）
M12 台灣鳳山無線電信所（設計提案）
M14 蘇聯颱風級核潛艇（設計提案）

It has been proposed, but has not been carried out. Examples:

A0 Tokyo Highway Market, Tokyo, Japan (proposal)
A8 Rio-Santos Highway Orangutan Jungle Park, Rio, Brazil (proposal)
C6 Cromarty Firth, Scotland, United Kingdom (proposal)
D4 Ronghua Dam, Taoyuan, Taiwan (proposal)
M12 Fongshan Wireless Communications Station, Taiwan (proposal)
M14 Typhoon class Nuclear Submarine (proposal)

瀕死｜Dying

該基礎設施目前已逐漸停用或退役中

It is gradually going obsolete and falling into disuse.

平地｜Above Ground

該基礎設施在地表面以上，例如：

B5　中國上海船廠 1862 藝術中心
D2　英國倫敦泰特現代美術館
M1　德國柏林飛船機庫室內熱帶度假村

It is located above ground level. Examples:

B5　Shipyard 1862, Shanghai, China
D2　Tate Modern, London, United Kingdom
M1　Tropical Islands Dome Resort, Krausnick, Germany

地下｜Underground

該基礎設施在地表面以下地下隧道，礦坑，地下碉堡，例如：

A3　英國倫敦地下隧道滑板萬斯之家
C5　中國上海佘山地下深坑酒店
D3　台灣高雄青埔垃圾發電景觀公園
B2　丹麥赫爾辛格海事博物館
M5　瑞典斯德哥爾摩核掩體數據資訊中心

It is an underground structure or located below ground level, like tunnels, mines and bunkers. Examples:

A3　House of Vans London, London, United Kingdom
C5　InterContinental Shanghai Wonderland, Shanghai, China
D3　Kaohsiung Metropolitan Park, Kaohsiung, Taiwan
B2　Maritime Museum of Denmark, Helsingør, Denmark
M5　Data Center Pionen White Mountain, Stockholm, Sweden

山間｜Mountains

該基礎設施在山間，例如：

A8　巴西里約叢林 Rio-Santos 高速公路未成段（設計提案）
D4　台灣桃園榮華壩（設計提案）

It is located in the mountains. Examples:

A8　Rio-Santos Highway Orangutan Jungle Park, Rio, Brazil (proposal)
D4　Ronghua Dam, Taoyuan, Taiwan (proposal)

可移動｜Mobile

該基礎設施是移動設施，尤其是軍事單位，例如：

B1　丹麥哥本哈根起重機飯店
M13　美國紐約無畏號航空母艦海空暨太空博物館
M14　蘇聯颱風級核潛艇（設計提案）

It is mobile, especially some military facilities. Examples:

B1　The Krane, Copenhagen, Denmark
M13　The Intrepid Sea, Air & Space Museum, New York, USA
M14　Typhoon class Nuclear Submarine (proposal)

水｜Body of Water

該基礎設施在水上，水底，水邊，港口設施，鑽油平台，軍事設施，例如：

B3　荷蘭阿姆斯特丹吊車梁辦公室
C4　馬來西亞西巴丹鑽油平台潛水旅館
M9　英國樸茨茅斯港海上堡壘旅館

it is on or below the water surface or near a body of water, like port facilities, rigs and military facilities. Examples:

B3　Kraanspoor, Amsterdam, The Netherlands
C4　Seaventures Dive Rig Resort, Sipadan, Malaysia
M9　Palmerston Sea Forts Hotel, Solent, United Kingdom

剩餘空間｜Urban Leftovers

該基礎設施在橋下或其他都市剩餘空間，此書中幾乎都在橋下，例如：

A1　法國巴黎綠蔭步道藝術高架橋
A4　荷蘭贊丹 A8ernA 高速公路公園
A5　日本東京秋葉原 2k540 商業藝廊
A6　瑞士蘇黎世翻新高架拱橋
A7　台灣台北環河南路五金街

it is located under a bridge or other kinds of leftover space. In this book, most examples that occupy urban leftovers are under bridges. Examples:

A1　Viaduc des Arts, Paris, France
A4　A8ernA, Zaanstad, the Netherlands
A5　2k540 Aki-Oka Artisan, Tokyo, Japan
A6　Refurbishment Viaduct Arches, Zurich, Switzerland
A7　Huanhe S. Rd Hardware street, Taipei, Taiwan

基礎工程剩下部分的形式，直接影響重生的空間型態
The form and shape of the infrastructure project that is to be reused. This directly affects the spatial feature of the reborn structure.

單棟建築：個體 | Single Building

該基礎設施為單棟建築，例如：
D2 英國倫敦泰特現代美術館
W2 中國瀋陽水塔展廊
M8 英國薩福克海岸拿破崙式海防塔樓住宅

It is a single building. Examples:
D2　Tate Modern, London, United Kingdom
W2　Water Tower Pavilion, Shenyang, China
M8　Martello Tower Y, Suffolk coast, United Kingdom

線性空間 | Linear Structure

該基礎設施屬於線性空間，多為橋梁上方空間利用，橋梁下方空間利用，地下隧道空間利用，例如：
A2 美國紐約高線公園
A3 英國倫敦地下隧道滑板萬斯之家
A5 日本東京秋葉原 2k540 商業藝廊

it is a linear structure, like the space above or under a bridge or a tunnel itself. Examples:
A2　High Line Park, New York City, USA
A3　House of Vans London, London, United Kingdom
A5　2k540 Aki-Oka Artisan, Tokyo, Japan

群體建築：聚落 | Multiple Buildings

該基礎設施為多棟建築，例如：
C1 西班牙巴塞隆納水泥廠辦公住宅
D1 德國卡爾卡爾核電廠仙境主題樂園
M12 台灣鳳山無線電信所（設計提案）

it belongs to a group of buildings. Examples:
C1　La Fábrica, Sant Just Desvern, Spain
D1　Wunderland Nuclear Reactor Amusement Park, Kalkar, Germany
M12　Fongshan Wireless Communications Station, Kaohsiung, Taiwan (proposal)

負空間：凹陷坑洞 | Depression

該基礎設施在地下或山裡的自然與人工形成的坑洞，例如：
C2 羅馬尼亞圖爾達地下鹽礦博物館
C3 英國康沃爾郡伊甸園植物園
M5 瑞典斯德哥爾摩核掩體數據資訊中心

it is a natural or artificial depression under the ground or in the mountains. Examples:
C2　Turda Salt Mine, Turda, Romania
C3　Eden Project, Cornwall, United Kingdom
M5　Data Center Pionen White Mountain, Stockholm, Sweden

機械裝置：物件 | Machinery

該基礎設施本身多由鋼鐵機械設備「可動」裝置構成，例如：
B1 丹麥哥本哈根起重機飯店
B4 荷蘭阿姆斯特丹法拉達起重機豪華酒店
M1 德國柏林飛船機庫室內熱帶度假村
M13 美國紐約無畏號航空母艦海空暨太空博物館

it is part of a machinery or equipment that is mobile. Examples:
B1　The Krane, Copenhagen, Denmark
B4　Faralda Crane Hotel, Amsterdam, the Netherlands
M1　Tropical Islands Dome Resort, Krausnick, Germany
M13　The Intrepid Sea, Air & Space Museum, New York, USA

筒形建築 | Cylinder

該基礎設施為單一或多個筒形構造，在本書中出現頻率很高，例如：
C1 西班牙巴塞隆納水泥廠辦公住宅
M11 台灣新竹建功國小大桶教室

it is a cylindrical structure or a group of cylinders. There are numerous examples in this book. Examples:
C1　La Fábrica, Sant Just Desvern, Spain
M11　The Oil Storage Classroom JianGong Primary School, Hsinchu, Taiwan

基礎工程原本的建造是為了服務的對象或它們的作用物，
例如運輸或儲存某種天然資源，運輸或產生能源，服務火車或汽車
The target the original infrastructure facility wishes to serve,
such as a certain natural resource (in its transport or storage), a type of energy (in its transport or production), or vehicles (trains or cars).

礦產天然資源 | Mine

服務標的為生產，儲存，或輸送礦產，例如：
C2 羅馬尼亞圖爾達地下鹽礦博物館
C3 英國康沃爾郡伊甸園植物園
C5 中國上海佘山地下深坑酒店

The infrastructure is used to produce, store or transport minerals. Examples:
C2 Turda Salt Mine, Turda, Romania
C3 Eden Project, Cornwall, United Kingdom
C5 InterContinental Shanghai Wonderland, Shanghai, China

軍事設施 | Military Facility

服務標的為軍事，例如：
M4 荷蘭屈倫博倫赫碉堡599水線紀念碑
M6 奧地利維也納高射炮塔水族館
M8 英國薩福克海岸拿破崙式海防塔樓住宅

It is used to fulfill military purposes. Examples:
M4 Bunker 599, Zijderveld, The Netherlands
M6 Haus des Meeres Flak Tower aquarium, Vienna, Austria
M8 Martello Tower Y, Suffolk coast, United Kingdom

能量 | Energy

服務標的為生產，儲存，或輸送能源，例如：
D1 德國卡爾卡爾核電廠仙境主題樂園
D2 英國倫敦泰特現代美術館
D3 台灣高雄青埔垃圾發電景觀公園

it is used to produce, store or transport energy. Examples:
D1 Wunderland Nuclear Reactor Amusement Park, Kalkar, Germany
D2 Tate Modern, London, United Kingdom
D3 Kaohsiung Metropolitan Park, Kaohsiung, Taiwan

運輸 | Transportation

服務標的為交通運具，多為火車或汽車，例如：
A2 美國紐約高線公園
A3 英國倫敦地下隧道滑板萬斯之家
A7 台灣台北環河南路五金街

It serves vehicles, mostly trains and cars. Examples:
A2 High Line Park, New York City, USA
A3 House of Vans London, London, United Kingdom
A7 Huanhe S. Road Hardware Street, Taipei, Taiwan

物品 | Product

服務標的為生產或儲存原料的半成品或成品，例如：
C1 西班牙巴塞隆納水泥廠辦公住宅：水泥
S1 丹麥哥本哈根 Frøsilos 集合住宅：大豆加工
S3 比利時韋訥海姆酒廠筒倉住宅：麥，釀酒原料
S6 南非開普敦非洲當代藝術博物館：糧食
S7 加拿大蒙特婁攀岩健身房 Allez-Up：糖加工
S8 中國上海民生碼頭八萬噸筒倉：糧食

it is used to produce or store semi-finished products or finished products. Examples:
C1 La Fábrica, Sant Just Desvern, Spain (cement)
S1 Frøsilos SILO apartment, Copenhagen, Denmark (soybeans)
S3 Kanaal, Wijnegem, Belgium (barley, for brewing)
S6 Zeitz MOCAA, Cape Town, South Africa (grains)
S7 Allez Up Rock Climbing Gym, Montreal, Canada (sugar)
S8 80,000-tonne Silo Warehouse, Shanghai, China (grains)

氣體 | Gas

服務標的為儲存或輸送氣體，例如：
S4 奧地利維也納瓦斯槽城市
S5 英國倫敦國王十字天然氣槽住宅

It is used to store or transport gas. Examples:
S4 Gasometer, Vienna, Austria
S5 King's Cross Gasholders, London, United Kingdom

液體 | Liquid

服務標的為生產，儲存，或輸送，多為水或燃料油，例如：
C4 馬來西亞西巴丹鑽油平台潛水旅館
D4 台灣桃園榮華壩（設計提案）
W1 荷蘭羊角村觀景塔
W3 比利時安特衛普森林水塔住宅
S2 芬蘭赫爾辛基 Silo 468 燈塔
M11 台灣新竹建功國小大桶教室

it is used to produce, store or transport liquid, mostly water and oil. Examples:
C4 Seaventures Dive Rig Resort, Sipadan, Malaysia
D4 Ronghua Dam, Taoyuan, Taiwan (proposal)
W1 Watch Tower Sint Jansklooster, Jansklooster, The Netherlands
W3 Water Tower Housing Brasschaat, Belgium
S2 Silo 468, Helsinki, Finland
M11 The Oil Storage Classroom JianGong Primary School, Hsinchu, Taiwan

6 個
時間屬性
6 Pieces of
Rebirth Information

A1

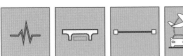

法國巴黎綠蔭
步道藝術高架橋

6 時間屬性 Rebirth Information

巴黎綠蔭步道（La Promenade Plantée 英譯 Green Walkway）大概是目前我們所能找到世界上最早的利用舊有高架鐵路線路改造成城市公共空間的案例。身為遺址再利用的先驅，故它當之無愧作為《重生之路》案例選集的第一例：A1。

位於巴黎市 12 區，4.5 公里長的空中花園，西起巴士底歌劇院，

選項一：拆，以便在街道上實施新的建築物。對公部門而言，拆掉應是最經濟與省事的方案，同時徹底解決南北區域被鐵路隔離的問題。但由於原本的鐵路高架橋寬度僅 12 公尺，拆掉後的地塊對於新建建築仍太過狹窄，難有效益。因此，加上其他各種考量之下，走選項二。

在時間上與它一生遭遇有關的6個演化時間軸上的「時間屬性」

a. 死亡原因
b. 重生利基
c. 重生週期
d. 原始機能
e. 改造手法
f. 重生物種

重生物種與死因是本書一大重點，基礎設施停用死亡，喪失其原設計目的主要原因：資金短缺、產業變遷、戰爭結束、政治影響、服務目標消失、天然資源耗盡，以及公安問題。

本書所列死因僅以我們所選出的54個案例為主，並不包含所有種類基礎設施的死因，尤其是年代更久遠的古代基礎設施死因。

例如：古羅馬人在西元一世紀時修築的位於法國南部普羅旺斯地區，跨越加東河谷，橋身高度49公尺的三層拱水道橋──嘉德水道橋輸水道，是古羅馬時期最高的橋樑。系統總長50公里，有17公尺的高度差。輸水道絕大部分位於地下，僅一小部分位於地面上，只有當橫跨河川時才以建造水道橋的方式供水。每天它由山區輸送流量4萬立方公尺的泉水至尼姆市內，供給5萬人口使用。然而因地質中富含石灰，逐漸在水道內部形成了厚厚的碳酸鈣沉澱，遭泉水水垢堵塞的輸水道供水能力逐漸減為每天流量1萬立方公尺。隨著羅馬帝國的衰亡與尼姆市的人口減少，城內既有的天然泉水足夠供應當時的城內需求，故逐漸取代了輸水道。嘉德輸水道的死因主要是當時的輸水道除垢技術的瓶頸，以及其他多項外部因素交互作用下的結果，難以總結出純粹的單一死因。由於至今仍保存的古代基礎設施多被各國列為文史資產，甚至世界遺產，所以對它們不可能有再利用計畫，自然也就無法在我們討論的範圍裡重生了。

基礎設施的死因多半為年代久遠，所以我們自己反而很容易去接受所能找到資料的說法或以結果推斷。然而相較於死因，當下重生的原因或許更重要，不過可惜的是，在現今對於大部分重生案例的介紹中，我們難以從公開的資料調查與理解其背後重生的原因，而對工業遺產的懷舊很少是其中的決定性關鍵，它們的背後包含了複雜的地產、經濟、政治等因素，最終呈現出來的是這些因素交互作用下錯綜複雜的結果。因此我們無法歸納出一套系統來解釋重生的原因，而是局部轉移至此討論到「物件／地點先決」這個簡單的二分法。況且，相較於重生原因，我們的精力更著重在重生的結果──重生物種。

6 Pieces of Rebirth Information

The six pieces of rebirth information regarding the life and evolution of the infrastructure projects are:
a. cause of death
b. decisive factor
c. years passed
d. before rebirth
e. purpose of rebirth
f. after rebirth

Among these, the two items with the greatest significance are the cause of death and the new "species" born after rebirth. The following are what may cause an infrastructure facility to go desolate and no longer serve its original purpose: a lack of funding, a change in the industry, the end of a war, a political issue, the disappearance of its target of service, the exhaustion of the natural resource, and a threat to public safety.

One thing must be clarified. The causes of death listed here are induced from the 54 case studies we selected for the composition of this book, which is to say, the list does not identify all possible causes of death of all types of infrastructure there are in the world, particularly those facilities from ancient times.

Take the Pont du Gard for instance. It is a three-tier aqueduct bridge the Romans built in southern France in the first century AD. The 49-meter high bridge that crosses the Gardon River is the highest of its kind. The whole aqueduct is 50 km long and it descends in height by 17 m. The majority of the aqueduct is underground while only a small portion is above ground to cross the river in the form of a bridge. Every day, it carried 40,000 cubic meters of spring water from the mountains to the 50,000 citizens of Nîmes. However, because the water is rich in calcium carbonate, a thick layer of deposit slowly formed inside the channel, reducing the daily water supply down to 10,000 cubic meters. Upon the collapse of the Roman Empire and the following decline in Nîmes' population, the natural springs within the city proper became sufficient to meet the demand and replaced the aqueduct. Therefore, the main cause of death of the Pont du Gard and the aqueduct is not having the technologies to remove the deposit at that time, but along with many other factors, it is hard to determine the exact cause. Since most of extant infrastructure facilities from ancient times have been designated as national heritage sites or even as world heritage sites, it is impossible to reuse them and naturally they are not included in our discussion.

Most of the time, the death in question can be traced back to a long time ago, so we tend to be quick to accept what is stated in the document or to deduce the cause of death from the result. However, it is not so with a more important piece of information – the decisive factor in its rebirth. Unfortunately, it is very difficult to inquire into the reason behind the rebirth of most of the reborn facilities discussed in this book from public sources and data. It is worth noting that nostalgia for industrial relics is seldom the decisive factor in the rebirth; it is often a result of much more complex factors put together, factors related to the real estate market, economy, and politics. Thus, I have difficulty developing a satisfactory system to explain the reasons behind those rebirths. As a compromise, I resort to a simple dichotomy, grouping the factors according to whether they are "object-oriented" or "location-oriented." Having said that, I am still going to put the most focus on the result of the rebirth – the new species that is reborn.

資金短缺 | Lack of Funding

蓋到一半沒錢繼續蓋，或營運到一半沒錢繼續營運下去只好關門收店。例如：
A8 巴西里約叢林 Rio-Santos 高速公路未成段（設計提案）

When financial issues appear during construction or operation. Examples:
A8 Rio-Santos Highway Orangutan Jungle Park, Rio, Brazil (proposal)

產業變遷 | Change in the Industry

產業因時代改變而退出歷史舞台，其相關基礎設施也就變得無用。
發電方式改變，供暖方式改變，例如：
D2 英國倫敦泰特現代美術館
S4 奧地利維也納瓦斯槽城市
S5 英國倫敦國王十字天然氣槽住宅

When there is a change in part of the industry – such as a change in the way electricity is generated or heat is provided – in the wake of a new era and the relevant infrastructure becomes useless. Examples:
D2 Tate Modern, London, United Kingdom
S4 Gasometer, Vienna, Austria
S5 King's Cross Gasholders, London, United Kingdom

服務目標消失 | Absence of Target

該基礎設施服務目標消失。例如不用的倉庫，鐵路改道或停駛，航班太少關門的機場。例如：
A3 英國倫敦地下隧道滑板萬斯之家
B7 美國紐約甘迺迪機場 TWA 飛行中心飯店
S1 丹麥哥本哈根 Frøsilos 集合住宅
S3 比利時韋訥海姆酒廠筒倉住宅

When the target of service disappears. This can be a warehouse no longer in use, a railway no trains are using because of the existence of an alternative route, or an airport that has to be closed down because there are just too few flights. Examples:
A3 House of Vans London, London, United Kingdom
B7 TWA Flight Center Hotel, New York, USA
S1 Frøsilos SILO apartment, Copenhagen, Denmark
S3 Kanaal, Wijnegem, Belgium

政治影響 | Political Issue

民眾抗議基礎設施造成的汙染或景觀問題等，或為選舉政見被政府下令停工或關閉。例如被迫停工的核電廠，政府審核通過也進行試運營但最後因為抗議，政府決定放棄這個電廠，設施服務的需求還在，卻因為公部門的決策而放棄，例如：
D1 德國卡爾卡爾核電廠仙境主題樂園

When citizens protest against construction of the facility due to potential pollution issues or blocking of view, or when during election the government is pressured into having the construction suspended (for example, nuclear plants that are shut down), or when the construction has been completed and started operation but the government decides to give it up in the end due to public opinions. Examples:
D1 Wunderland Nuclear Reactor Amusement Park, Kalkar, Germany

天然資源耗盡 | Exhaustion of Natural Resource

礦挖完了油不開採了，或存量不足造成經濟規模難以支持繼續開採。例如：
C2 羅馬尼亞圖爾達地下鹽礦博物館
C3 英國康沃爾郡伊甸園植物園
C5 中國上海佘山地下深坑酒店

When there is no more mineral to be extracted from a mine or no more petroleum to be pumped from an oil well, or when what remains is just too little to be of any profit. Examples:
C2 Turda Salt Mine, Turda, Romania
C3 Eden Project, Cornwall, United Kingdom
C5 InterContinental Shanghai Wonderland, Shanghai, China

公共安全問題 | Public Safety

礦開採過於危險，意外頻傳，不得不關閉結束開採。或是基礎設施年久失修過於危險而放棄繼續運作。

When working in the mine becomes too much of a hazardous venture and there are too many accidents, so that the extraction has to be stopped, or when the infrastructure falls into disrepair and becomes too dangerous to operate.

戰爭結束 | End of War

這世界沒有戰爭了，那軍事設施也就無用武之地，或是軍事科技進步只有退役一途，願它們都重生為和平用途，例如：
M系列

When there are no more wars and the military facilities become useless or there is no need to encourage advance in military technology. May all of these facilities be reused in a way to promote world peace! Examples:
The M series

共生 | Coexistence

不是死亡，反而共生。對「正在使用」狀態的基礎設施其所造成的剩餘空間加以利用，書中只有橋下空間的案例中會出現這種「共生」的狀態，例如：
A4 荷蘭贊丹 A8ernA 高速公路公園
A5 日本東京秋葉原 2k540 商業藝廊
A6 瑞士蘇黎世翻新高架拱橋
A7 台灣台北環河南路五金街

When the infrastructure is not dead but actually quite alive and coexists with a new structure taking up its leftover space. This usually refers to the space under a bridge. Examples:
A4 A8ernA, Zaanstad, the Netherlands
A5 2k540 Aki-Oka Artisan, Tokyo, Japan
A6 Refurbishment Viaduct Arches, Zurich, Switzerland
A7 Huanhe S. Road Hardware Street, Taipei, Taiwan

對工業遺產的懷舊難以成為重生決定性原因，而是背後複雜的地產、經濟、政治等因素交互作用的結果，
因此我們對重生原因並沒有歸納出一套系統，而是局部轉移到討論「物件／地點先決」這個簡單的二分法並配合「都市縫合」的類別進行討論。

As mentioned above, nostalgia for industrial relics rarely is a decisive factor in the rebirth; rather, it is often a result of much more complex factors put together,
including the real estate market, economy, and politics. Thus, I decided against a system to explain the reasons behind those rebirths.
As a compromise, I resort to a simple dichotomy, grouping the factors according to whether they are "object-oriented" or "location-oriented." I also created another distinctive category: urban stitching.

物件先決（建築先決）｜ Object-oriented

因為建築物本身的特殊或趣味性或具歷史重要性所以被進行重生改造，重生前後被移動到新基地作為項目大亮點的，例如：
B1 丹麥哥本哈根起重機飯店
C4 馬來西亞西巴丹鑽油平台潛水旅館
W2 中國瀋陽水塔展廊
或是特殊的建築改造潛力者，多為住宅或旅館例如：
W4 德國布蘭登堡水塔住宅
M8 英國薩福克海岸拿破崙式海防塔樓住宅
M9 英國樸茨茅斯港海上堡壘旅館

The infrastructure facility is renovated because it is special or interesting in some aspect or has historical importance. Sometimes, this kind of structures are moved to another place and become places of interest, such as:
B1 The Krane, Copenhagen, Denmark
C4 Seaventures Dive Rig Resort, Sipadan, Malaysia
W2 Water Tower Pavilion (Renovation), Shenyang, China
Sometimes, they have great potential to be converted to a certain type of buildings, mostly private houses or hotels, such as:
W4 BIORAMA-Projekt, Brandenburg, Germany
M8 Martello Tower Y, Suffolk coast, United Kingdom
M9 Palmerston Sea Forts Hotel, Solent, United Kingdom

地點先決（環境先決）｜ Location-oriented

因為建築物所處的位置由於社會變遷、城市發展，所以過去不重要的邊陲地點，城外變成了城內，現在變成有價值的精華地點，所以被進行重生改造，例如：
B6 德國漢堡易北愛樂廳
D2 英國倫敦泰特現代美術館
S5 英國倫敦國王十字天然氣槽住宅
S8 中國上海民生碼頭八萬噸筒倉

the infrastructure facility is renovated because the area which used to be marginal and unimportant is turned into an expensive neighborhood. Examples:
B6 Elbe Philharmonic Hall, Hamburg, Germany
D2 Tate Modern, London, United Kingdom
S5 King's Cross Gasholders, London, United Kingdom
S8 80,000-tonne Silo Warehouse, Shanghai, China

都市縫合｜ Urban Stitching

因高架橋造成都市空間的割裂，對無論是「已經死亡」或「正在使用」的高架橋下的空間加以利用以縫合割裂，例如：
A1 法國巴黎綠蔭步道藝術高架橋
A4 荷蘭贊丹 A8ernA 高速公路公園
A5 日本東京秋葉原 2k540 商業藝廊
A6 瑞士蘇黎世翻新高架拱橋

Elevated highways can impair the wholeness of urban space. In order to reverse the trend and "stitch" the cut, the space under elevated bridges (no matter they are out of use or still being used) can be put into good use. Examples:
A1 Viaduc des Arts, Paris, France
A4 A8ernA, Zaanstad, the Netherlands
A5 2k540 Aki-Oka Artisan, Tokyo, Japan
A6 Refurbishment Viaduct Arches, Zurich, Switzerland

演化周期 / 重生時間｜ Years Passed

已重生者才有所謂演化或重生時間，花多少週期從第一次死亡（停用）到第一次再利用，有第二次再利用就有兩個時間週期。

The length of time for the infrastructure to be reborn. Needless to say, this piece of information is only available when the infrastructure in question has undergone rebirth. The number of years passed is calculated from its death to its rebirth; if it has been reborn twice, there will be two numbers.

高架橋／路｜Elevated Highway/Roadway

一種以高架結構支撐，突出地面的橋。使汽機車與火車可以利用橋面上的車道或軌道快速便利跨越地形障礙如山谷、河川或越過其他道路，特別是在都市區域。

An elevated bridge structure that is raised above ground to enable cars or trains to make crossings over obstacles like valleys, rivers or surface roads (particularly in urban areas).

地下鐵｜Subway

鐵路運輸的一種形式，指在地下運行為主的城市軌道交通系統，本書縮小定義範圍指的就是「專有路權的地下鐵道」，與其他運具線路無平交，不限定主要以電力驅動的軌道交通，但必為地下隧道。

A form of railway transportation that is mainly underground and in urban areas. In this book, I give out a stricter definition: an underground railroad with right-of-way that does not intersect with other roads and is not necessarily powered by electricity.

礦場｜Mine

地底下或地表面開採有經濟價值的礦物或其他物質集中的礦床，例如鑽石、煤、石灰石、岩鹽、鐵、各類金屬、稀有金屬等人類無法透過農業而取得的原始物質。

An excavation in an underground tunnel or on the ground surface that is used to extract profitable minerals and materials, such as diamond, coal, lime, rock salt, iron and other metals, that cannot be grown or harvested through agriculture.

鑽油平台｜Oil Rig

位於海上用於鑽井提取石油和天然氣的大型設施，平台包含開採設施，暫時儲存，以及容納勞工的海上居住。石油、天然氣的開採也能算作採礦業範疇，但鑽油平台因其特殊的工程構造被本書獨立出來。

A large structure located offshore that is used to extract petroleum and natural gas. On the platform, there are facilities for drilling and temporary storing, and living quarters for the workers. Although the extraction of petroleum and natural gas is also a type of mining, here I separate it from the other category due to the special form of the rig.

發電廠｜Power Station

將熱能或動能轉換為電能的設施，是電力供應的核心部分。根據能量來源區分，人類目前有過核能，火力 ，水力，風力，地熱，太陽光電，太陽能加熱發電廠。

A facility that converts thermal energy or kinetic energy into electric energy, forming the core of power supply. Depending on the source of energy, there are nuclear, fossil fuel, hydroelectric, wind, geothermal, photovoltaic, and concentrated solar power stations.

起重機｜Crane

指用吊鉤或其他取物裝置吊掛重物，一定範圍內垂直提升和水平搬運重物的循環性作業起重機械，工作特點是做間歇性運動。

A type of machine that uses a hook to vertically lift up heavy objects and horizontally move them to the designated place. Intermittent motion is one of its characteristics.

造船廠｜Shipyard

從事建造及修理船隻的工廠。包括建造船體用的船台、船塢及艤裝工廠、碼頭等相關設施。

A place where ships are built and repaired. This includes the building slips, the docks and the outfitting factories.

軍事設施｜Military Facility

為維護軍事區域其硬體設施所設定的範圍或裝備，都屬於軍事設施。塑造作戰環境的工程活動，整合了對機動和整體力量的支持，包括支持的部隊防護、搜索、反爆炸、環境保護，只要是支援軍事用途的建築物與構造物都算。

This includes equipment that protects the hardware of a military zone, constructions that supports mobilization and military efforts (force protection, scouting, anti-explosion, environmental protection, etc.), and all the buildings and structures that can fulfill military purposes.

儲存設施｜Storage Facility

儲存一般物品，大多靠近港口，讓貨物進行臨時或短期存放保管的建築物。其主要作用是便利貨物貯存、集運、加速車船周轉。例如貯存散裝穀物的建築物。穀物在倉內可以進行淨化、乾燥、灌包、計量和裝車等工作。穀倉結構形式主要有兩種：樓層式和圓筒式，而港口多圓筒式。其他資源的儲存設施有地表上架高的儲水塔，也有儲存易燃、易爆的天然氣能源與儲存原油及石化產品的建築物。

For the storage of common goods, facilities are usually near the port. They are buildings that store goods temporarily or in the short term, and they can fulfill the functions of convenient storage, mass transport, and quick circulation of ships and trucks. Take grain storage buildings for example. In a building like this, the grains are cleansed, dried, bagged, counted and loaded. There are two forms of grain storage buildings: the multi-story form and the cylindrical form. Those near the port are usually of the latter. Storage facilities that store other stuff include water towers elevated high above the ground and oil depots that store combustible natural gas, petroleum and other petrochemical products.

單純只做建築單體改造，將基礎工程改造為博物館，注重歷史價值的保存和重現，將基礎工程改造為工業博物館、文化藝術展區等。例如：

W 系列的改造為水塔住宅

C1　西班牙巴塞隆納水泥廠辦公住宅
C4　馬來西亞西巴丹鑽油平台潛水旅館
M9　英國樸茨茅斯港海上堡壘旅館

One of the strategies is to preserve and reconstruct the historical importance of the infrastructure; for instance, to convert an infrastructure facility to an industrial museum or an art and culture park. Examples:

The W Series
C1　La Fábrica, Sant Just Desvern, Spain
C4　Seaventures Dive Rig Resort, Sipadan, Malaysia
M9　Palmerston Sea Forts Hotel, Solent, United Kingdom

擴大範圍到整體都市設計。借助基礎工程的物件本身外部造型提供多元化的設計和改造，進行空間重塑和意義重置。例如：

A2　美國紐約高線公園
D2　英國倫敦泰特現代美術館
S8　中國上海民生碼頭八萬噸筒倉

The third strategy is to utilize the exterior shape of the facility to design and renovate it in a diversity of ways, ultimately reshaping the space and giving it new meanings. This elevates the rebirth project to the level of urban planning. Examples:

A2　High Line Park, New York City, USA
D2　Tate Modern, London, United Kingdom
S8　80,000-tonne Silo Warehouse, Shanghai, China

景觀開放空間，公共領域的打造。注重景觀營造和重塑，多結合基礎工程本身的建構和周圍的生態環境，改造為一種城市公園和生態景觀。例如：

C3　英國康沃爾郡伊甸園植物園
C5　中國上海佘山地下深坑酒店
D3　台灣高雄青埔垃圾發電景觀公園

Another strategy focuses on landscaping and reshaping the environment. Most of the time, the original structure of the facility is integrated with the nearby ecosystem to convert it to an urban park or visitor attraction. Examples:

C3　Eden Project, Cornwall, United Kingdom
C5　InterContinental Shanghai Wonderland, Shanghai, China
D3　Kaohsiung Metropolitan Park, Kaohsiung, Taiwan

綜合的設計，對於人工或天然形成的地下空間，或對具備移動能力的軍事武器。例如：

A3　英國倫敦地下隧道滑板萬斯之家
C2　羅馬尼亞圖爾達地下鹽礦博物館
M5　瑞典斯德哥爾摩核掩體數據資訊中心
M14 蘇聯颱風級核潛艇（設計提案）

Finally, there is a comprehensive strategy that mixes all of or only two of the above three; for instance, recently a few infrastructure facilities were converted to housing units. Examples:

A3　House of Vans London, London, United Kingdom
C2　Turda Salt Mine, Turda, Romania
M5　Data Center Pionen White Mountain, Stockholm, Sweden
M14 Typhoon class Nuclear Submarine (proposal)

基礎工程死亡（停用）到目前由於符合當代再利用計畫而重生創造出來的新物種，基礎工程與建築物的混合體，既不再是基礎工程也不完全是建築物，透過再一次的人擇而產生，與最初基礎工程建造目的迥然不同的當代功能使用，視為基礎工程的演化。

The new species born from a dead infrastructure facility that has found a new use and purpose. It is a hybrid of infrastructure and architecture, but does not belong to either one of them. It is reborn by artificial selection, rather than natural selection, and its function is entirely different from that of the original species.

開放空間｜Public Open Space

開放意味著必須對公眾開放，供市民休閒；是戶外而非室內的並包含植被。都會公園、廣場、綠園道、人行道都可以算是開放空間。

The word "open" implies that it is open for public access, so that people can go there in their pastime. It is outdoor rather than indoor and is vegetated. Parks, squares, greenways and promenades are all part of the public open space.

眺望台｜Observation Deck

高層建築上用作觀賞景致的平台。周圍一般有欄杆防護，常伴隨著付費望遠鏡。

A sightseeing platform situated upon a tall building. There are usually protective railings around it and coin-operated telescopes.

辦公室｜Office Building

一種讓人們在其中工作的場所，通常是房間型態。建築內部由辦公室所組成的出租或銷售的商業建築則稱為辦公大樓。

An office is a place where people can work. It is usually a room. Commercial buildings that rent out or sell offices are called office buildings.

展覽空間／博物館｜Gallery / Museum

一個非營利性、對外開放、永久經營的機構，以服務人群、促進社會發展為宗旨，主要從事蒐集、保存、研究、傳播、展示等活動。

A permanent non-profit institution that opens to the public and the goal of which is to serve the general public and promote developments of the society. Specifically, the purpose of a gallery or museum is to collect, preserve, study, and exhibit items of importance.

遊樂園｜Theme Park

以某個主題為核心而建立的人造公園。「主題」意味由一個或幾個相關主題所主導，再配合不同的人工設計景觀和設施，讓遊客體驗到主題感覺。從旅遊業、服務業角度，主題樂園是提供旅遊體驗的旅遊產品，並且能帶動周邊相關餐飲或者購物設施的經濟。

A type of amusement park that centers on a specific theme. There can be one or more themes that are related to each other, with various man-made designs and facilities to create the expected atmosphere. For the tourism and service industry, it is a tourism product that provides a unique experience and can help boost the economy of surrounding areas.

運動設施｜Sport Facility

室內室外提供人參與活動的空間或場地，如各種球類運動場館、田徑場、游泳池、射箭場、射擊場、攀岩場、划船場、極限運動場、溜冰場等及其附屬設施。

An indoor or outdoor space for people to engage in sports activities. This can be a general gymnasium, a track and field stadium, a swimming pool, an archery range, a shooting range, a rock-climbing gym, a rowing club, an extreme sports facility, an ice rink and so on.

住宅｜Private Home

只有一戶家戶為單位的獨立建築物。

A building that houses only one family.

集合住宅｜Multi-dwelling Unit

有多個住宅單元，具有共同基地和共同空間或設備，或共同持分土地產權，設施公共的多戶建築物。

A building with multiple housing units where there are public space and common facilities for all residents to use. Sometimes, residents share the ownership of the property.

旅館 | Hotel

有多個居住單元，通常有共用的公共空間或餐廳，以提供非居民住宿空間及服務為主要目的的建築物。

An establishment with multiple lodging rooms and usually one or more public areas and restaurants for people who are not locals to have a place to stay.

資訊中心 | Data Center

網路時代，用於安置電腦系統設施、儲存並傳遞資訊的空間。在全球協同運作的特定裝置網路中，有在網路基礎設施上傳遞、加速、展示、計算、儲存資料和資訊的功能。

A building used to house computer systems and associated equipment and to transfer, speed up the transmission of, display, calculate and store data and information.

倉儲 | Warehouse

工業時代大量生產的商品或大量開採之資源暫時儲放的地方，常位於交通運輸設施或是商品／資源生產地旁。

A building that temporarily stores manufactured goods or extracted resources. It is usually located somewhere with easy access to transportation or the raw material.

學校 | School

一個容納以教育為主要目的之機構／組織的空間，為教師及學生提供學習環境，使學生能系統化的增進智識和價值體系。

An educational institution that provides a learning environment for the teaching of students, with the aim of systematically enhancing their level of knowledge.

購物中心 | Shopping Mall

是現代城市或郊區市鎮中常見的建築物，是結合購物、休閒、文化、娛樂、飲食、展示及資訊等設施於一體的商業設施。或著說：給人花錢的地方。

A type of building very common in modern cities and towns which integrates shopping, leisure, culture, entertainment, food, display and information; a place where one spends money!

發電廠 | Power Station

自工業革命以降，電力成為民生必需品而且隨科技進步需求持續擴張。發電廠係為以火力、水力、核能、再生能源等方式，產生電力的地方。

Since the Industrial Evolution, electricity has become an essential commodity, the demand of which ever increasing due to the advancement of technology. A power station is a facility where electric power is generated by means of thermal, hydro-, nuclear power or renewable energy.

動物保育所 | Animal Shelter

是專門收留受傷、棲地受破壞的動物，與安置走失、被遺棄或流浪動物的場所。

A place that shelters wounded, lost, abandoned or stray animals, or animals whose habitat is destroyed.

紀念碑 | Monument

一種紀念性建築物，用以紀念人或事。

A structure erected to commemorate a person or event.

水族館 | Aquarium

收集和繁殖生活在水中的各種動物。設有大規模水池、水箱，供人觀賞的地方。

A place where all kinds of aquatic animals are collected and bred, with large tanks and pools to display the creatures for public viewing.

電影院 | Cinema

一個提供大眾觀影、播映電影的場所。電影大多由放映機投射在觀眾席前的銀幕上，故電影院通常是室內空間。

A place where movies are played on a screen at the front of the auditorium. It is usually indoor.

海上城市 | Floating Town

城的發展形式與人類聚落型態的一種，在海上建設能自給自足的城市，提供就業與居住，是提供人類生活型態的可能途徑。

A plausible type of self-sufficient settlement where the inhabitants live and work on the sea.

疫苗研究室 | Vaccine Laboratory

感染症是威脅人類健康的疾病之一，而透過對疫苗接種是預防感染症最直接有效的方法，並能降低醫療成本。因其研究性質的高危險性，研究疫苗的空間與選址需要與外界有高度的隔離。

Infectious diseases threaten the existence of human beings. One of the most effective ways to prevent infection and to lower healthcare cost is vaccination. However, due to the inherent hazard of vaccine development, the laboratory is often located in a highly isolated area.

表演廳 | Theater

特定的、通常是永久性的建築構造圍塑出之室內／半戶外表演場所。大多分為兩個區域，一為表演者進行展演的區域，另一為觀者觀看之區域。

A dedicated, usually permanent place where performances are produced. It can be indoor or semi-outdoor and is divided into two main areas, one for the performers and the other for the audience.

溫室 | Greenhouse

專用作種植植物的建築物。可透光的建造物料是玻璃或塑料，溫室會因太陽發出的電磁輻射而加熱，使溫室內的植物、泥土、空氣等變暖，因為可以提早種植也比較不受氣候影響，在亞熱帶和溫帶國家很流行，在乾燥地區還有防止水分過度蒸發的效果。溫室的興建就是可以抓住電磁輻射和防止對流現象。

A structure that is built specifically for the purpose of growing plants. It is built with glass, plastics or other transparent material, so that the air, the soil and the plants in the interior all warm up via solar radiation. For this reason, the plants can be planted sooner or later in the year, disregarding the weather. Greenhouses are very common in subtropical and temperate regions, and can also prevent water loss through evaporation in dry places. Overall, the construction of greenhouses is to aid radiation and prevent convection.

3

重生案例紀錄

The Collection of Reborn Infrastructure

書中很多內容並非所謂的研究，而是資訊的組織方式。

資訊也有它的形狀。同樣的資訊，就跟水一樣，在不同容器中，可以有不同的形狀。當然，形式上，它們更可以跟水一樣產生——「相變」。

編輯本書的過程中，更多時候我們在做的是調整資訊的組織方式、整合方式與轉換方式。一直以來無論《寄生之廟》或《重生之路》，真正優先的核心重點並不是把我們搜集到的內容去做更深層次的提煉與消化去，而是「演繹」與「再詮釋」，研究是附帶的成果，「必要」卻不是最重要。

把一段文字變成另外一種說法的文字，
把一段文字變成一個符號，
把一段文字變成圖。

統計學從數學分離出來，走上獨立發展的道路；建築學從藝術分離出來，走上獨立發展的道路；都市學從景觀分離出來，也走上了獨立發展的道路。圖學做為上述各項專業的工具，早已獨立發展分離出來。這些案例的等角圖與剖透圖可以視為一種建築與都市與環境再現的形式。《寄生之廟》或《重生之路》這兩本書原本都可以開啟一個章節，對建築與都市圖學本身展開討論，不過這類偉大著作已經太多，此不贅述。

就算只是單純地、深度無改變地把資訊從一種形式轉換成另外一種形式，便已產生「溝通」與「被解讀」的不同，而無論是哪一種結果，都算是資訊的演繹與再詮釋。內容本身題材固然重要，然而當今時代中能夠決定資訊傳播影響力的，卻是「呈現內容的形式」。

雖然資訊的形式轉換工作並非限定於有建築或都市背景的人，但唯獨此背景的人偏向挑選這個主題，並透過書中的方式來呈現。這種翻譯資訊的方式和表達方式，雖可由工程師或別的專業領域來處理，但轉換結果必然與我們不同。

精緻殘骸的發現，
一場工程師與建築師的世紀接力

發掘並收集那些「非建築師主導設計而完成」的基礎設施再利用的案例與紀錄，包括高架鐵路、高速公路、地下隧道、軍事掩體、礦坑、筒倉、發電廠、鑽油平台、垃圾掩埋場，乃至核潛艦等。對廢棄基礎設施再利用的模式，目前都為實體空間再利用，源自於以下三種情形：

1

對「已停止使用」、「未完工」、「未啟用」狀態的基礎設施殘骸本身加以利用。

Most of the content in this book cannot be called a study; rather, it is just a representation of the way I organize the information.

Information can be turned into a certain shape. Like liquid, the same piece of information can assume different shapes depending on what container you put it into. And, like liquid, information also experiences phase transition.

Most of the time while compiling this book, I was only fine-tuning the way the information was organized, integrated and converted. The truth is, no matter we are talking about *Parasitic Temples* or *The Road to Rebirth*, the whole point is never to analyze what I gathered and bring it to another level, but to deduce and interpret it. The so-called study is only one of the outcomes; it is inevitable but not of the utmost importance.

What I have done is:
To turn a group of words into words expressed in another way.
To turn a group of words into a set of symbols.
To turn a group of words into drawings.

Statistics branched out from mathematics and became a separate field of study; architecture branched out from art and became a separate field of study; urban planning branched out from landscaping and became a separate field of study. As for drafting, it had become a separate discipline long before the others, being the essential technique of all the above fields. The isometric drawings (sometimes with sectional perspectives) of the case studies in this book can be seen as a representation of the architecture, the city, as well as the surrounding environment. For *Parasitic Temples* and *The Road to Rebirth*, I could have written a whole chapter to discuss the art and science of drafting. In the end, I decided against this, since there were already numerous treatises on the subject.

However, even though it is only a matter of simply changing the "shape" of the information without deeper probing, a different dialogue has been presented. In other words, the information has been reinterpreted. As important as the subject matter is, it is depending on how it is presented that the information can wield any amount of influence nowadays.

Of course, this task of reshaping information about infrastructure can be performed by those who don't know much about architecture or urban planning. However, only those who do know something about these fields tend to choose this topic and present the information the way I do. Other professionals like the engineers can also interpret and present the same information, but the end product will not be the same.

The Discovery of These Marvelous Ruins

In this book, I have discovered and collected case studies of infrastructure facilities that were designed and built by non-architects and then reused at a later time, from elevated railways, elevated highways, underground tunnels, military facilities, and mines, to silos, power stations, rigs, landfills, and even nuclear submarines. So far, reuse of infrastructure facilities is limited to the tangible kind, and these reborn projects can be grouped into three categories:

1

The rebirth of infrastructure facilities that had been derelict, that had not finished construction at all, and that had finished construction but never been used. The majority of the cases studied in this book belong

書中各系列的案例多為此情形，例如：
「已停止使用」——
C2　羅馬尼亞圖爾達地下鹽礦博物館；
「未完工」——
A8　巴西里約叢林 Rio-Santos 高速公路未成段；
「未啟用」——
D1　德國卡爾卡爾核電廠仙境主題樂園。

2

對「正在使用」狀態的基礎設施之剩餘空間加以利用。例如 A 系列橋梁的案例中，在橋下空間出現此種「共生」狀態——
A4　荷蘭贊丹 A8ernA 高速公路公園；
A5　日本東京秋葉原 2k540 商業藝廊；
A6　瑞士蘇黎世翻新高架拱橋；
A7　台灣台北環河南路五金街。

3

對「已退役的軍事設施與軍事單位」加以利用。軍事單位退役後的命運不是只有出售拆解，現實中總找得到實案且不斷有公開的徵選再利用方案，只不過更多是停留在想像中，例如：

「已退役軍事設施」——
M1　德國柏林飛船機庫室內熱帶度假村；
M9　英國樸茨茅斯港海上堡壘旅館。
「已退役軍事單位」——
M13　美國紐約無畏號航空母艦海空暨太空博物館；
M14　蘇聯颱風級核潛艇。

雖然很期待有廢棄太空站的再利用案例，但似乎對人類而言，目前還沒有再利用效益。前面提過本書中的基礎設施重生案例中，我們並未把文化與休閒建設列入基礎設施的範圍，美術館、博物館乃至圖書館等文化建築雖對城市至關重要，但在認定上，我們偏向選擇那些維持城市基本運作與基本維生功能，與能構成這些系統關鍵部分的建築。

然而我們篩選案例的標準是什麼？很難給一個全面性的標準。以電影選角為例，即使在同一部電影中的不同角色，選角的標準會隨著角色改變而改變，無法用一個統一標準去選擇所有的角色，對於角色 A 會有一個選角標準，而角色 B 則是另一個標準。選擇的標準是無常且不斷變動的。

另外，重生成果的精彩，或多或少也影響了我們的選擇，我們永遠無法保持客觀性，也不否認自己的選擇多少帶著戀物癖的情懷。

書中的 54 個精選案例可以歸納成下面七個系列：

A系列——交通（9 個案例）

B系列——港口（7 個案例）

C系列——資源開採（6 個案例）

D系列——供電（4 個案例）

W系列——供水（6 個案例）

S系列——倉儲（8 個案例）

M系列——軍事（14 個案例）

這七大類 54 個案例並未完全涵蓋所有的基礎設施，但大多是由我們所找到的 300 多個「再利用案例」分類歸納而來。像七種不同類型的電影，或是同一部電影中不同的七個角色，它們有各自獨立的選擇或取捨標準。若要總結這七類中的共同性或通則，那就是該案例是否開啟了再利用內容與功能上的另一個可能性。

---

to this category. For example:
Those that had been derelict –
C2 Turda Salt Mine, Turda, Romania
Those that had not finished construction –
A8 Rio-Santos Highway Orangutan Jungle Park, Rio, Brazil (proposal)
Those that had never been used –
D1 Wunderland Nuclear Reactor Amusement Park, Kalkar, Germany

2

The renovation of infrastructure facilities that are still in use. Type A contains good examples where the space underneath a bridge coexists with the infrastructure itself. Examples:
A4 A8ernA, Zaanstad, the Netherlands
A5 2k540 Aki-Oka Artisan, Tokyo, Japan
A6 Refurbishment Viaduct Arches, Zurich, Switzerland
A7 Huanhe S. Road Hardware Street, Taipei, Taiwan

3

The rebirth of military facilities and vessels. After a military facility or vessel fulfills its duty, it doesn't have to be sold out or torn apart. The reality is, there are quite a few cases when a military facility or vessel is reused and once in a while we also hear about renovation projects calling for candidates to carry out the task, although sometimes they are only proposals, waiting to be realized.
The reuse of military facilities –
M1 Tropical Islands Dome Resort, Krausnick, Germany
M9 Palmerston Sea Forts Hotel, Solent, United Kingdom
The reuse of military vessels –

M13 The Intrepid Sea, Air & Space Museum, New York, USA
M14 Typhoon-class Submarine (proposal)

Although I am looking forward to the rebirth of space stations, currently these facilities will not benefit us if reused. I have mentioned above that I didn't include in this book the type of infrastructure facilities that have cultural or leisure values, such as museums and libraries. These buildings are without doubt crucial to a city, but I decided to pick only the systems or their key parts that support the basic operation and function of a city.

Nevertheless, if you ask me about the criteria I use to choose the infrastructure facilities, I will have a hard time giving an absolute answer. It is like choosing the cast for a movie. In the same movie, the director will have different criteria for different characters; it is impossible to have an absolute standard when choosing actors and actresses. For character A, there is one set of requirements; for character B, there is another one. The criteria are unfixed and ever-changing.

Admittedly, I am more or less influenced by how ingenious the facilities turned out to be; after all, we can never be entirely objective or rational. I will not deny that my choice does reflect to a certain degree my obsession with obsolete infrastructure works.

The 54 case studies are divided into seven types, as follows:

A——Transportation (9)
B——Port Facilities (7)
C——Mines (6)
D——Power Supply (4)
W——Water (6)
S——Storage (8)
M——Military Structures (14)

These types by no means cover all infrastructure buildings in the world; they are only a classification of over three hundred examples I have found around the world. Like seven types of movies, or seven characters in the same movie, each category (there are six plus one) has its own set of criteria. If there is one general principle for all the seven types, that would be –

The new possibilities they open up regarding the nature and function of infrastructure.

A

交通 Transportation

A0
日本東京首都
高速公路市場
（設計提案）
**Tokyo Highway Market,
Tokyo, Japan** (Proposal)

A1
法國巴黎綠蔭
步道藝術高架橋
**Viaduc des Arts,
Paris, France**

A2
美國紐約高線公園
**New York High Line Park,
New York, USA**

A3
英國倫敦地下隧道
滑板萬斯之家
**House of Vans London,
London, United Kingdom**

A4
荷蘭贊丹 A8ernA
高速公路公園
**A8ernA, Zaanstad,
the Netherlands**

A5
日本東京秋葉原
2k540商業藝廊
2k540 Aki-Oka Artisan, Tokyo, Japan

A6
瑞士蘇黎世
翻新高架拱橋
Refurbishment Viaduct Arches, Zurich, Switzerland

A7
台灣台北
環河南路五金街
Huanhe S. Rd Hardware street, Taipei, Taiwan

A8
巴西里約叢林
Rio-Santos
高速公路未成段
（設計提案）
Rio-Santos Highway Orangutan Jungle Park, Rio, Brazil (Proposal)

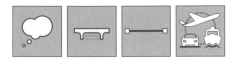

想像

1964   proposal   市場

日本東京首都
高速公路市場
（設計提案）

Tokyo Highway Market, Tokyo, Japan (Proposal)

DESIGNER 賴伯威｜Po Wei Lai

因應 1964 年東京奧運的時代需要，東京市政府為了把來自世界各地的旅客從成田機場帶到東京市中心，興建首都高速公路，造成了在日本橋 Nihonbashi 這個日本道路元標（日本道路網的起點）所在周邊的景觀改變。過去日本橋川上的許多歷史名橋，包含日本橋本身，如今都籠罩在基礎設施鋼鐵怪獸的陰影下。

1991 年所啟動的美國波士頓 Big Dig 大挖掘（永恆之掘），將市中心主要的高架道路系統地下化以釋放道路上空，把城市平面還給步行者，也還給波士頓更好的市容。後來受此影響的重大工程與開放空間案例，著名的有韓國首爾的清溪川，2003 年開始拆除其上建於 1968 年的高架道路。江戶時代的浮世繪，常描繪日本橋及富士山的景色，居民們也期待日本橋能類似波士頓 Big Dig 或首爾清溪川的計畫，重現往日風華。回應老市民期待的做法是將首都高速公路經過日本橋周邊這段埋入日本橋川河岸的地底下，埋入部分確實能避開複雜的地下鐵與公共管道。然而此計畫雖然技術上可行，但因其耗費鉅資，東京都廳一直未通過此工程。

本設計案不去辯證世界上這類都市交通改善與美化工程是否值得，直接選擇在「首都高速公路日本橋段地下化」的假設前提下進行，作為 2006-2007 哈佛建築研究所設計課尋找東京市場的自提基地。首都高速公路服務東京超過半世紀，即使拆除後理應留下點什麼。設計師選擇保留首都高速公路距離日本橋已遠且對景觀影響小的部分，利用與山手線結構共構的這一段，將其改造為市場。也因共構重疊的這一段首都高速公路拆除難度較高，且會影響東京車站的火車進出，因此建議保留。

最大的挑戰是如何把人帶上高架的市場，這一段高速公路高度超過 18 公尺。一共有六個垂直動線，位置與都市地面層的人行動線結合。一整年裡，人們都看得到木板搭建的維修站出現在高架橋的各位置，這是首都高的常態，因此設計師便利用這種維修站的木板外皮來包覆電扶梯與樓梯，把人們從地面帶上高速公路市場。在拆除過程中，路燈、信號燈、交通標誌牌等都成為廢棄物，設計師直接原地把它們在高架的市場上再利用──變成市場照明與市場標識系統。由於高速公路是為汽車量身打造的基礎設施，因此汽車的尺度可說是最適合的尺度，汽車被改造成市場各種不同物業的商店，以市場商店的形式重回高速公路。

高架的開放空間：在市場層上加了一層觀景台，讓人們散步中欣賞都市風景。過去是基礎設施功能時由汽車獨占的領域，現在人可以登上高速公路，體驗前所未有的高速公路逛街與市場購物。市場層仍有由汽車改造的商店存在，其上的都市觀景台是專屬於人的空間，宛如一座水平的艾菲爾鐵塔。

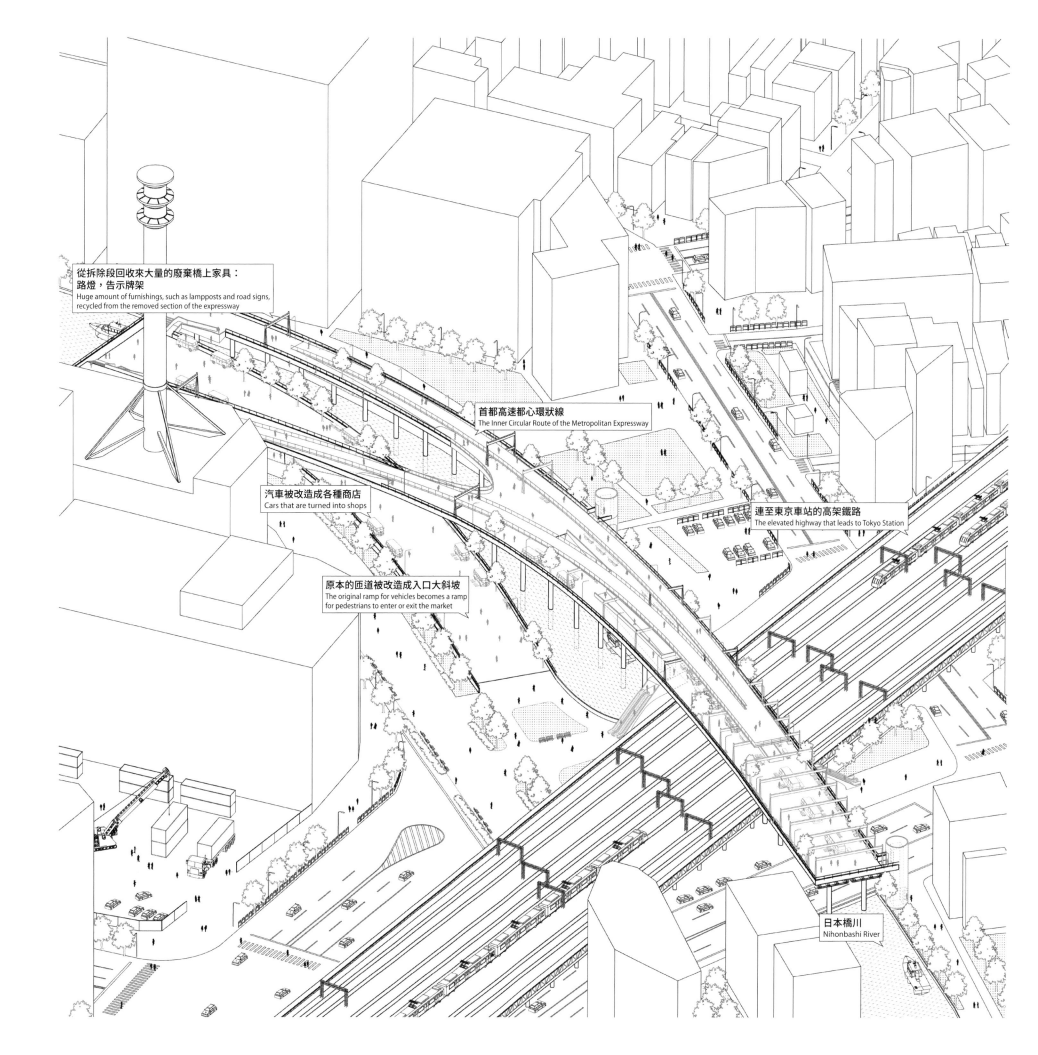

從拆除段回收來大量的廢棄橋上家具：
路燈，告示牌架
Huge amount of furnishings, such as lampposts and road signs,
recycled from the removed section of the expressway

首都高速都心環状線
The Inner Circular Route of the Metropolitan Expressway

汽車被改造成各種商店
Cars that are turned into shops

連至東京車站的高架鐵路
The elevated highway that leads to Tokyo Station

原本的匝道被改造成入口大斜坡
The original ramp for vehicles becomes a ramp
for pedestrians to enter or exit the market

日本橋川
Nihonbashi River

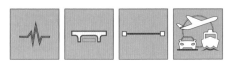

A1

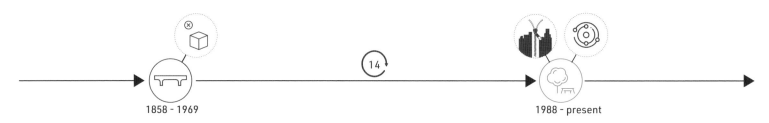

1858 - 1969　　　　　　　　　14　　　　　　　　　1988 - present

法國巴黎綠蔭
步道藝術高架橋

Viaduc des Arts,
Paris, France

DESIGNER Patrick Berger

巴黎綠蔭步道（La Promenade Plantée 英譯 Green Walkway）大概是目前我們所能找到世界上最早的利用舊有高架鐵路線路改造成城市公共空間的案例。身為遺址再利用的先驅，故它當之無愧作為《重生之路》案例選集的第一例：A1。

位於巴黎市12區，4.5公里長的空中花園，西起巴士底歌劇院，東至於巴黎環線外的文森綠地。主要包括一、藝術高架橋（Viaduc des arts）；二、新高架橋；三、勒伊；四、勒伊—比克布斯；五、比克布斯—環城大道等五段。過程中經歷多種基礎設施重生的都市空間型態，而其景觀的處理方式，讓多年後出現的紐約 High Line 與首爾路7017，相形之下都太過人工化。

綠蔭步道在重生之前，是一條建於1858年的巴黎—文森鐵路線。鐵路線建設的目的是當時巴黎為了應付城市擴張所衍生的交通運輸需求，日間以客運為主，夜間則以貨運為輔。1929年巴黎城市地鐵開始興建，使得這條火車線路走向衰落（死因：替代線路或運輸產業迭代），1939-1969間各段逐漸停駛至鐵路全線停用。

面對這座當時已經無用的鐵路高架橋，政府有兩種選擇：拆與不拆（再利用）。

選項一：拆，以便在街道上實施新的建築物。對公部門而言，拆掉應是最經濟與省事的方案，同時徹底解決南北區域被鐵路隔離的問題。但由於原本的鐵路高架橋寬度僅12公尺，拆掉後的地塊對於新建建築仍太過狹窄，難有效益。因此，加上其他各種考量之下，走選項二。

選項二：不拆，而是將鐵路高架橋改造成空中花園、景觀長廊，並利用拱形的橋下空間。修復後的高架橋可以產生許多活動，並為其所在的多梅尼大道增添許多人氣。

右圖所選擇描繪的，是藝術高架橋（Viaduc des arts）這段。藝術高架橋是整個綠蔭步道的第一段，也是最知名的部分。全長1.4公里，由71個拱所構成的拱廊，1990年由帕特里克·伯傑（Patrick Berger）規劃的「高架橋藝術與創意商街（Le Viaduc des Arts）」中標，將拱廊改造成一個藝術畫廊。修復工程的設計主要是清理高架橋，並在兩側用玻璃牆封閉拱頂。利用大小在150至450平方公尺之間的64個拱形的橋下空間，創造出一條藝術畫廊和咖啡館。改造帶來了新的透明性，拱形的店鋪吸引了以前不願意到北邊來的南邊有錢有閒階級。因此，即使保留了基礎設施不拆除，從前的邊界依然被跨越與打破，使城市在不同向度與高度上進行了縫合。

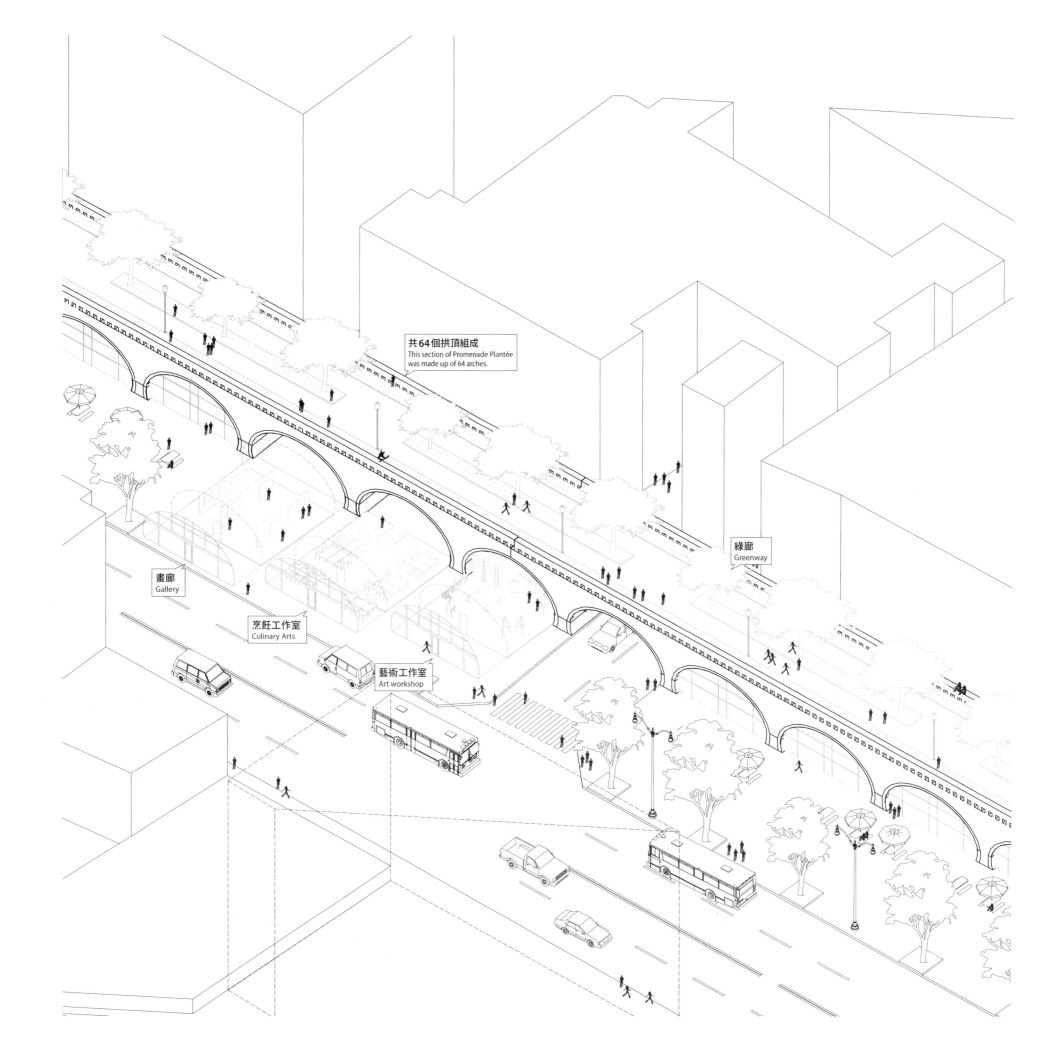

共64個拱頂組成
This section of Promenade Plantée
was made up of 64 arches.

綠廊
Greenway

畫廊
Gallery

烹飪工作室
Culinary Arts

藝術工作室
Art workshop

A2

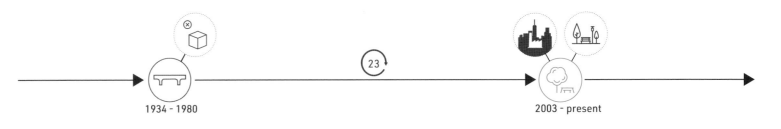

1934 - 1980　　　　　　　23　　　　　　　2003 - present

美國紐約
高線公園
New York High Line
Park, New York, USA

DESIGNER Diller Scofidio +
Renfro, James Corner

1934年，「高線鐵路 High Line」開始為鐵路貨運專用線使用，運送生鮮肉品、牛奶等，從港口直通廠房，減少地面的交通干擾。1950年代由於州際公路的發展，貨物逐漸轉為由卡車運送，鐵路面臨衰退的命運，高線鐵路在1980年功成身退（死因）。沿途的加工區日漸荒廢，鐵路上雜草叢生，也漸漸變成犯罪的搖籃。原先市政府認為這個與城市發展格格不入的舊廠區必須進行拆除，如此曼哈頓下西區才能再發展。存與廢的爭吵持續了20多年，期間，在周圍社區的抗議之下，拆除了其最南段的5個街區內的結構。1999年，由兩位附近居民創立的非營利機構「高線之友」（Friends of the High Line），呼籲要把它作為「紐約的祕密花園」保存。為了集思廣益，高線之友在2003年舉辦一個設計競賽，本來預設的規模僅為社區性的概念徵集，最後卻收到來自36個國家共720個規劃設計作品。最後，由紐約市建築師 Diller Scofidio + Renfro 與紐約景觀建築師柯納（James Corner）聯手的設計方案勝出。

總長約2.4公里，距離地面約9.1公尺高，南北串聯了22個街區。「高線公園（High Line Park）」規劃結合了彎曲混凝土鐵道與自然植被的景觀，並運用固定和可動式的座椅、燈光等設計手法塑造這座公園調性。在多個街道的入口處設置樓梯連接，有些入口更可以搭乘電梯，為行動不便的人開創了無障礙的空間。

2006年，一期景觀工程開工。至2014年，全線對外開放。公園總體設計和建設投資為1.53億美元，由紐約市政府所擁有，管轄權則歸於紐約市公園管理局（New York City Department of Parks & Recreation）。在90年代，公園管理局以「公私合營（Public Private Partnership, PPP）」的管理模式，委託高線之友負責公園的保育與營運，以私人基金應付基本開支，其中70%經費來自捐款。

而今，這條在高架鐵路遺蹟上建立的空中綠色步道，不僅提供市民多樣化的戶外休閒空間，更吸引外地遊客參觀。人潮的流入帶動城市的更新與開發，也直接創造了許多就業機會及巨量的經濟效益。高線公園應可算是自紐約市中央公園以來，最重要也最具影響力的開放空間計畫。高線公園的建成，證明以「高線之友」這樣一個在地草根組織之力，也能達到對政府城市規劃的影響，以及公共資金投資於公共空間，帶動城市產生巨大的提升效應。

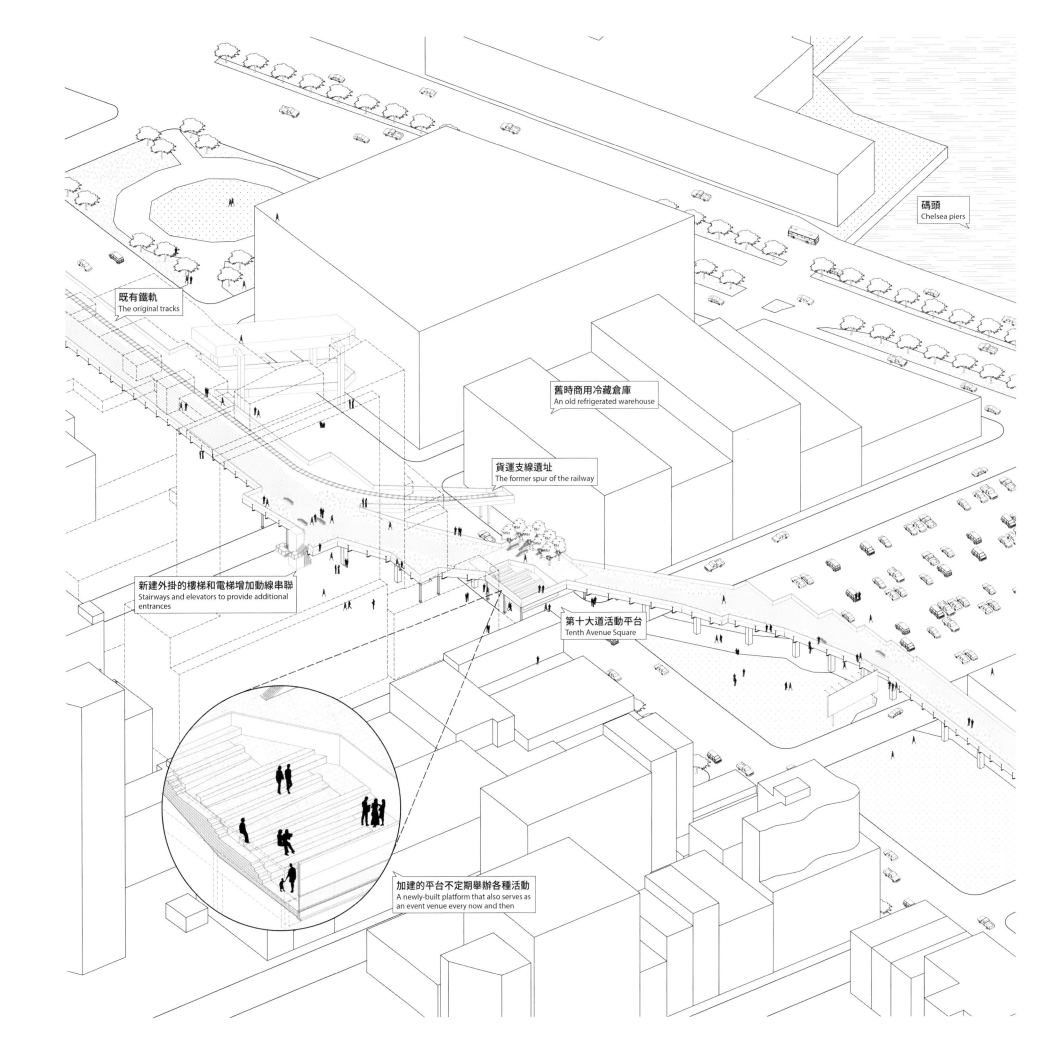

碼頭
Chelsea piers

既有鐵軌
The original tracks

舊時商用冷藏倉庫
An old refrigerated warehouse

貨運支線遺址
The former spur of the railway

新建外掛的樓梯和電梯增加動線串聯
Stairways and elevators to provide additional entrances

第十大道活動平台
Tenth Avenue Square

加建的平台不定期舉辦各種活動
A newly-built platform that also serves as an event venue every now and then

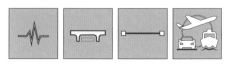

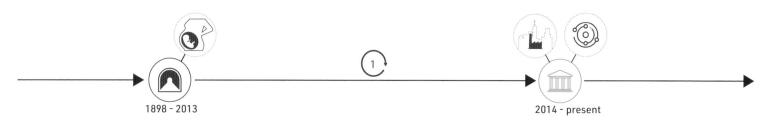
英國倫敦地下隧道滑板萬斯之家

House of Vans London, London, United Kingdom

DESIGNER Tim Greatrex

舊維克隧道（The Old Vic Tunnels/ London House of Vans）是19世紀工程的幽靈遺跡，隱藏在滑鐵盧車站和開膛手傑克之下，是倫敦滑鐵盧火車站地底下逾2,780平方公尺未被使用的鐵路隧道。在第二次世界大戰期間，滑鐵盧隧道被用作太平間，當時被稱為大墓地站。後來The Old Vic舊維克隧道藝術空間向英國鐵路有限公司買下了228-232隧道這段，改造隧道用作地下展覽和表演空間。它於2009年首次開放，在幾年內舉辦了許多活動，很可惜2013年3月15日舊維克隧道藝術空間宣布關閉。

其後2014年Vans贏得了運營228-232隧道的競標，並再次將其改造成了一個名為House of Vans的地下滑板公園。整個空間禁止任何形式的結構破壞以及磚牆破壞，所有牆面都被保留，地板則全面翻修，往磚牆照射的燈光和指標系統凸顯了磚牆的歷史痕跡，在舊有的磚牆上展示著Vans品牌的故事。

現在這個由街頭品牌Vans經營的複合式空間，位於倫敦市中心滑鐵盧車站下方的隧道裡，不特意尋找很難發現在都市中隱身這樣有強烈個性的空間。循著地圖一開始是先找到塗鴉街Leake street，它原本是鐵道下方的人行隧道，慢慢地，藝術家開始在這裡創作，成為舉辦小型演唱會、街拍、布滿藝術作品的塗鴉街。在塗鴉街裡有一個往上的窄小樓梯，抱著懷疑走上去才會發現House of Vans——這個如今主要開放給市民免費使用的滑板空間。

Vans的要求是為滑板、藝術、電影和音樂提供文化中心。利用隧道的布局，該場地被劃分為四個主要功能，以便每個都安置在特定的隧道內。它們分為以下幾個部分：一個藝術隧道；一個藝術家實驗室的畫廊，用於創作和展示藝術展覽；電影隧道，包含電影院和放映室；一個音樂隧道，是一個可以容納850人的演出場地；還有一個滑板隧道，適合各種滑板玩家在此大展身手。總體目標是創造一個能孵化創造力的空間，引發滑板和建築的聯繫，特別是滑板運動員與其滑行的環境。

滑板場的計畫是它可以容納三個隧道空間，以允許不同的布局。主要空間是用於專業用途的混凝土「碗」，第二個區域是中等能力的「街道場景」，第三個區域是給初學者練習的「迷你坡道」。該場地的五個獨立隧道用橡膠地板統一。首先來到的是提供初學者或小朋友的平面滑板空間，往後走才進入專業的滑板空間，途中會經過一些可以先看到大滑板空間的牆洞，透過這些牆洞可以瞥見使用者展現華麗的滑板技巧。走進最裡面來到由兩個大碗公組成的滑板空間，你會發現這個拱型隧道空間的最高處接近兩層半，足夠提供高強度的滑板運動使用。建築師在歷史隧道中置入了這些街道活動，也讓這個再利用的空間保留了原本的特質。此案除了討論了滑板和室內建築之間的聯繫，也是滑板和舊有隧道拱形空間的結合。

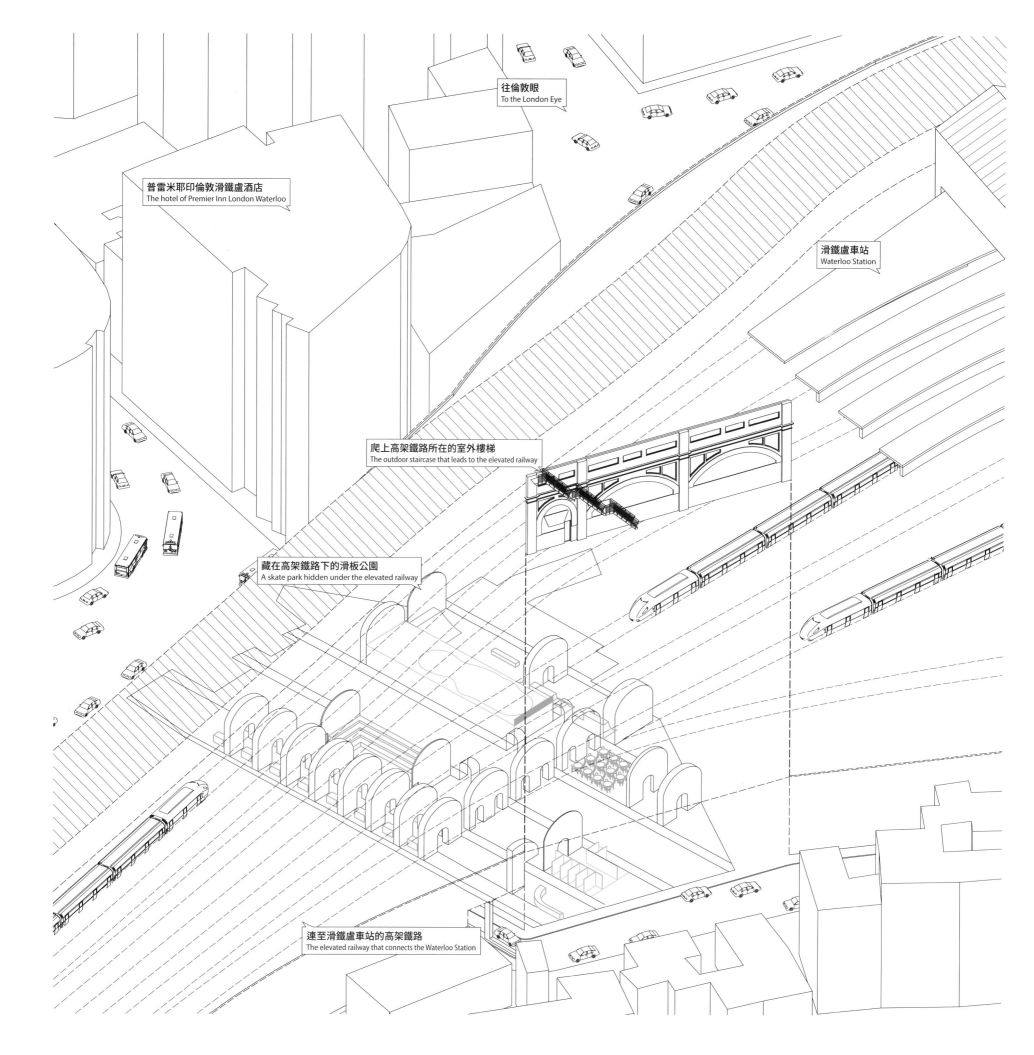

往倫敦眼
To the London Eye

普雷米耶印倫敦滑鐵盧酒店
The hotel of Premier Inn London Waterloo

滑鐵盧車站
Waterloo Station

爬上高架鐵路所在的室外樓梯
The outdoor staircase that leads to the elevated railway

藏在高架鐵路下的滑板公園
A skate park hidden under the elevated railway

連至滑鐵盧車站的高架鐵路
The elevated railway that connects the Waterloo Station

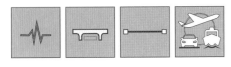
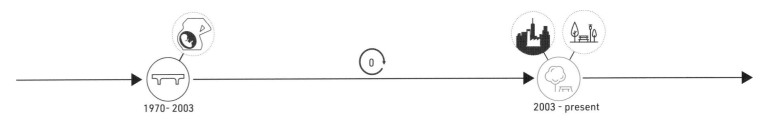

荷蘭贊丹 A8ernA 高速公路公園
A8ernA, Zaanstad, the Netherlands

DESIGNER NL Architects

由 NL Architects 所設計的荷蘭 A8 高速公路公園，位於 Koog aan de Zaan，這是阿姆斯特丹附近的一個可愛的小村莊，坐落於贊河（Zaan）河畔。在 20 世紀 70 年代早期，為了跨過 Zaan 河，在這裡興建了一條 A8 高速路。這條新的高速路就這樣地穿過小鎮的城市肌理，殘酷地劃分了城鎮的結構。A8 高架路下的橋墩約有 7 公尺高，在基礎設施的樓板下面，一條寬約 40 公尺長約 400 公尺的主要用於停車的地帶被長期忽視，更重要的是這個斷裂帶導致了小鎮的教會和政府的分離：高架路的一側是一個小教堂，而另一側則是曾經的市政廳。

2003 年 Zaanstad 市議會決定進行城市規劃，以重新連接 Koog aan de Zaan 的兩側，並活化高架路下的空間，為居民提供休閒空間和服務設施。項目基於樂觀的態度，將基礎設施的巨大存在視為機遇而非障礙。機遇在於：由於其型態和靠近河流中心位置，高架路下的空間可被視為一個大型的公民拱廊，便於滿足所有公民的要求。高架路下方的空間置入了超市、花店、魚店和娛樂設施，包括公園、滑板公園和水上運動碼頭，不但重新連接城市的兩側，並將它們連接到附近的河流。橋下的空間變得非常具有發展潛力：它可以成為一個教堂空間的延伸。

兩條與之交叉的道路將這個大型拱廊（橋下空間）劃分為三段不同的區域。

中央部分是一個有蓋的廣場，那裡有超市、花卉和寵物商店，以及一個發光的噴泉。在東端，穿過高街，有一個「雕塑」巴士站和一個帶全景平台的小港口。港口將水帶到橋下空間。全景平台為市民提供了一個特殊的窗口。

在西端有一個兒童和青少年遊樂場，包括一個塗鴉畫廊、一個滑板公園、一個嘻哈舞台、一些手足球和乒乓球桌、一個七人制足球場、一個籃球場和「戀人長椅」。

除了高速公路覆蓋的區域外，在市政廳和教堂周邊也有公共空間，沿著高速公路連結起來產生了一系列公共空間。教堂前方，相關單位開放了現有的綠地廣場，用於舉辦露天博覽會和慶祝活動。市政廳一側則有一座新的公園，容納了一個保齡球場、一個種植了白樺樹的小山坡、一個烤肉區和一個由金屬柵欄環繞的足球場，還有一座射擊場位於高速公路出口坡道下方。

當地政府集思廣益，邀請當地居民與開發商共同協作參與，最終建造出 A8ernA 的公共空間。成功的關鍵在於政治意圖加上地方政府與社區之間的合作，使得一個被忽視了 30 多年的空間的轉變成為可能。

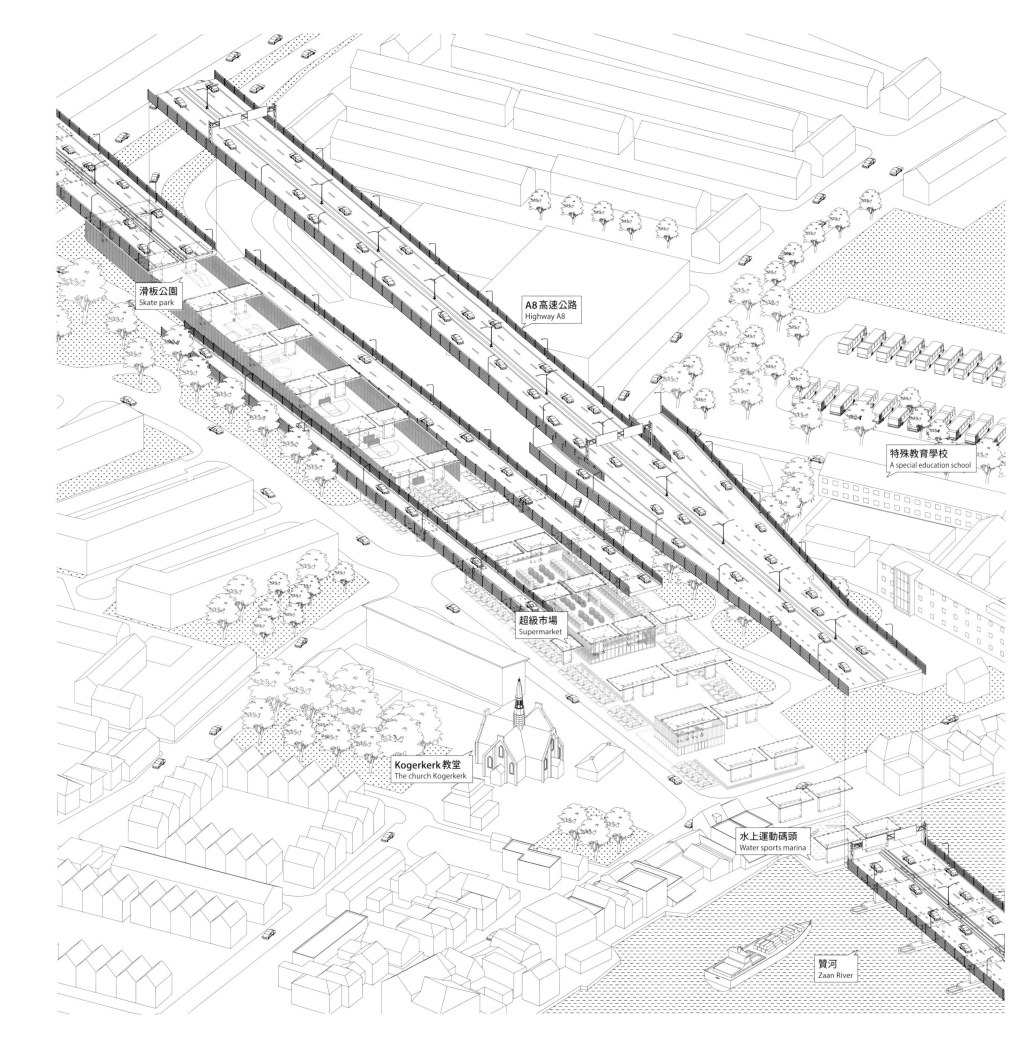

滑板公園
Skate park

A8高速公路
Highway A8

特殊教育學校
A special education school

超級市場
Supermarket

Kogerkerk教堂
The church Kogerkerk

水上運動碼頭
Water sports marina

贊河
Zaan River

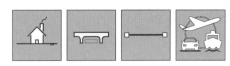

A5

日本東京秋葉原 2k540 商業藝廊
2k540 Aki-Oka Artisan, Tokyo, Japan

DESIGNER 東日本都市開發公司

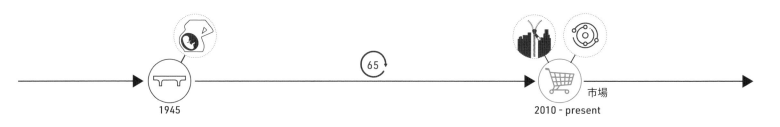

1945

65

2010 - present 市場

2010年12月誕生的「2k540 AKI-OKA ARTISAN」，是日本近幾年來成功讓老舊橋下空間重生的例子之一。從東京車站出發，到御徒町站和秋葉原站高架橋下的距離為2公里540公尺，所以此區域簡稱為2k540，是日本JR東日本鐵道集團旗下的JR東日本都市開發公司，基於地域開發而推動的一項都市更新案。以往電車的高架橋下，常給人冰冷、昏暗甚至髒亂之感，御徒町站和秋葉原站之間原本亦是如此。

其實像JR、metro、京王、東急等等不同的電鐵公司，不論公營私營，沿線的車站與路段皆是電鐵公司的地產。所以不難想像，為了增加營收，與車站站體一體成形的商場百貨（像是澀谷站或新宿站）或是沿線下的空間（2k540、中目黑車站沿線下面的商店──東急東橫線等等）都會成為商業活動聚集的地方。

如今的2k540整體以黑與白為基調，純白簡潔的小鋪量體順著橋柱整齊排列，柏油鋪面質感由外側馬路延伸進橋下的遊逛動線，空間巧妙利用了橋柱搭配地嵌式照明設計，將原本陰暗雜亂的橋下閒置空間修建為明亮清爽且具有設計感的挑高場所。店鋪所需要的各式線路及中央照明管線，經過整理及設計後，也成為藝術裝置般展示於一隅。這裡除了職人街的功能外還有著藝廊般的氛圍：內部聚集了近50間店鋪，從布藝、紙藝、陶藝、皮革、手工飾品、銀匠，

到咖啡烘豆鋪、餐館等，將相距886公尺的秋葉原與御徒町在慢速悠遊的動線中連接起來。

在江戶時代，御徒町是日本武士聚居的小城鎮，階級較低的武士通常只有非常微薄的薪水，這些武士常藉著販賣自製的手工藝品給當地居民以維持生計，隨著時間推移，尤其到西元1860年代，明治天皇開始推行工業化和現代化運動後，這些武士家族拋棄原本的工作，轉而開始製作珠寶和首飾等手工藝。時至今日，仍有許多年邁的職人居住在這裡，街道上也看得到不少珠寶店、皮件批發商，由於店面大多已隨歲月失去當年的朝氣及繁華風貌，而逐漸難以吸引年輕族群到此一訪，所以2k540的出現，讓傳統老工藝職人與年輕的手工創作者們有了可以重新釋放能量的地方，每一間小店都充分利用微型創作空間，盡情散發專屬於日本文化的職人精神。

2k540作為地區共生型的傳統技藝創新基地，不只是成為僅供消費的商業設施，更注重在地居民的參與，店鋪之間所空出的巷道空地和中央廣場，成為舉辦各種活動的場域。店家們甚至聯合鄰近市場籌辦地區性活動，帶進外部人潮活絡地方產業。這裡完整地展示了如何利用因城市發展所餘下的空間，承繼歷史涵構並帶入社區設計，進而再造區域繁華。

多條高架橋並排產生多樣的橋下柱梁結構
Multiple pillar structures under multiple elevated bridges

傳統老工藝職人與年輕的手工創作者商店
Shops owned by artisans of the traditional crafts and craftsmen of the younger generation

秋葉原站至御徒町站的高架鐵路
The elevated railway between Okachimachi Station and Akihabara Station

A6

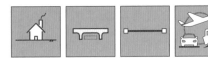

瑞士蘇黎世
翻新高架拱橋

Refurbishment
Viaduct Arches,
Zurich, Switzerland

DESIGNER EM2N

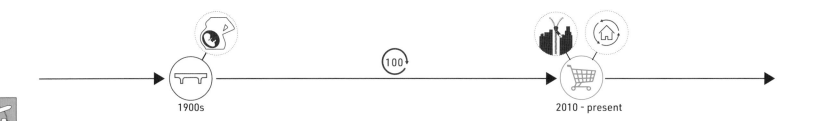

1900s

100

2010 - present

歷史悠久的高架橋拱門轉變為蘇黎世的時尚購物區。

　　全世界正以大膽的方式將橋梁和高架橋下的剩餘空間再利用。瑞士設計公司 EM2N 和景觀公司 Zulauf Seippel Schweingruber 將蘇黎世一系列19世紀的高架橋拱門翻新成為「最令人興奮的購物街」，在橋下拱門空間引進了一系列的現代商店和餐廳，同時保持了高架橋的建築完整性。

　　高架橋成為大型連接器和線性建築，過去用作鐵路的基礎設施將形成一個線性公園，成為文化、工作和休閒的一部分。過去隔斷空間的障礙將轉變成為新的連接。建築師和景觀設計師在這裡置入了由黑色鋼質結構組成、附有泡泡天窗的「容器」——但不至於填滿整個拱門以保持拱門上方視線通透，也讓天光能進到下方的「容器」中。高架橋下的空間一共重新置入了36個「容器」，這群黑色容器將藝術畫廊、商店、餐館等新功能組合在一起。帶有泡泡天窗的市場大廳位於此項目的中心，為20個當地農民和食品供應商提供產品展銷空間。這些新功能與高架橋結構形成共生連接的關係。

建築採用當代和極簡主義的材料設計，以便讓人將注意力集中在既有的拱門和外露的 Cyclopean 磚石上。為了使 Cyclopean 磚石成為主要元素，建築師對裝修制定了準則：一定程度地限制新加入的元素，以強調凸顯出現有的拱門。在裝修「容器」的室內時，未來的租戶可以從一系列預設好的設計元素中進行選擇或自行設計空間。因此，舊的石牆完全保留下來，高架橋本身則幾乎沒有被改動。

　　這個案例將城市歷史建築的橋下空間轉變為可用的文化和商業景點。首先於 2004 年提出建築設計競賽方案，並於 2010 年完工。在此案例中，EM2N 建築師宣稱研究了兩個基本問題：作為受保護如紀念碑一般的基礎設施元素如何被重新編程，並成為新的城市系統的一部分？在一個法規越來越多（能源、衛生、消防等）和更高的舒適度要求的時代，如何實施低預算項目？19世紀城市早期的鐵路線的高架橋結構，過去將蘇黎世的城市部分和市中心分開，現在透過 EM2N 的設計，實現了將這兩個區域透過文化和商業空間結合在一起。

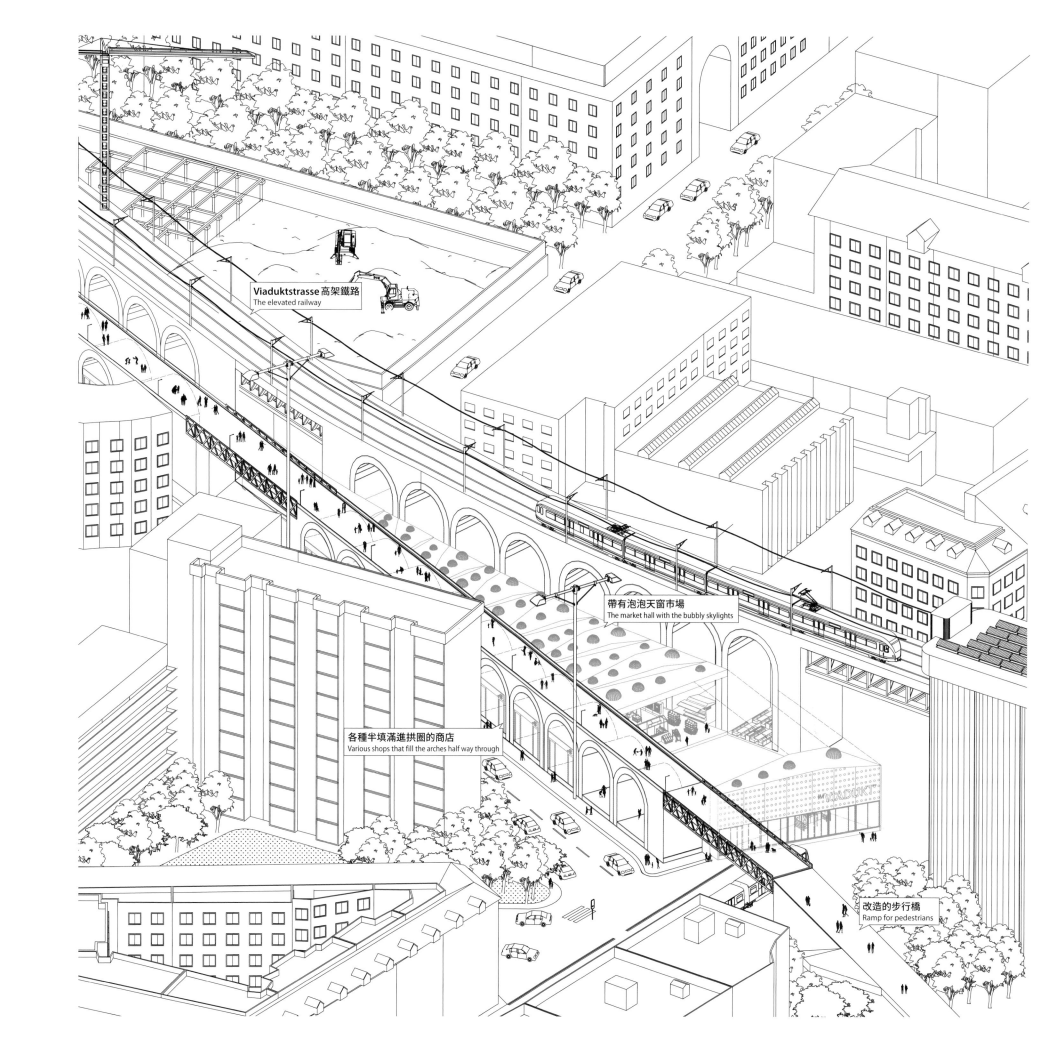

Viaduktstrasse 高架鐵路
The elevated railway

帶有泡泡天窗市場
The market hall with the bubbly skylights

各種半填滿進拱圈的商店
Various shops that fill the arches half way through

改造的步行橋
Ramp for pedestrians

A7

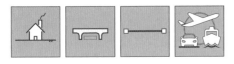

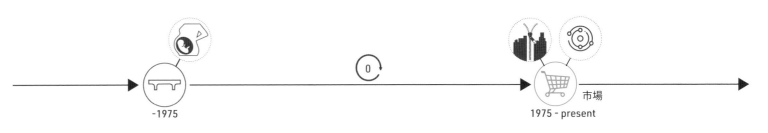

-1975

市場

1975 - present

台灣台北
環河南路五金街
Huanhe S. Rd Hardware street, Taipei, Taiwan

DESIGNER 不詳

路過台北環河南路五金街，應該很少人會不訝異於這裡的街景：高架橋上方來往不息的車輛與下方靜止不動的店鋪共存。

這樣的河濱商圈，早從日治時代就出現雛形，當時旁邊是艋舺碼頭，物流方便，故出現了這樣的商場聚落。這裡的店面高達135間，有上萬件商品，雖然不像量販店那樣整齊，但有很多人就是因為喜歡挖寶的感覺，特地慕名而來。大大的雙喜燈、霓虹燈、香腸攤的打彈珠台，小到各種稀奇玩具，大到在夜市裡常見的攤車，在這裡都買得到。

回溯其歷史，早在日治時代民國12年，這裡還只是兩排平房，地上都是黃土，後來蓋了高架橋，橋下變成一間一間的店面，都在賣這些五金零件。

這個商場，是民國64年8月22日那天成立的。五金街歷史悠久，不少人愛到這裡來體驗挖寶的感覺，甚至是那股街頭雜亂呈現出的人情味。

環河南路五金街至今仍存在代表台北都市的異質性與包容性，難以否認的活力與特色，使它成為台北都市橋下空間的代表。

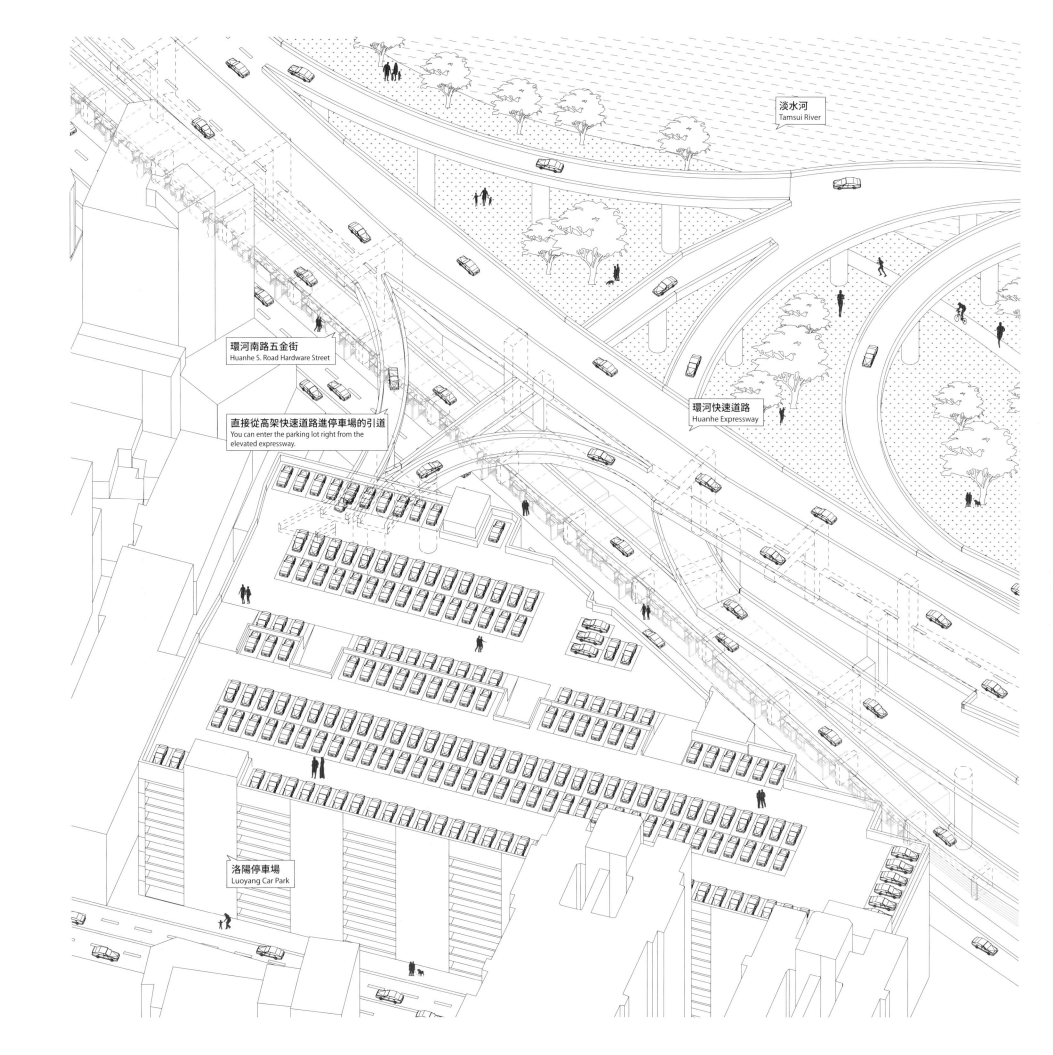

淡水河
Tamsui River

環河南路五金街
Huanhe S. Road Hardware Street

直接從高架快速道路進停車場的引道
You can enter the parking lot right from the elevated expressway.

環河快速道路
Huanhe Expressway

洛陽停車場
Luoyang Car Park

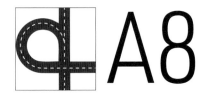

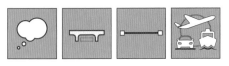

1960s-1976　　　　　想像　　　　　proposal

巴西里約叢林
Rio-Santos
高速公路未成段
（設計提案）

Rio-Santos Highway Orangutan Jungle Park, Rio, Brazil (Proposal)

DESIGNER 侯雅齡

Rio-Santos 高速公路的建設始於20世紀60年代，到1976年，該路段將與現有高速公路相連，是里約至桑托斯高速公路的一部分。但是，這些計畫在最後一分鐘由於預算不足被改變了，以致現有道路與沿海路線相連，而新建但未使用的高架橋則被迫放棄。柏油高架橋位於叢林之上40公尺處，長300公尺，高40公尺的高架道路以擋土牆和一個巨大的混凝土基礎為特色。

這座高架橋所處的環境是在一整片自然山瘠中，由於從未被使用，因此已經被周遭的自然逐漸蔓延攀附，處在這樣的環境而顯得突兀的人造物，試圖重新思考人為建造與廢棄的反轉價值，透過設計讓被遺棄的混凝土人造物，成為周遭環境的守護者與動物的中介站。

基地規劃部分，以高架橋作為動物保育的核心，逐漸從人到自然的層次來界定場域，最後透過水及綠的軟性隔離形成整個園區的邊界。

高架橋作為園區核心，必須容納人與猩猩不同機能，設計概念以橋面作為新的界面，透過橋面橋墩水平垂直的不同形式特性，劃分人與猩猩及共享的空間。

當人與猩猩不同物種，透過新舊空間的交疊及置入，讓自然之中的人造物，轉變為生態共生的媒介，讓人在這場域中不再扮演占據者的角色，反思建造的意義及扭轉價值，在自然環境中達到平衡。

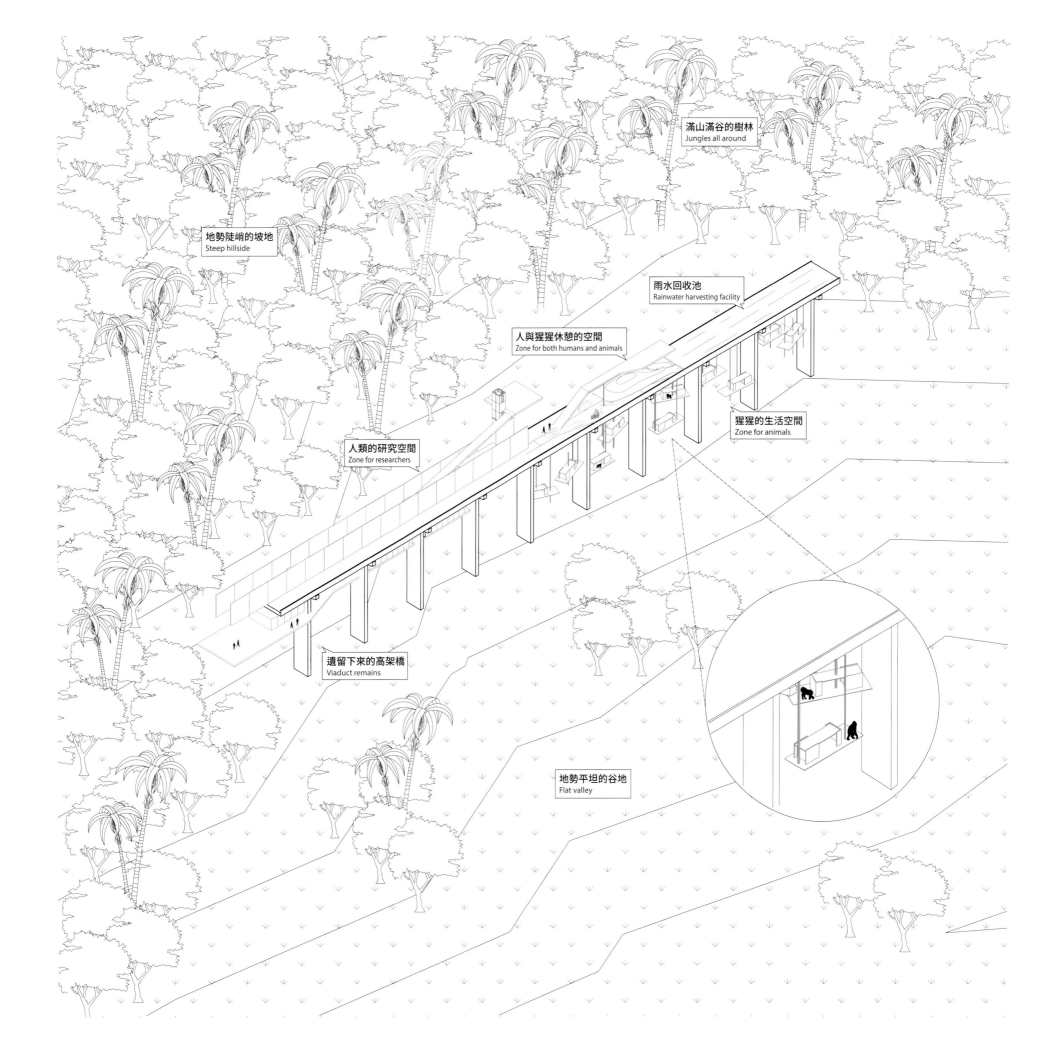

滿山滿谷的樹林
Jungles all around

地勢陡峭的坡地
Steep hillside

雨水回收池
Rainwater harvesting facility

人與猩猩休憩的空間
Zone for both humans and animals

猩猩的生活空間
Zone for animals

人類的研究空間
Zone for researchers

遺留下來的高架橋
Viaduct remains

地勢平坦的谷地
Flat valley

B

港口 Port

B1
丹麥哥本哈根起重機飯店
The Krane, Copenhagen, Denmark

B2
丹麥赫爾辛格海事博物館
Maritime Museum of Denmark, Helsingør, Denmark

B3
荷蘭阿姆斯特丹
吊車梁辦公室
Kraanspoor, Amsterdam, The Netherlands

B4
荷蘭阿姆斯特丹
法拉達起重機豪華酒店
Crane Hotel Faralda NDSM, Amsterdam, The Netherlands

B5
中國上海船廠
1862藝術中心
Shipyard 1862, Shanghai, China

B6
德國漢堡易北愛樂廳
Elbe Philharmonic Hall, Hamburg, Germany

B7
美國紐約甘迺迪機場
TWA飛行中心飯店
TWA Flight Center Hotel, New York, USA

丹麥哥本哈根
起重機飯店

The Krane,
Copenhagen, Denmark

DESIGNER Arcgency-Mads Møller

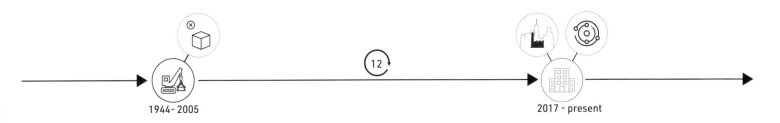

1944 - 2005   12   2017 - present

建築公司 Arcgency 將哥本哈根港口的工業煤炭起重機變成了一個豪華的招待所 The Krane，提供沉浸式的多感官體驗，營造出如世外桃源般的獨特豪華度假場所。室內裝潢刻意設計以黑色為主——以回應過去用來裝載煤炭的起重機。

此案例是由具遠見的開發商 Klaus Kastbjerg 和 Arcgency 首席建築師 Mads Møller 所帶領。The Krane 目前的擁有者 Klaus Kastbjerg 是哥本哈根市許多其他濱水項目的開發商，其中包括一個曾是糧倉的住宅大樓。此港口起重機為 Klaus Kastbjerg 在周邊的項目 Harbour House 帶來亮點，這也是起重機能重生的契機。

事實上，起重機原本不在目前這個位置，而是在設計雪梨歌劇院的建築師 Jørn Utzon 建議下，將起重機移到景觀更美麗的位置——位於 Nordhavn 的邊緣，是丹麥首都最後一個正在裝修的港口之一。The Krane 坐落在海濱，可通過吊橋式樓梯抵達。

其多層結構包括一樓的接待區和玻璃牆會議室。二樓設有 Spa 區和露台，頂樓設有一間帶休息室和露台的客房／起居區。每個區域都可單獨租用。建築頂部是一個 50 平方公尺的 Krane Room 生活空間，設有環繞式窗戶，可欣賞大海、天空、海港和哥本哈根市全景。在子宮般的 Krane Room 下面，Spa 區從地板到天花板都是灰色的石頭，巨大的玻璃牆可以欣賞海港和大海的廣闊景色，住客可以在 Spa 區的兩個浴缸中放鬆身心。

水的重要性可能占了整個起重機住宿體驗的 80％，憑藉其歷史和如此美麗的水景，此改建案例成功地將退役的港口起重機轉化為具現代感且有靈魂的空間。

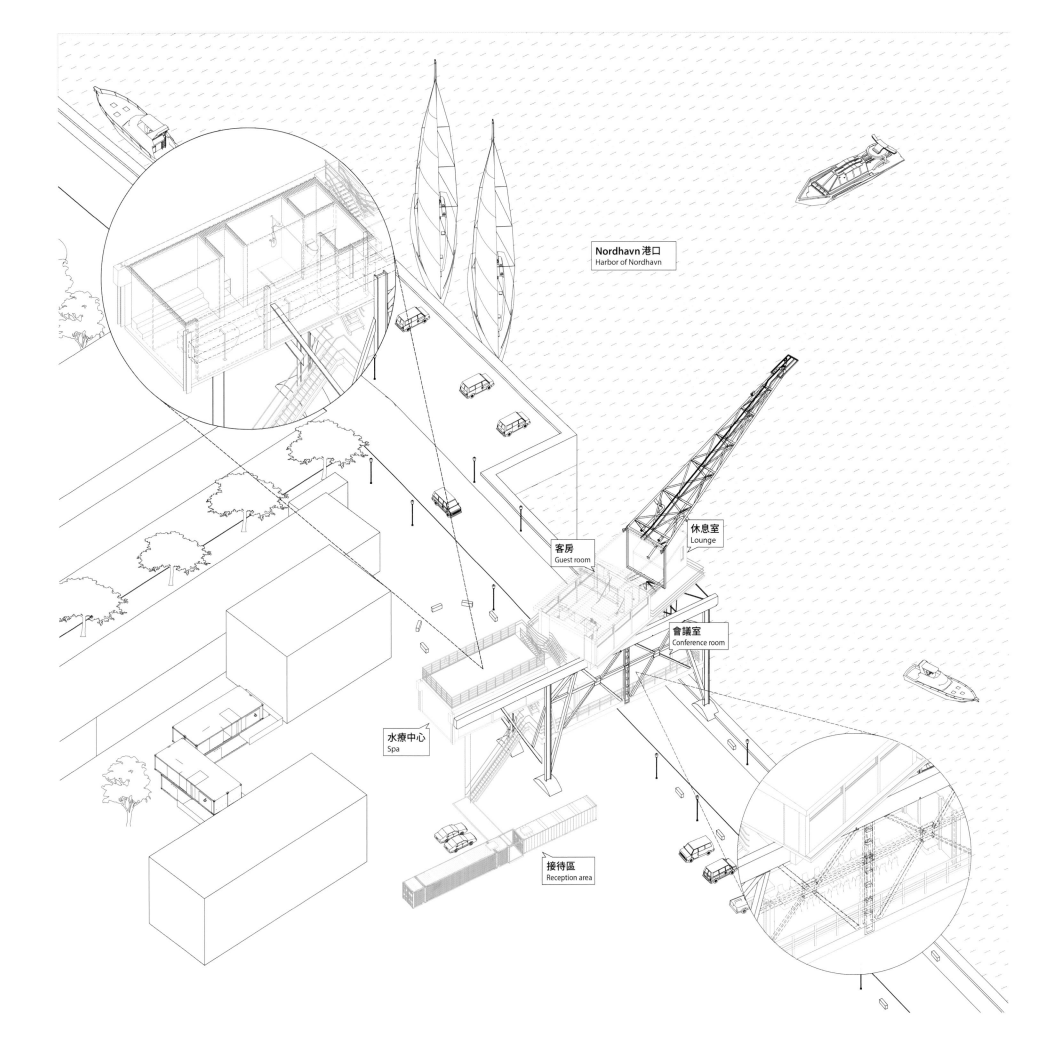

Nordhavn 港口
Harbor of Nordhavn

客房
Guest room

休息室
Lounge

會議室
Conference room

水療中心
Spa

接待區
Reception area

B2

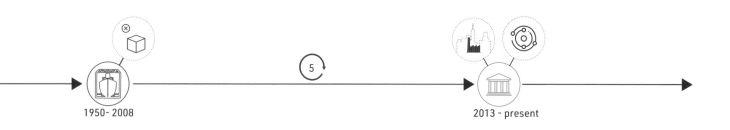

丹麥赫爾辛格
海事博物館
Maritime Museum
of Denmark,
Helsingør, Denmark

DESIGNER BIG

離哥本哈根不遠的赫爾辛格Helsingør，是一個優美古老的海港小鎮，為了推動觀光，相關單位發起「文化海港中心計畫（Kulturhavn Kronborg）」，以1420年建造的克隆城堡Kronborg為主角——這裡正是莎士比亞名劇《哈姆雷特》的故事背景——改建鄰近的Elsinore造船廠，並於2010年成立文化碼頭Kulturværftet。而被夾在這兩者之間60年的老船塢，歷時5年的改造後，變身為海事博物館，於2013年10月開幕。

出於對克隆城堡的尊重，建築師將海事博物館保持幾乎完全隱形於地下，將舊的混凝土乾船塢改造成現代化的博物館，採用了之前未曾在丹麥使用的建築技術，在海平面以下建造一座博物館。舊船塢本身1.5公尺厚的混凝土牆壁和2.5公尺厚的混凝土地板被切割開來，重新組裝成精確如現代鋼橋般的博物館。一系列複層橋跨越舊船塢，既可以作為城市連接，也可以為遊客提供短途停留在博物館的不同區域。海港大橋在作為港口長廊的同時關閉了碼頭；博物館的大廳是連接相鄰文化庭院和克隆城堡的橋梁；傾斜曲折的橋梁將遊客引導至正門。當遊客下降到博物館空間時，這座橋將新舊結合在一起，俯瞰舊船塢船形空間的雄偉，將展覽空間與大廳、教室、辦公室、咖啡廳和博物館內的舊船塢樓層連接起來。

地下畫廊沿著舊船塢牆體周圍，連續配置在一個兩層長方形結構中，展示了丹麥航海歷史至今的故事。將舊船塢向內翻轉，它如船形坑的空間成為展覽的核心——一個開放的室外區域，遊客可以在船形坑體驗到造船的規模，同時保留充滿故事的舊船塢牆體改造成一個中庭院，將日光和空氣帶入地下博物館的中心。

海事博物館是由國家核准，11個基金會聯合出資的獨立營運機構。地下面積7,600平方公尺，建造費用5,400萬美元。丹麥成功打造了吸引國際目光的海事博物館和文化中心，更反映了丹麥作為世界領先的海洋國家之一的歷史和當代角色。

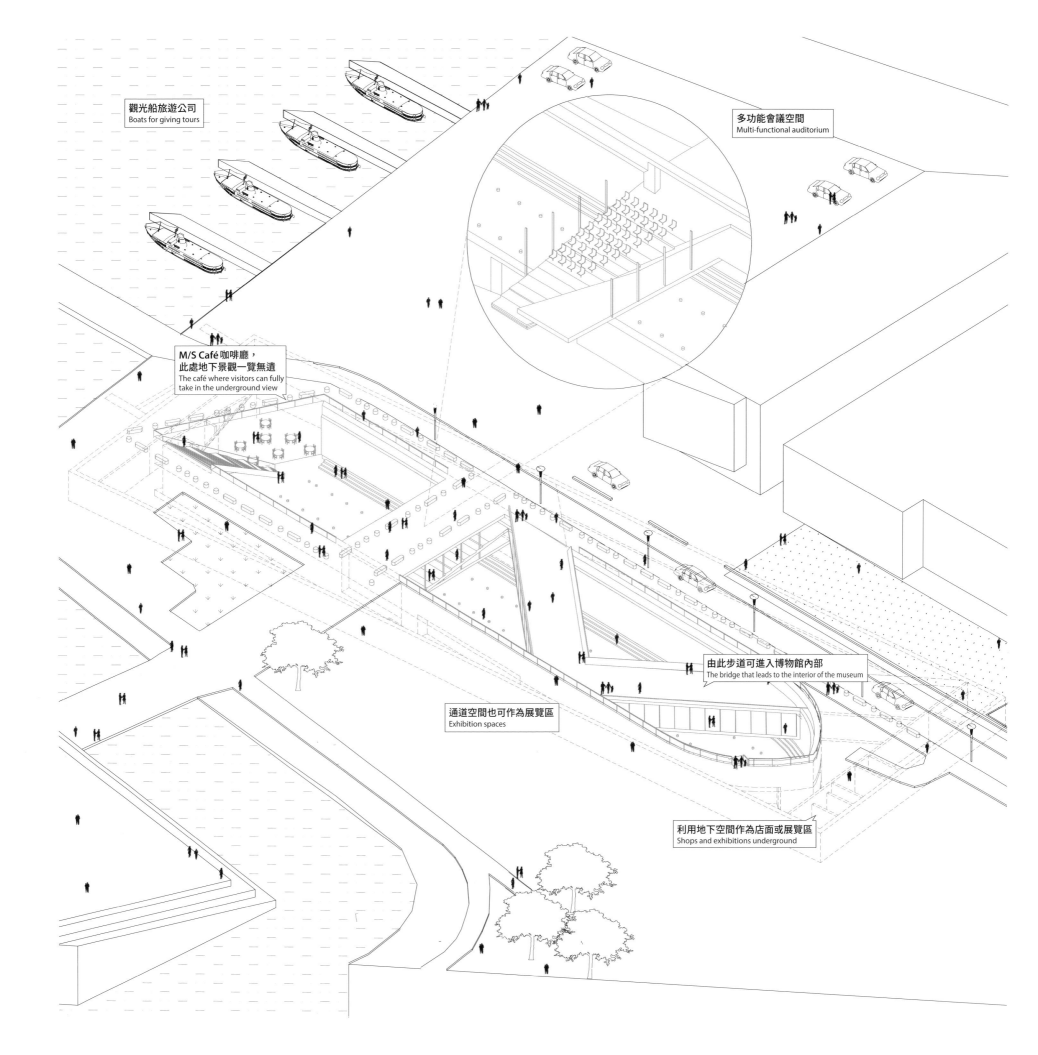

觀光船旅遊公司
Boats for giving tours

多功能會議空間
Multi-functional auditorium

M/S Café 咖啡廳，
此處地下景觀一覽無遺
The café where visitors can fully
take in the underground view

由此步道可進入博物館內部
The bridge that leads to the interior of the museum

通道空間也可作為展覽區
Exhibition spaces

利用地下空間作為店面或展覽區
Shops and exhibitions underground

B3

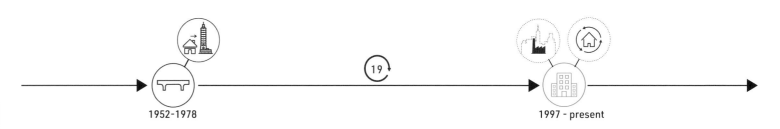

1952-1978
19
1997 - present

荷蘭阿姆斯特丹吊車梁辦公室

**Kraanspoor,
Amsterdam,
The Netherlands**

DESIGNER OTH Architecten

位於阿姆斯特丹海濱西側的碼頭和工業區原先屬於「荷蘭碼頭和造船公司（Nederlandsche Dok en Scheepsbouw Maatschappij, NDSM）」。這個巨大的場地包含一條1952年建造的混凝土結構的起重機軌道，由JD Potsma設計，用作兩部起重機的移動平臺和船舶工程的碼頭。1970年代，造船業務開始下降（死因），許多碼頭消失。1978年，NDSM造船廠也關閉，起重機軌道也發了拆遷許可證。

1997年，OTH的建築師Trude Hooykaas在舊的起重機軌道上看見了破舊的空曠景觀，並構想了在舊混凝土軌道上疊加現代化辦公大樓的計畫。結果當地政府決定不要拆毀這座吊車軌，而是調整了總體規劃，也預見了黃金海岸那頭的可能性。這座吊車軌象徵著阿姆斯特丹航運業的遺跡，由兩個水平桁架組成，由四個位於水中的雙主要入口門架支撐，由建築師Trude Hooykaas設計了一幢3層樓高、浮在現有吊車軌結構上方OTH的透明辦公室，可以從內部看到

舊日的起重機軌道，擁有超過10,000平方公尺的獨特工作空間，並且坐擁阿姆斯特丹最開闊的水路視野。

設計挑戰在於舊基礎的承載能力有限，新建築選擇自重較小的鋼骨結構執行，這有利於建築物的透明度和輕量化。玻璃外殼的新建築遵循起重機軌道的整個長度和寬度，並由3公尺高的細長鋼柱支撐。一樣是270公尺長，13.8公尺寬，高度為10公尺。這座建築裝配有一套可移動的玻璃百葉窗表皮，將自身完全的透明性對外展示。入口安置樓梯和全景升降機，通過這些升降機可以進入上面的樓層。

如今，吊車梁辦公室Kraanspoor因作為工業遺產而受保護，擁有了新的功能。它不僅是一座壯觀的建築，也是一座綠色建築。這是一套新舊之間的無縫結合，以工業遺產和現代建築之名，保存歷史，卻又不失活力。

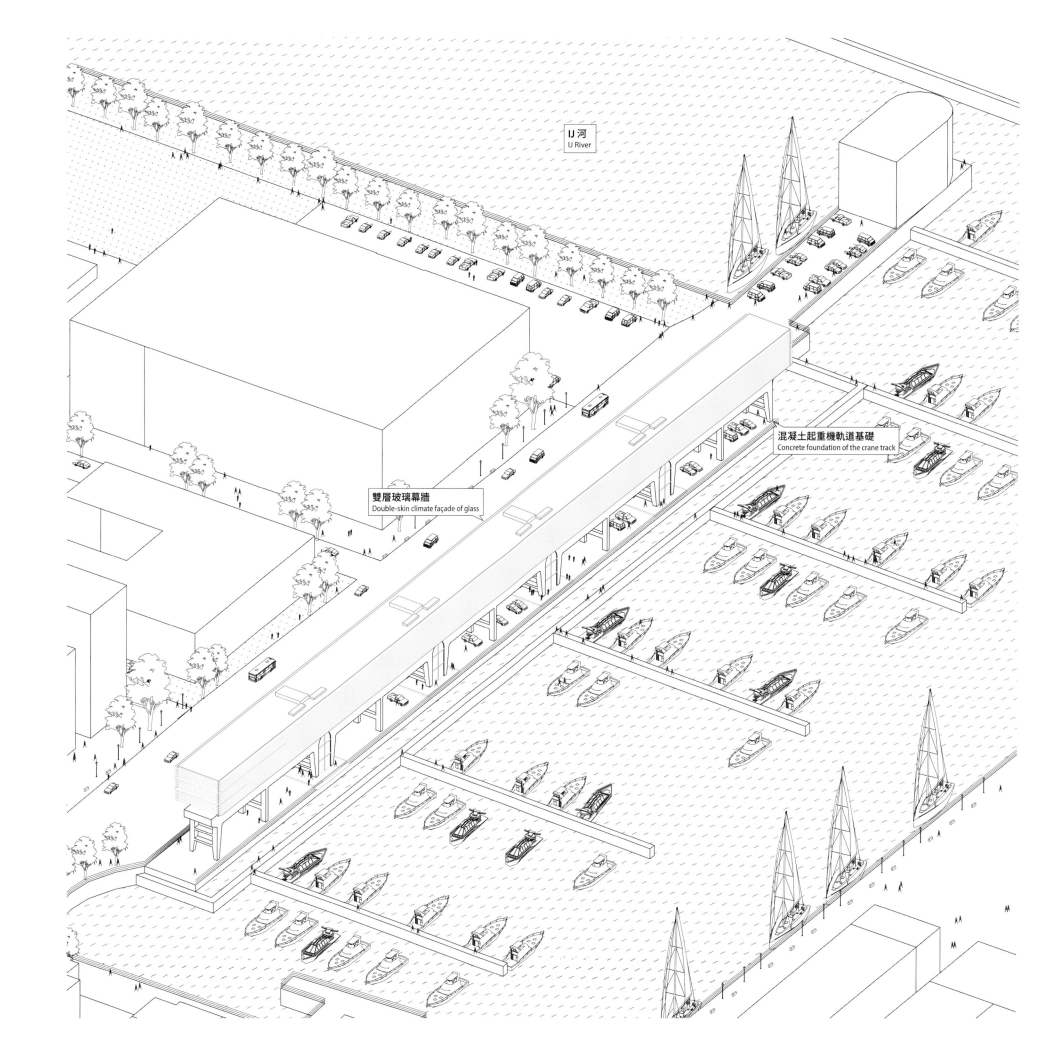

IJ 河
IJ River

混凝土起重機軌道基礎
Concrete foundation of the crane track

雙層玻璃幕牆
Double-skin climate façade of glass

B4

荷蘭阿姆斯特丹
法拉達起重機
豪華酒店

**Crane Hotel Faralda
NDSM, Amsterdam,
The Netherlands**

DESIGNER IAA Architecten
Amsterdam

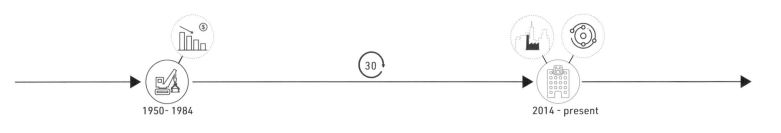

1950 - 1984   30   2014 - present

Faralda NDSM Crane Hotel 是全球首間以起重機作為建築結構骨架的酒店，在高 50 公尺、重 250 公噸的起重機骨架裡裝入貨櫃作為客房。狹小且難以做太多更動的框架下，僅提供有限的三間房間，但每一間房型都附有小閣樓，以三種不同的風格裝修：豪華、摩登、工業風。外觀上鋼結構恢復原來的色調，所有新設計的元素都以紅色為主，因此新舊的對比亦可從顏色上明顯識別。

編號 13 號的前 NDSM 起重機，由 Hensen 在 20 世紀 50 年代建造。1984 年後，在「荷蘭碼頭和造船公司（NDSM）」破產後，起重機便不再使用（死因）。25 年後，破舊的起重機 13 號原本將像其他的起重機一樣，因缺乏維護以致危及公共安全而被炸掉或拆除，還好開發人 Edwin Kornmann Rudi 及時地將這座殘破的起重機從毀滅中拯救出來，並賦與重生。2011 年 8 月，他取得阿姆斯特丹市政府的許可，開始制定再利用計畫，以重建這個鋼鐵巨人。2013 年 6 月 7 日，國家重建基金會與 Edwin Kornmann Rudi 之間進行合作簽約。同時，阿姆斯特丹市政府也批准他提出的重建計畫並搬移到新的基地。

該項目因針對眾多實際建築與工程問題所提出的獨特解決方案，而受到大力讚揚。包含三間套房的起重機塔架繼續圍繞樞軸軸承旋轉。在這個幾公分直徑的非常小的旋轉軸中，所有管道，阻力聯軸器和防火安全措施都能順暢運行。起重機的拆卸與搬移工作於 2013 年 7 月 22 日開始，運往弗拉內克，在一座 100 公尺長的浮橋上進行修復。2013 年 10 月 22 日，返回的起重機恢復正常。2014 年 4 月 4 日，起重機被正式用作起重機酒店 Faralda。

一樓處，起重機旁邊增加了一個入口，樓梯和電梯分別通向一個 10 公尺高的平台和通往樓上套房的轉換層。10 公尺高的平台位於起重機基腳之間，為工作室及辦公區域。從轉換層，人們可以乘坐外掛的電梯到三個套房。按摩浴缸位於酒店的最頂樓，喜愛極限運動的旅客甚至可以在起重機懸臂末端體驗刺激的高空彈跳。房客只要從客房或頂樓的按摩浴缸就可以遙望市區最繁忙的地段與阿姆斯特丹著名的河景，相信對任何人而言，這都會是獨特難忘的住宿體驗。

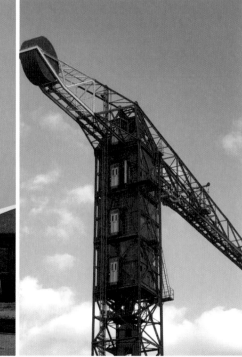

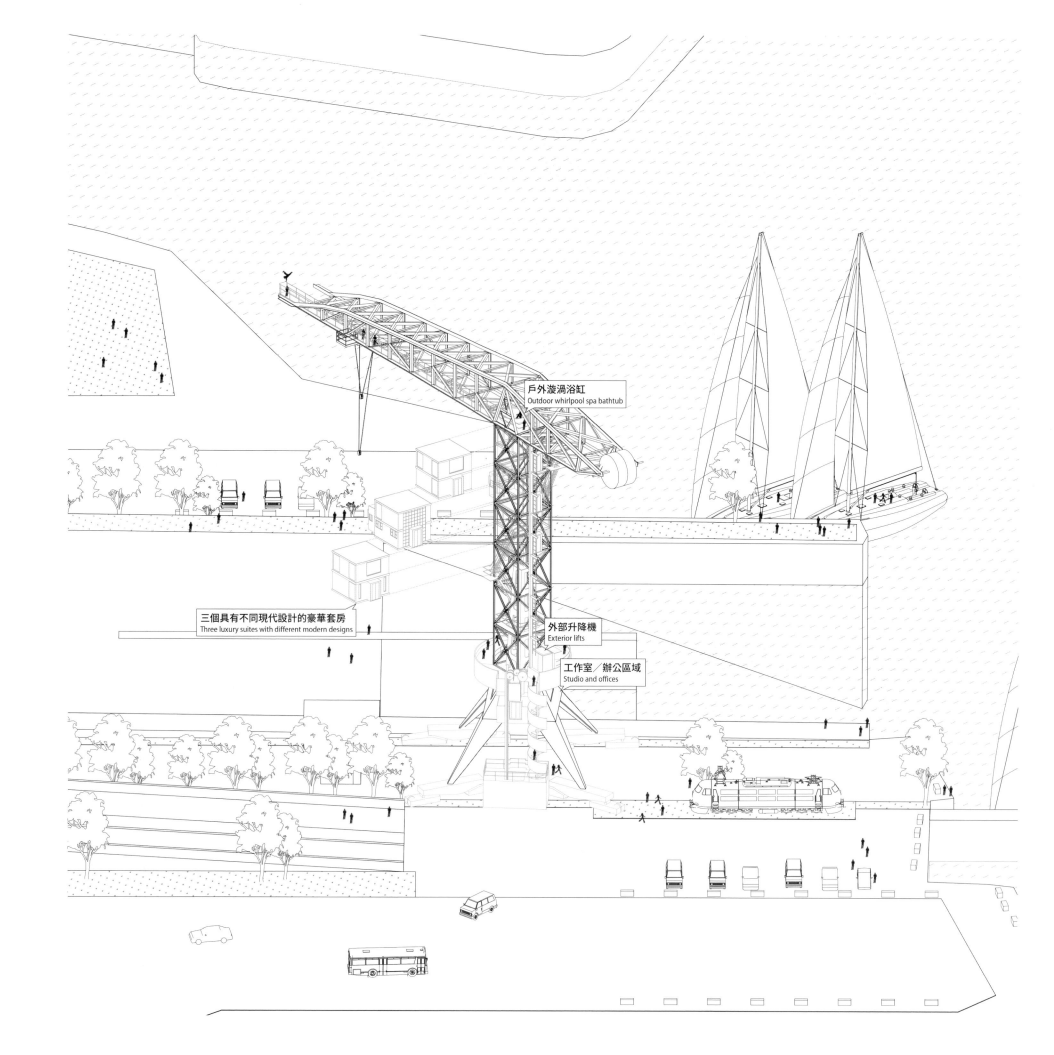

戶外漩渦浴缸
Outdoor whirlpool spa bathtub

三個具有不同現代設計的豪華套房
Three luxury suites with different modern designs

外部升降機
Exterior lifts

工作室／辦公區域
Studio and offices

B5

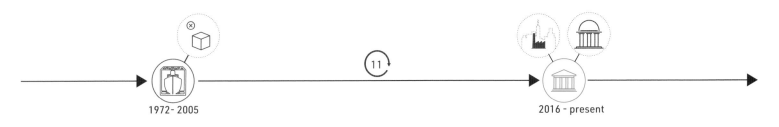

1972- 2005 　　　　11 　　　　2016 - present

中國上海船廠
1862藝術中心
Shipyard 1862,
Shanghai, China

DESIGNER 隈研吾建築都市設計事務所
| Kengo Kuma
and Associates

「船廠1862」建於1972年，位於上海浦東陸家嘴，是由上海船廠鍛造機間舊址改造而成的藝術商業綜合體。1862的數字則指出其真正在歷史上展現的重大意義。清末，西方列強迫使清朝開港通商，1862年由英商在浦東陸家嘴開設的祥生船廠，拉開了上海灘輝煌的造船序幕。祥生船廠初期製造軍火，後來修造船舶，通過兼併浦西的幾家造船廠，以及一戰時期接收了德國的瑞鎔船廠，形成當時英聯船廠的極盛版圖。太平洋戰爭爆發之後，船廠的東家也由英國、日本，最後落入國民政府手裡，成為上海船廠的一部分。

2005年，上海船廠整體搬遷至崇明後（死因），開發商與中國船舶工業集團合作，在船廠原址上開發占地136萬平方公尺的陸家嘴濱江金融城，而最靠近江邊的這座機間舊址被納入改造計畫的核心，期待透過其轉變重生的能量能重新激發陸家嘴一區的活力。

改造計畫委託隈研吾建築都市設計事務所與Arup合作，開啟了船廠1862充滿詩意的重生。船廠被改造為一棟以劇院為主體的藝術中心，26,000平方公尺的室內商場保留老船廠原本的工業感與歷史感，

地上五層地下一層的建築體，包含了一座有800座位席可以欣賞到浦西全景的劇院，以及商場、餐廳、展覽等配套空間，總建築面積達31,600平方公尺。

由於船廠保存良好，結構系統以新舊分離的策略進行設計，舊結構獨立設計的方式，更讓它們具備展品般的效果。因為被賦予多元的新機能，機電設備的介入是痛苦複雜的工程，既要符合現代化的需求，又要避免在視覺上影響歷史感。除了空間之外，建築師也重新利用船廠已廢棄的設備，例如舊蒸氣管改作空調送風管等手法，都在劇場專家的諮詢下完成，以確保符合劇場空間功能上的獨特性。

外部立面保留了既有的梁柱、桁架及局部磚牆，隈研吾事務所在充滿記憶的元素中引入輕盈、充滿現代感的玻璃帷幕，外側再疊加上以不鏽鋼繩懸掛著如圖元化的磚塊系統，以創新卻不失歷史感的表皮設計為新舊介面做銜接，這種特殊的立面設計也成為此案最吸睛的一道風景。

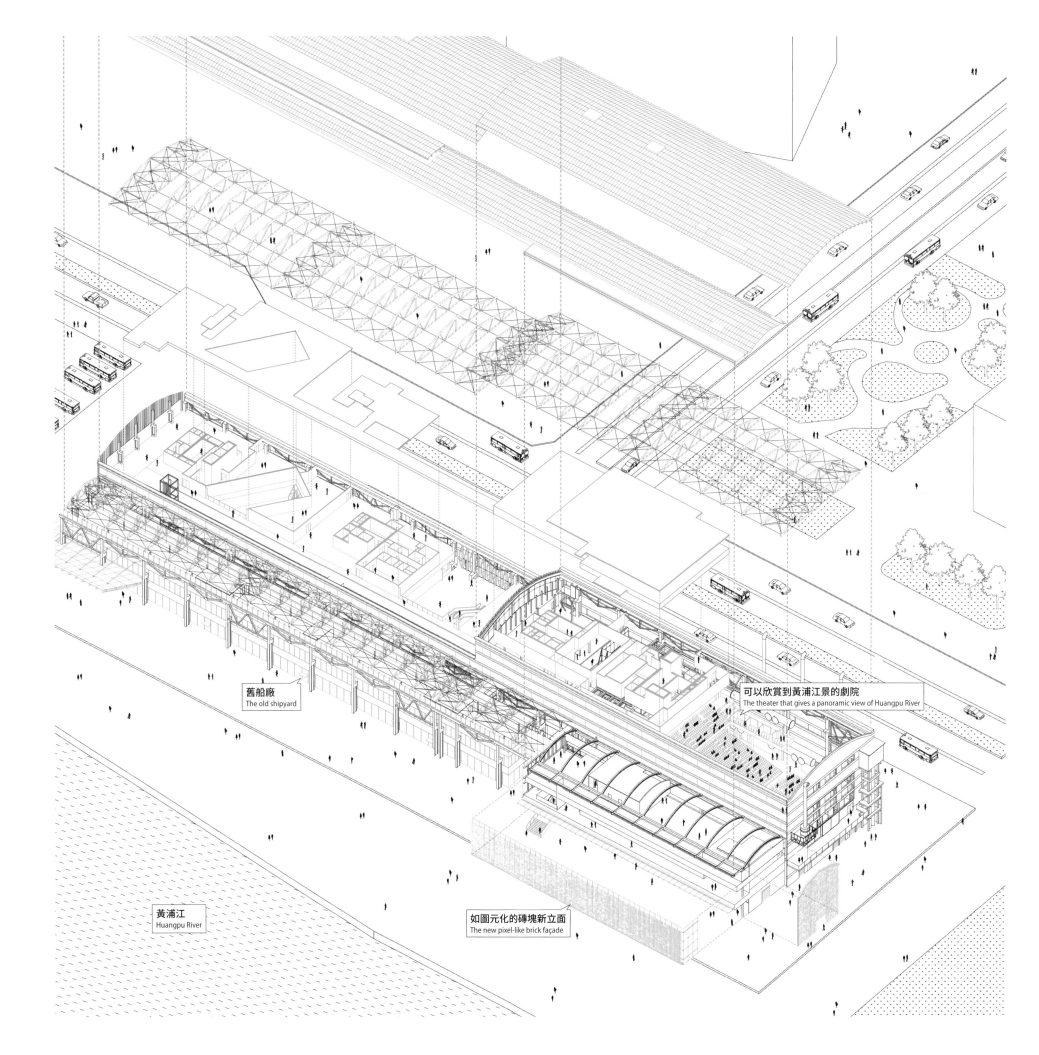

可以欣賞到黃浦江景的劇院
The theater that gives a panoramic view of Huangpu River

舊船廠
The old shipyard

黃浦江
Huangpu River

如圖元化的磚塊新立面
The new pixel-like brick façade

B6

德國漢堡
易北愛樂廳
Elbe Philharmonic Hall,
Hamburg, Germany

DESIGNER Herzog & de Meuron

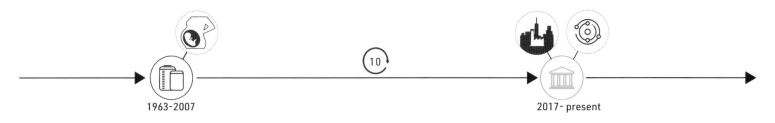

1963-2007 ⟲10 2017- present

史上最貴的音樂廳?史上最貴的工業遺產改造!

2001年歐洲規模最大的都更計畫「港城」HafenCity,在漢堡市中心的易北河港展開,計畫將舊港口改頭換面。UNESCO評為世界遺產的「倉庫城」Speicherstadt,便是港城中保存完善並持續使用的古蹟。Kaispeicher A的紅磚倉庫由Werner Kallmorgen設計,建於1963年至1966年,近50年的歷史裡一直儲存著可可豆和茶葉。它最初是為了承受成千上萬袋可可豆的重量而建造的。Kaispeicher A的易北愛樂廳標誌著原本漢堡大多數人都知道但從未真正注意到的位置,現在搖身一變成為漢堡市民與世界遊客的日常生活與社會文化的新中心。

建築師Herzog & de Meuron提出「新舊垂直疊加」往天空爭取空間:留下整座舊倉庫,就讓新音樂廳漂浮在舊倉庫的頂上,而將舊倉庫改造成立體停車場。下半部分既有紅磚倉庫的牆面和上半部分的鋼結構形成對比。舊倉庫結構承載能力極強,新建鋼結構部分透過一個結構轉換系統,將垂直力轉換到既有倉庫的結構系統上。結構技師在底部增加了1,761支水下鋼筋混凝土支撐柱,每支承載能力超過200噸。既有紅磚倉庫高度約37公尺,這是漢堡港平均天際線的高度。建築師在舊倉庫的上空設計了一個現代的、輕盈的、通透的、如同浪潮又如同冰山一般,最高高度超過70公尺的巨大虹彩晶體。其外觀因天空、水和城市的反射而不斷變化。整棟建築最高點26層,總高度110公尺,建築面積12萬平方公尺,總重量20萬噸。

為了解決易北愛樂廳地處城市末端的交通不便,漢堡市政府建造了一座連接老城區和Sandtorhafen的吊橋,以及將未來地鐵4號線(U4)延伸至此。此外還有多條公車路線和渡船停靠易北愛樂廳前。易北愛樂廳不僅是一個音樂場所,它更是一個完整的住宅和文化綜合體。2,100人座位的音樂廳,以及能容納550名聽眾的室內音樂廳,嵌入在豪華公寓和五星級酒店之間,內置服務設施,如餐廳、健身中心、會議設施等。為了使新的愛樂樂團成為真正的公共景點,這裡還設有一個享有漢堡景色的全景露台。這些不同的功能組合在一個建築物中,就像一個微型城市。

易北愛樂廳已不只是漢堡市,甚至是整個德國的標誌性建築,但這個項目最出名的並非它的設計成果,而是其冗長的工期與驚人的費用:完工時間延遲7年,預算增加10倍,不斷的工程安全和技術問題等等,過程中遭受諸多爭議和訴訟。2007年公開招標時,漢堡市僅批准7,700萬歐元建造費用,但在Herzog & de Meuron的方案中標後,建造費用提升至1.14億歐元。2012年做的最終決算定格於5.75億歐元,但到最後整個項目花費了7.89億歐元!建造時間從3年延長至10年。這個多年來一直被巨型吊車圍繞的工地,工程後期甚至有市民提議保留幾座鷹架,因為它們占地數年,早已成為市容的一部分!2017年1月11日音樂廳終於正式開幕,舉辦了第一場演奏會。

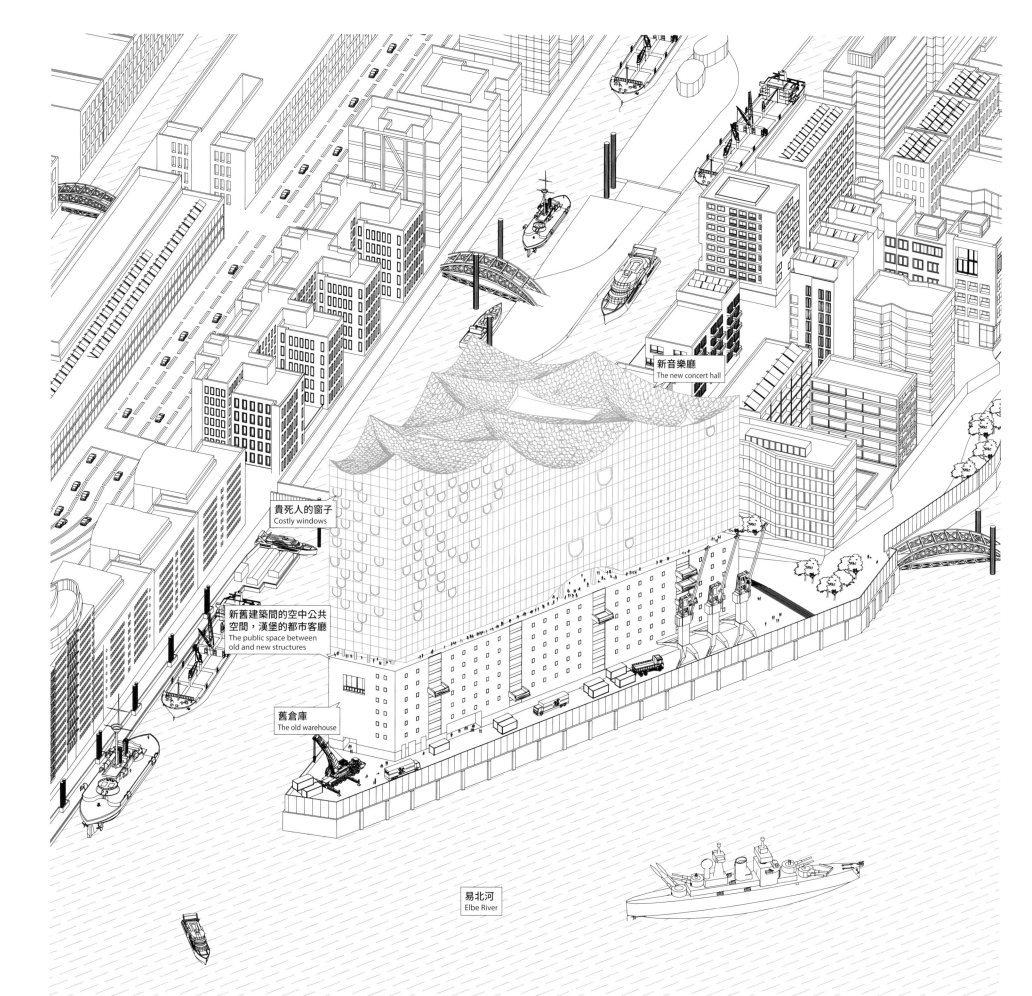

新音樂廳
The new concert hall

貴死人的窗子
Costly windows

新舊建築間的空中公共
空間，漢堡的都市客廳
The public space between
old and new structures

舊倉庫
The old warehouse

易北河
Elbe River

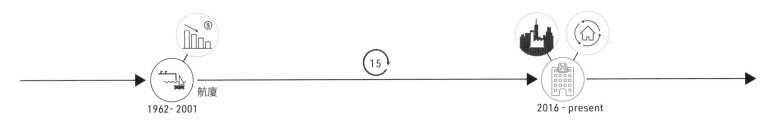

航廈
1962 - 2001
15
2016 - present

美國紐約甘迺迪機場TWA飛行中心飯店
TWA Flight Center Hotel, New York, USA

DESIGNER Beyer Blinder Belle & Lubrano Ciavarra Architects

位於美國紐約市約翰·甘迺迪國際機場的「環球航空飛行中心（TWA Flight Center）」於1962年啟用，是建築師埃羅·薩里寧（Eero Saarinen）為環球航空（Trans World Airlines）所設計。其向兩側揚起的翼形薄殼混凝土屋頂、從室內延續到室外的流線型結構、非比尋常的管狀紅地毯走廊、鑲嵌在屋頂結構間隙的天窗，以及可看到飛機起降的廣角視野玻璃立面，有機形式形成密不可分的整體，經由細節堆疊，使環球航空飛行中心成為體現20世紀航空精神的代表建築。

環球航空飛行中心於1994年被定為紐約市歷史古蹟，2005年登上美國國家史蹟名錄，然而，由於環球航空於1990年代的財務困難，以及近代航空業營運模式轉變，於2001年停止營運。2008年竣工的第五航廈增建案，拆除了原航空站建築群的外圍結構，並新建了部分環繞原址的半月形量體，作為捷藍航空（JetBlue Airways）實際營運的第五航廈。此時環球航空飛行中心主建築仍在修復中，用途未定。

在關閉15年後，紐約州州長安德魯·古莫（Andrew Cuomo）和開發商MCR/ MORSE、建築師事務所Beyer Blinder Belle及Lubrano Ciavarra Architects展開了備受矚目的整建計畫，旨在保護和恢復環球航空飛行中心，反映其1962年的設計美學。此計畫包括於兩側新建酒店量體，修復建築本體及查爾斯·埃姆斯（Charles Eames）、雷蒙德·洛威（Raymond Loewy）和華倫·普拉特納（Warren Platner）的室內設計，並設置一個博物館展示紐約作為噴射機時代的誕生地、環球航空公司的歷史，以及現代設計運動。

環球航空飛行中心於2019年開始新的生命，修復後航空站作為大廳、商業空間、餐廳與酒吧。在大廳外的廣場上，一架60年的環球航空洛克希德星座L-1649A客機，將改造為酒吧使用。每個航廈的旅客都可以搭乘甘迺迪國際機場捷運，通過薩里寧設計的管狀走廊，到達TWA酒店，參訪逾半世紀的經典之作。

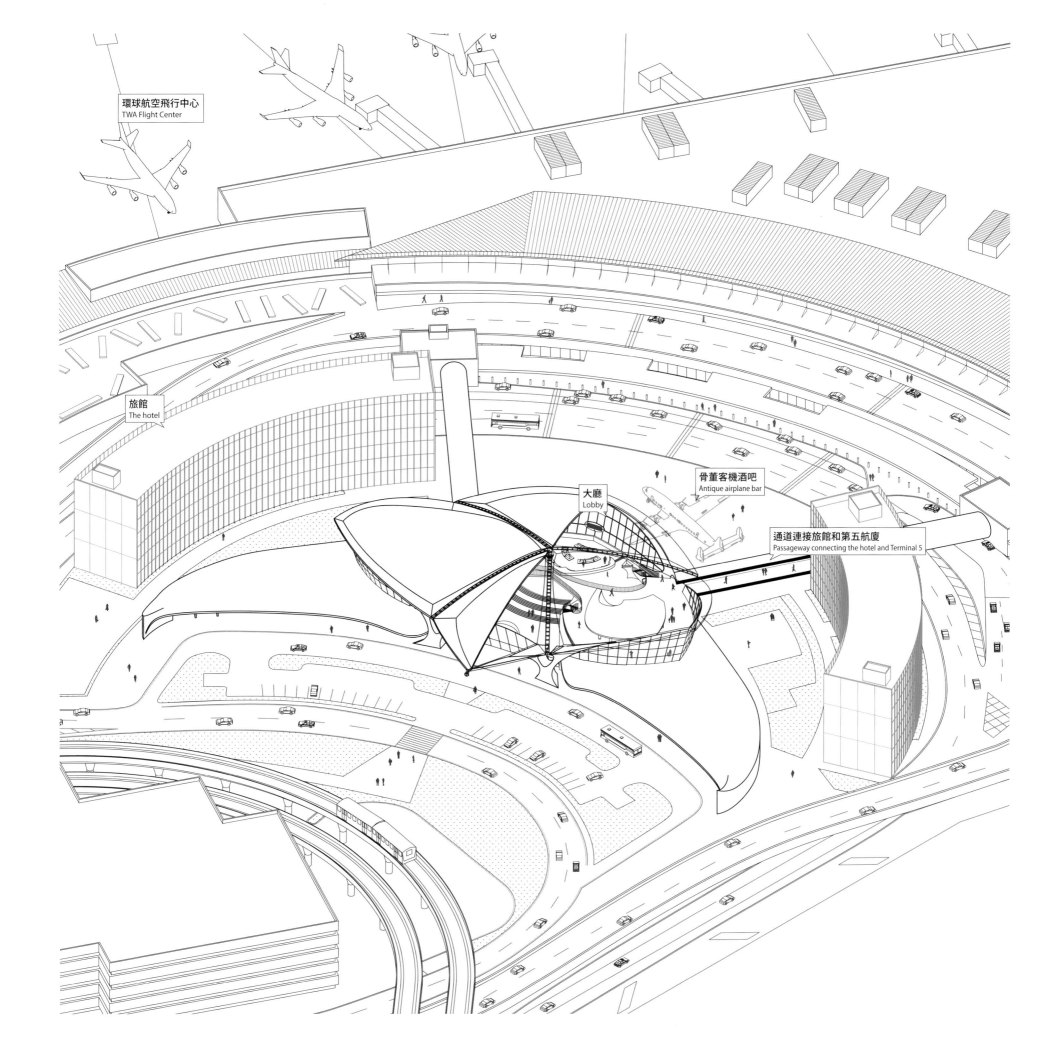

環球航空飛行中心
TWA Flight Center

旅館
The hotel

大廳
Lobby

骨董客機酒吧
Antique airplane bar

通道連接旅館和第五航廈
Passageway connecting the hotel and Terminal 5

C

資源開採 Mines

C1
西班牙巴塞隆納
水泥廠辦公住宅
La Fábrica, Sant Just Desvern, Spain

C2
羅馬尼亞圖爾達
地下鹽礦博物館
Turda Salt Mine, Turda, Romania

C3
英國康沃爾郡伊甸園植物園
Eden Project, Cornwall,
United Kingdom

C4
馬來西亞西巴丹
鑽油平台潛水旅館
**Seaventures Dive Rig Resort,
Sipadan, Malaysia**

C5
中國上海佘山
地下深坑酒店
**InterContinental Shanghai
Wonderland, Shanghai, China**

C6
英國蘇格蘭克羅馬帝灣
鑽油平台（設計提案）
Cromarty Firth, Scotland (Proposal)

C1

西班牙巴塞隆納水泥廠辦公住宅

La Fábrica, Sant Just Desvern, Spain

DESIGNER Ricardo Bofill

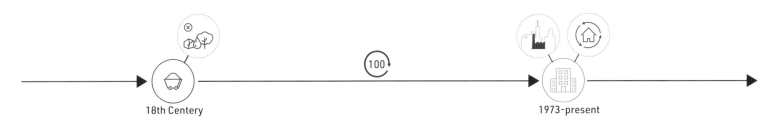

18th Centery 　　　　100　　　　 1973-present

這大概是所有建築師的夢想：像西班牙建築師里卡多·波菲爾（Ricardo Bofill）這樣，將舊水泥廠轉變為家庭和工作場所。

1973 年，里卡多·波菲爾和他的團隊來到了巴塞隆納郊外的一家水泥廠，對他而言這家工廠的狀況似乎相當不錯。廢棄的工廠可能對大部分的人而言是不值得再看一眼的巨大混凝土廢棄物，但對於里卡多·波菲爾和他的團隊來說，它充滿了潛力。這座被遺棄的、部分廢墟的工廠是一個超現實主義元素的彙編：爬上無處不在的強大的鋼筋混凝土結構，懸掛在空中的鐵塊，還有充滿魔力的巨大空間。

建築師和他的團隊決定購買和改造舊工廠，將其變成建築師的家和團隊的工作室。第一次的重塑工作歷時兩年。改造過程始於拆除舊結構的一部分，留下迄今為止我們所看到的形式，就好像混凝土已經雕刻一樣。一旦空間被定義，清除了舊有多餘的部分並被新生的綠化包圍，就開始啟動了「舊建物適應新功能」的過程。8 個筒倉仍然存在，拆除一些內牆以獲得可以作為設計工作室的開放空間。工廠的巨大開口得以維持，為寬敞的內部空間提供了理想的自然光線。這些筒倉成為辦公室、模型實驗室、檔案館、圖書館、投影室和被稱為「大教堂」的巨大空間，具有展覽、音樂會以及與專業活動相關的一系列文化功能。新的景觀綠化依照建築師的意志開始與舊建物的新功能結合，矗立在花園中。

總共經過 45 年的不斷翻新，La fábrica 已從魔幻成了現實。它的屋頂和周圍環境完全覆蓋著茂密的新植被，工廠的不同部分保留了原有的形式和建築風格，承載不同的新功能。這個項目證明了這樣一個事實：一個富有想像力的建築師可以將任何空間適應一個新的功能，不管它與原來的功能有多大的不同。

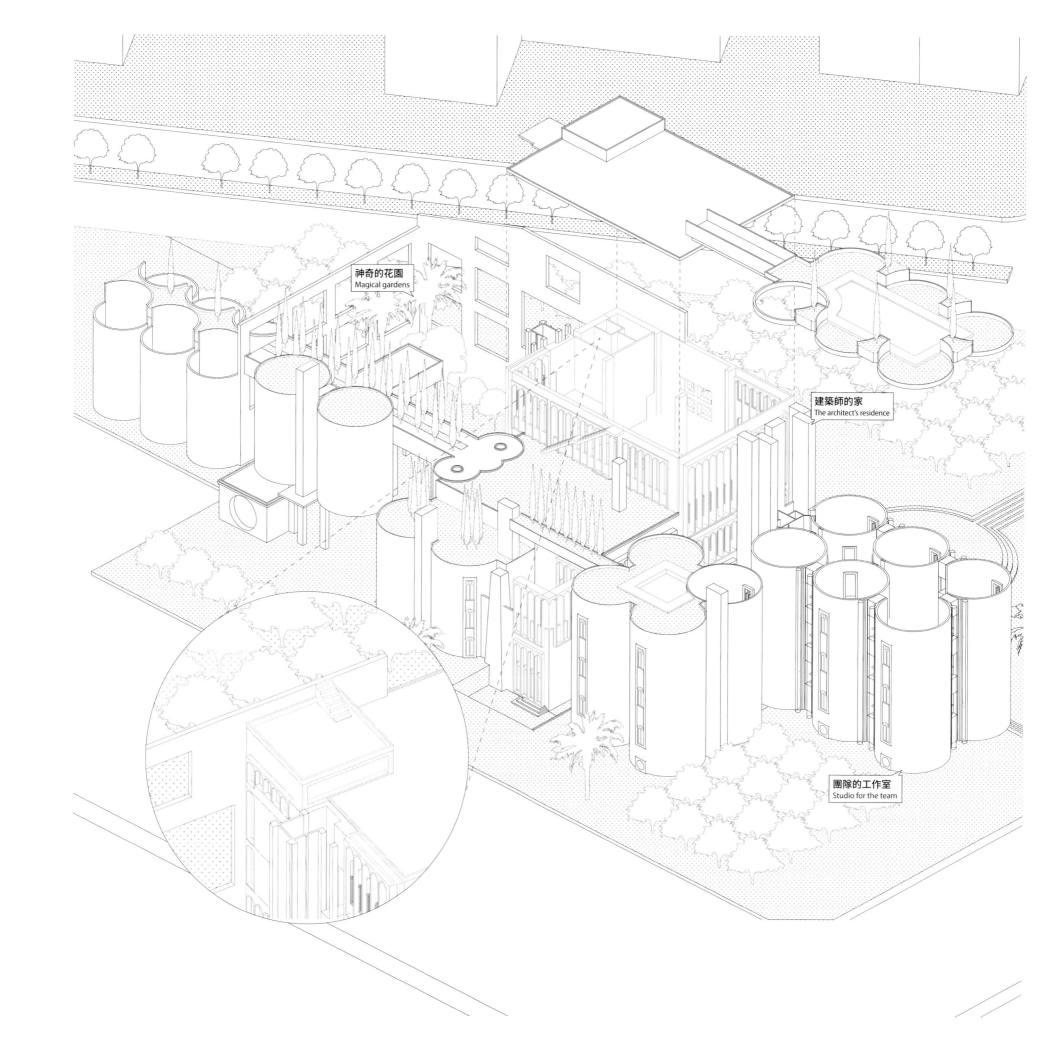

神奇的花園
Magical gardens

建築師的家
The architect's residence

團隊的工作室
Studio for the team

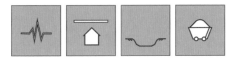

C2

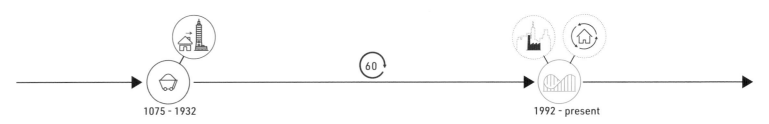

羅馬尼亞圖爾達地下鹽礦博物館
Turda Salt Mine, Turda, Romania

DESIGNER Ecopolis

這個令人嘆為觀止的地下空間過去曾經是羅馬尼亞圖爾達的一個巨大鹽礦坑，目前已經被精心翻修成為世界上最大的地下鹽礦遊樂場與鹽礦歷史博物館。從中世紀（該礦區在1075年首次出現於歷史文獻）到1932年間，這個礦場不斷生產食鹽。這裡鹽礦純度達到80%，但由於技術設備簡陋，產量下降，不敵另一個特蘭西瓦尼亞鹽礦的競爭，圖爾達鹽礦終於在1932年停止開採（死因）。

圖爾達鹽礦於1992年重新全年開放，用於旅遊和治療目的。值得一看的是保存完好的中世紀機器和鹽磨。在2009年歐盟的財政支持下，為了刺激當地旅遊，圖爾達鹽礦展開了大量的改造工程。歷經兩年的工程和600萬歐元的投資後，圖爾達改頭換面，目前此地設有帶天然氣溶膠的溫泉治療室、露天劇場、健身房，和一個「全景小型摩天輪」，旅客還可以在這裡欣賞鹽風化物的沉積和鐘乳石的形成。

礦區地表面的Durgau湖是受歡迎的旅遊景點，全年吸引著大量的遊客。遊客沿著曾經運輸了上千噸鹽的垂直動線向下行進，過程中會漸漸感受到過去挖掘此礦坑的巨大規模。在到達礦井底部時有一個地底湖，覆蓋著沙子般的鹽層。湖中小島上的設施在設計上似乎借鑑了某種深海生物的美學，礦井的底部幾乎是由木材構件製成的外加人工結構，並用懸掛的管燈照亮。礦井內部保持穩定的攝氏恆溫12度和80%的濕度，礦坑內幾乎完全沒有任何過敏原和任何細菌。獨特的微氣候成為呼吸道過敏患者的舒適目的地。

博物館實際上包括三個礦井：Terezia礦井在120公尺處達到最深，然後是108公尺處的安東礦井和42公尺處的魯道夫礦井。從每個礦井中可以看到數百個不同層疊出的如凝固流體般的岩層，這些岩層歷經極長的時間才形成。每個礦井內的構造物都重修改造後放回各自原本所屬的地點，以回復這些用數百年艱苦勞動所雕刻出的地下地形。動線由保存多年來運作礦井的路徑決定。

遊樂場包含一系列場景，採用經過修復的既有設備或至少統一用相同的設計元素表達。圓形劇場、運動場、摩天輪、迷你高爾夫球場和保齡球道等等，讓遊客可以體驗到礦山的不同面向。鐘乳石和鹽風化完成了巨鐘Terezia礦井的惰性平衡，礦井底部的地下湖深0.5至8公尺，在湖的中心有一個島嶼，由1880年後在這裡沉積的殘留鹽形成。最好玩的是可以在島上租用小船，遊覽這個被挖成圓錐鐘形空洞最低點的地下湖。

在2012年至2014年間，Terezia礦與Losif礦之間建立了一條連接隧道，長50公尺，包括在旅遊線路和Losif礦山中，尚未向大眾開放。最深處為地下120公尺。圖爾達地下鹽礦的重生案例巧妙利用了這個人工挖掘的地下地形，原汁原味地呈現。雖曰人工，豈非天然？

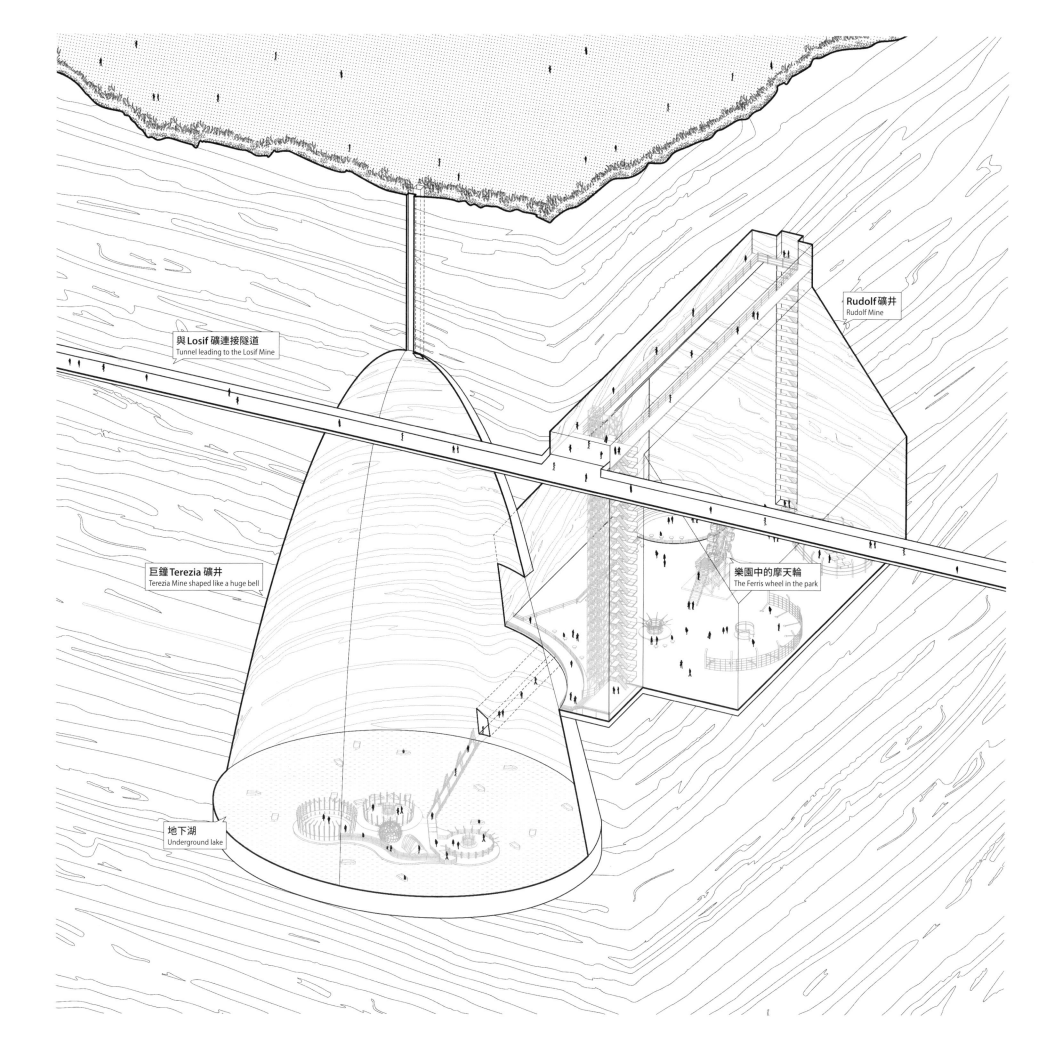

與 Losif 礦連接隧道
Tunnel leading to the Losif Mine

Rudolf 礦井
Rudolf Mine

巨鐘 Terezia 礦井
Terezia Mine shaped like a huge bell

樂園中的摩天輪
The Ferris wheel in the park

地下湖
Underground lake

C3

英國康沃爾郡
伊甸園植物園

Eden Project, Cornwall, United Kingdom

DESIGNER Nicholas Grimshaw

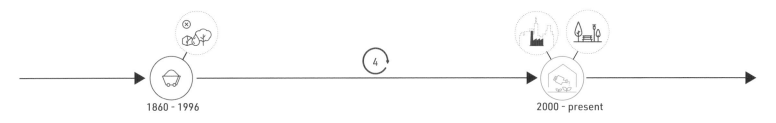

1860 - 1996　　　　4　　　　2000 - present

「伊甸園計畫（Eden Project）」位於英國康沃爾郡，是全世界最大的室內植物園，也是英國最受歡迎的休閒旅遊景點之一。這個人稱為世界第八大奇蹟的花園，坐落在一個30個足球場大小的礦坑中。伊甸園不僅僅是一個熱帶花園，它還是通往植物與人類之間關係的門戶，呈現人類對植物世界的迷戀，也保存了相關教育和知識的永續資源。

伊甸園計畫所在的製陶用黏土礦場在開採了160多年後，到了1990年代中期黏土礦場終被消耗殆盡（死因）。廢棄之後這裡留下了一個無用的大坑，建築師格里姆肯（Nicholas Grimshaw）面對著不斷變化的地形景觀，想到了肥皂泡。它們可以適應所處的任何表面，並且，當兩個或多個氣泡連接時，連接線始終完全垂直（直線向上和向下）。在肥皂泡上設置「傾斜」的生物群落結構是在不平坦的坑洞上建造溫室的完美方式。

於是，伊甸園計畫就由多個像肥皂泡的巨大溫室組成，每個溫室都模仿一個天然生物群。溫室的圓頂（Dome）是由數百個鋼框架支撐的六邊形和五邊形所組成。為了減重，每個六邊形和五邊形中的

透明「窗戶」採用乙烯—四氟乙烯共聚物（ETFE），充氣以形成一個個2公尺深的透明枕頭。雖然ETFE窗戶比起玻璃來說非常輕，但強度足以承受汽車的重量，可以傳輸紫外線並且容易清潔，使用壽命超過25年。

三大生物群落包括戶外花園、雨林生物群系和地中海生物群系，共占地約13公頃。其中最大的溫室中模擬了熱帶雨林環境，在雨林生態園遊客可領略到全球四大熱帶雨林環境：熱帶島嶼、東南亞、西非和南美洲。遊客可以漫步在「林冠走廊（Treetop Walkway）」，從上而下俯瞰整個園區，感受熱帶雨林的壯麗景觀。伊甸園項目的第一部分，遊客中心於2000年5月向大眾開放，整個園區則於2001年3月17日開放。

除了為所在的康沃爾郡經濟貢獻了超過10億英鎊，伊甸園計畫園區率先用綠色電力——能源來自康沃爾郡的風力與地熱發電機。它之所以稱之為「計畫」，是Project而不是Garden，與許多景觀項目的本質一樣，因為它是不斷生長、持續改變，與一次次新的階段目標，持續象徵著土地的「重生」（regeneration）。

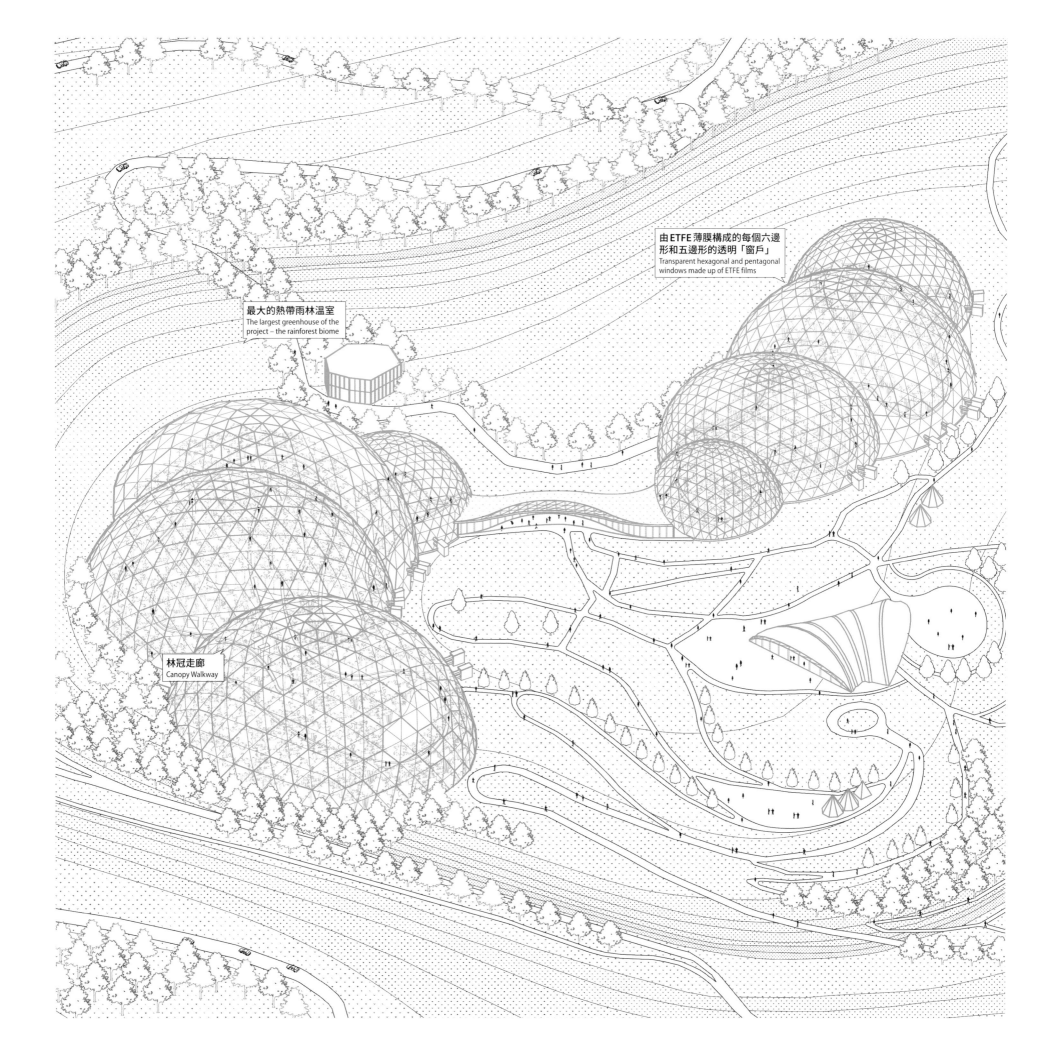

由ETFE薄膜構成的每個六邊形和五邊形的透明「窗戶」
Transparent hexagonal and pentagonal windows made up of ETFE films

最大的熱帶雨林溫室
The largest greenhouse of the project – the rainforest biome

林冠走廊
Canopy Walkway

C4

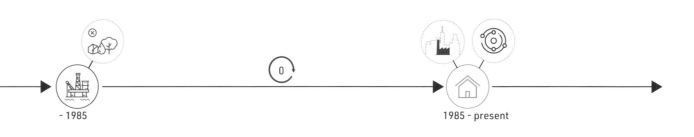

- 1985

1985 - present

馬來西亞西巴丹鑽油平台潛水旅館

Seaventures Dive Rig Resort, Sipadan, Malaysia

DESIGNER Morris Architects

Seaventures 潛水度假村以前是座鑽油平台，石油產業將其拖拽至不同地點開採石油和天然氣，早時稱為 The Rig。它在巴拿馬建造並由馬來西亞國家石油公司（Petronas Oil Corporation）所使用，1985年退役後在新加坡出售。

退役的鑽油平台改造成旅館或住宿功能的潛力，在於不需再砍樹或不需在海洋中進行土方工程，可以在不破壞海洋生態系統的情況下建造。挑戰是將工業石油平台改造成舒適的酒店──空調客房配有相鄰的洗手間設施和熱水淋浴，中央休息室可以享受清新的海風，空間寬敞到足以容納不定期的現場娛樂表演。最初它是一家以釣魚為導向的旅館，但後來轉型成潛水導向。Suzette Harris 家族將它從納閩 Labuan 拖走，遷移距離長達 800 公里，目的是前往世界上最好的潛水點。新的基地位於馬布島附近，靠近西巴丹島，是一個要申請許可證才能到達的國家公園，世界著名的潛水者天堂。

自1997年以來，這個鑽油平台一直留在目前的位置，其中的建築和庇護所為魚類創造了一個天堂。雖說 Seaventures 潛水度假村不算是奢華的地方，但這家擁有 25 個房間的工業旅館確實提供了充足的基本住宿條件，所有餐飲、船隻運輸、潛水和設備租賃均包含在套裝行程費用中。Suzette 已經成功地將一家小型新穎的潛水旅館擴展成為一處熱門的知名度假勝地。

這個案例雖然前所未有地再利用了退役的鑽油平台，但旅館以及旅遊本身並不完全環保。撇開動力來源是充滿臭味的柴油發電機不說，光是到達那裡的旅程就很耗能：搭機抵達馬來西亞首都吉隆坡後，接著得搭乘飛往斗湖的航班，到達馬來西亞婆羅洲沙巴州東海岸，再經過一小時的車程來到小港口城市仙本那過夜，隔天乘船一小時後才能到達鑽油平台。也許，有朝一日，另一座鑽油平台再利用的旅館會考慮增設太陽能板或風力發電機。

儘管如此，世界上出現這種極端的回收利用案例仍然是件很酷的事。

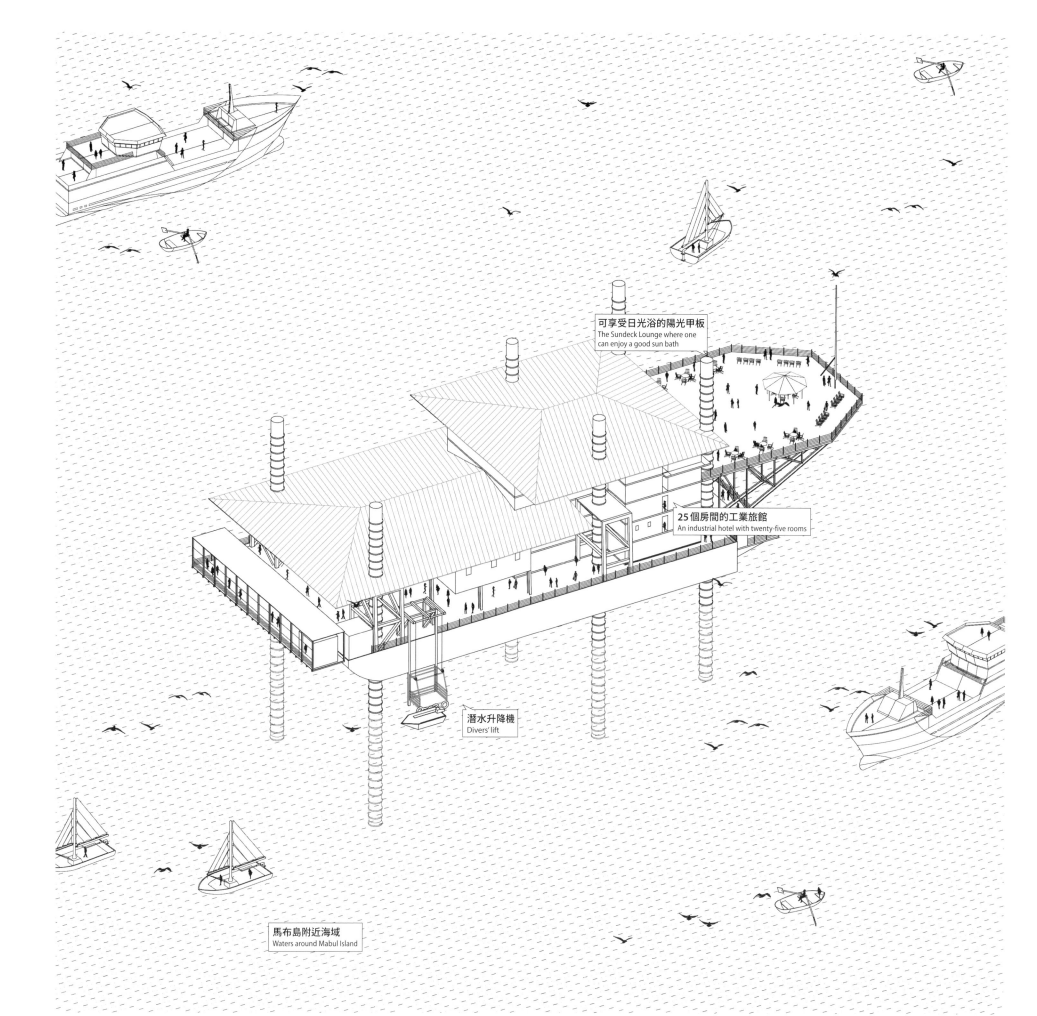

可享受日光浴的陽光甲板
The Sundeck Lounge where one
can enjoy a good sun bath

25個房間的工業旅館
An industrial hotel with twenty-five rooms

潛水升降機
Divers' lift

馬布島附近海域
Waters around Mabul Island

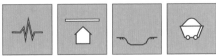
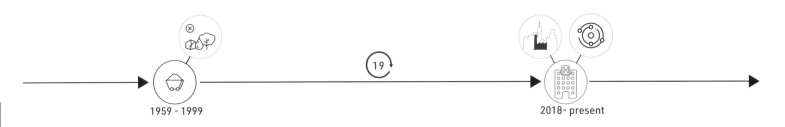

1959 - 1999   19   2018- present

中國上海佘山
地下深坑酒店

InterContinental
Shanghai Wonderland,
Shanghai, China

DESIGNER Martin Jochman

上海缺山少石，但天馬山深坑原本其實是一座山，東北遠望佘山，早前被炸山採石，至1950年代末，整個山丘已蕩然無存。1959年，天馬人民公社在小橫山設立採石場，隨著城鎮建設和經濟發展的需要，石材需求量與日俱增，開挖面積不斷擴大，石坑深度逐漸加深，終形成了80多公尺的深坑。2005年，世茂集團買下了佘山地區七個礦坑中最大的一個，他們選擇不抹去這一段記憶，而是承認了人類對地表的破壞是人類發展正常的一部分，世茂決定充分利用深坑的自然環境，發揮充分的想像力，建造了一座和這個深坑融為一體的五星級酒店。

2006年，深坑酒店正式成立，英國 Atkins 事務所的提案在國際設計競賽中獲勝。與傳統酒店往天空發展相反，深坑酒店扎根地下，然而因其「深度」同時也帶來了無數建築技術難題，例如消防、防水、防震等。經過7年的科學論證，深坑酒店在2013年開工。

深坑酒店總體高度約為70公尺，共19層樓，分為坑外和坑內兩部分，坑外為3層，坑內16層，其中有2層在水下，包含客房、會議室和餐廳。每間客房都是依山而建，主體的建築材料均採用金屬板材和玻璃，入住者可以在其中全視野觀看深坑的整體景色。設計師在這個廢棄的礦坑中建造了人工景觀湖，為了維持安全水位，設置了抽水泵，以確保每天湖水的水位變化不超過5公分的安全區間。火災預防方面，由於消防車進不了礦坑，所以要確保酒店的每一個陽台都和消防通道相連。

主體建築內共有337 間客房和一間總統套房。其中湖面以上的14層為標準客房，每一間都設有觀景露台，和走廊連接，形成「空中花園」。入住者可以在這裡看到西北側坑壁落差近百公尺的瀑布，瀑布是此開發案中最引人注目的特色之一。喜歡冒險的旅客還可以盡情享受攀岩的樂趣。最低的兩層是套房和總統套房，位於觀景湖水平面以下，擁有最好的水下景觀視線，試圖營造置身海洋中的新奇感。

建設酒店的時候，工程人員遇到了很多問題，研究人員為酒店研發了兩大向下深逾77公尺的混凝土輸送系統：簡稱「三級接力」和「一溜到底」。整個建案的工時超過10年，調派人力超過5,000名建築師、工程師、設計師和工人，獲得了39項專利，取得了數百項技術突破。深坑酒店的建築設計師馬丁認為，能夠將一個廢棄的工業疤痕重生推出成為全新的建築，賦予全新的功能，是人類變廢為寶的里程碑。它是中國對工業遺留問題的再一次成功案例，同時也是全球海拔最低的酒店。

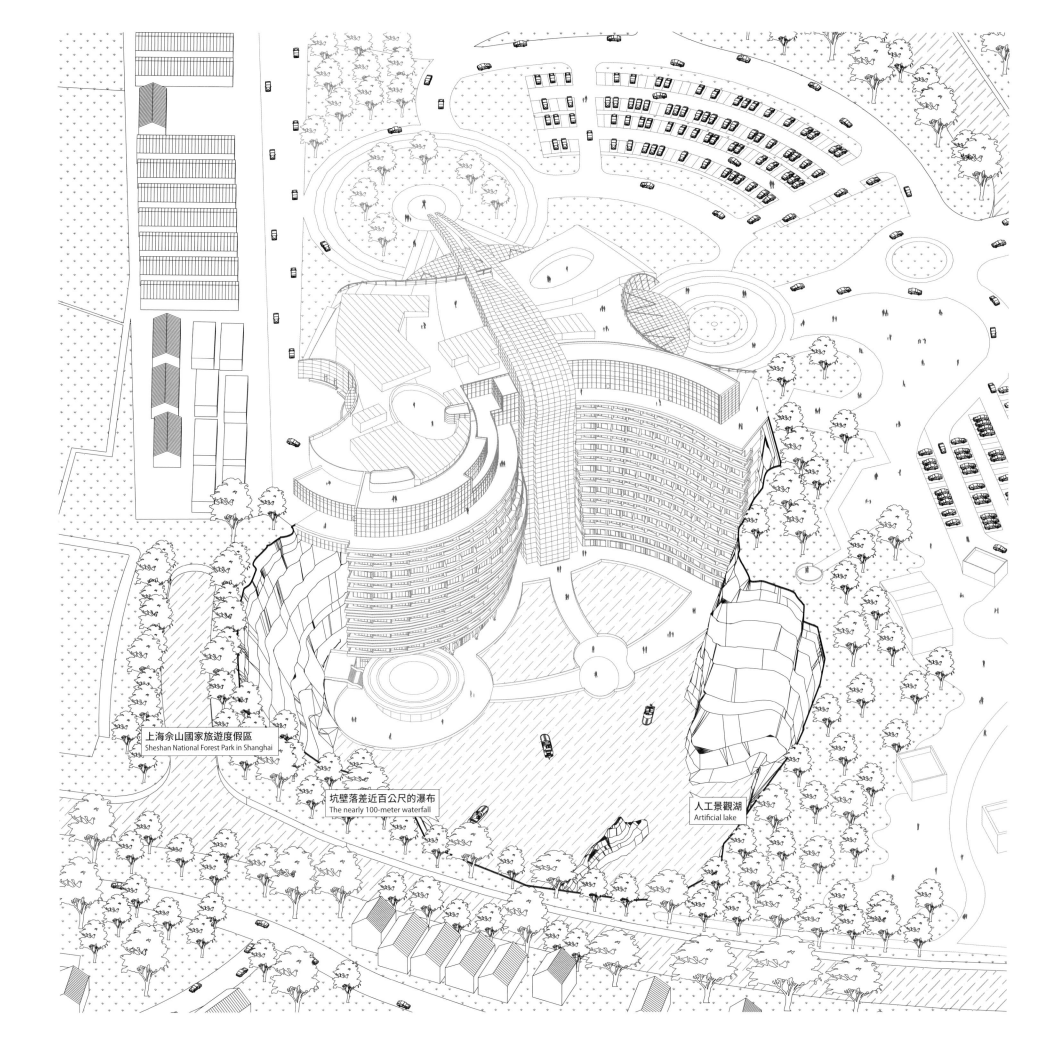

上海佘山國家旅遊度假區
Sheshan National Forest Park in Shanghai

坑壁落差近百公尺的瀑布
The nearly 100-meter waterfall

人工景觀湖
Artificial lake

C6

1980-2015　　　　想像　　　　proposal

英國蘇格蘭
克羅馬帝灣
鑽油平台
（設計提案）

**Cromarty Firth,
Scotland, United
Kingdom (Proposal)**

DESIGNER 王俐雯

鑽油平台為位於海上的大型結構設施，主要用於鑽井提取石油及天然氣，並能夠暫時儲存，直至石油被運送到陸地上的煉油廠去裂解原油成為產品。

鑽油平台上包含開採設施以及容納勞動力人口的居住空間，根據不同的需求也能夠產生出不同類型的鑽油平台，例如可移動式鑽油平台或是可以被固定在海底的固定式平台，而這些平台也可以相互連結，形成鑽油平台群。

這裡運用來設計改造的平台為半潛式海上鑽井平台，這種類型的平台有一個浮動鑽井單元，其中包含柱和浮筒，當被水淹沒時會將浮筒下沉到預定的深度，而通過其下部船體的「膨脹」和「縮小」使浮筒沉入水面以下，鑽機部分雖然會被淹沒，但仍然漂浮在鑽井現場上方。此類型鑽油平台下部船體充滿水時可在鑽井時提供穩定性，因此可以在海上穩定作業。

這個設計主要以蘇格蘭海灣的廢棄鑽油平台群作為發想。鑽油平台產業曾經處於高峰，但近期由於環境意識抬頭以及國際油價漸漸下降，鑽油平台正在面臨大量退役的問題。鑽油平台本身除了建造費用高昂之外，拆除費用也相當可觀，因此鑽油平台退役後，大部分的處置方式是將其閒置，因此在思考關於鑽油平台的再利用可能

時，以可以大量取用其原有設備及物件為主。

鑽油平台本身具有能在海中移動的特性，因此思考將退役的鑽油平台以單元式分類設計成不同的機能，例如住宅單元、研究單元以及生活機能單元，這些單元彼此可以透過海中的生質能以及太陽能板、潮汐能等自給自足、獨立運作，亦可以透過由設備改造而成的連接橋相互串聯，進而有機會可以形成海上都市，為人類提供一個有別於陸地的新住所。

設計的起頭是先分析原有鑽油平台的功能以及設備，因鑽油平台本身屬於機械設施，有許多可以重製或再利用的機會，因此在仔細分析之後，取用其中可以保留或是重新設計賦予其新機能的物件以及設備、空間，將其整合規劃後重新設計來繼續使用，如此一來除了可以有效利用現有的設備及空間，也能夠降低其物件報廢的成本。

面臨拆除價格過高以及汙染海洋的議題，重新設計後的鑽油平台將不再汙染環境，而是利用其原有設備重新設計及利用，轉而成為自行產生能源的單元，並且能夠作為研究及居住使用。期望未來若是可以將改造計畫落實在即將退役的鑽油平台上，將有機會可以產生出一種新的生活模式。

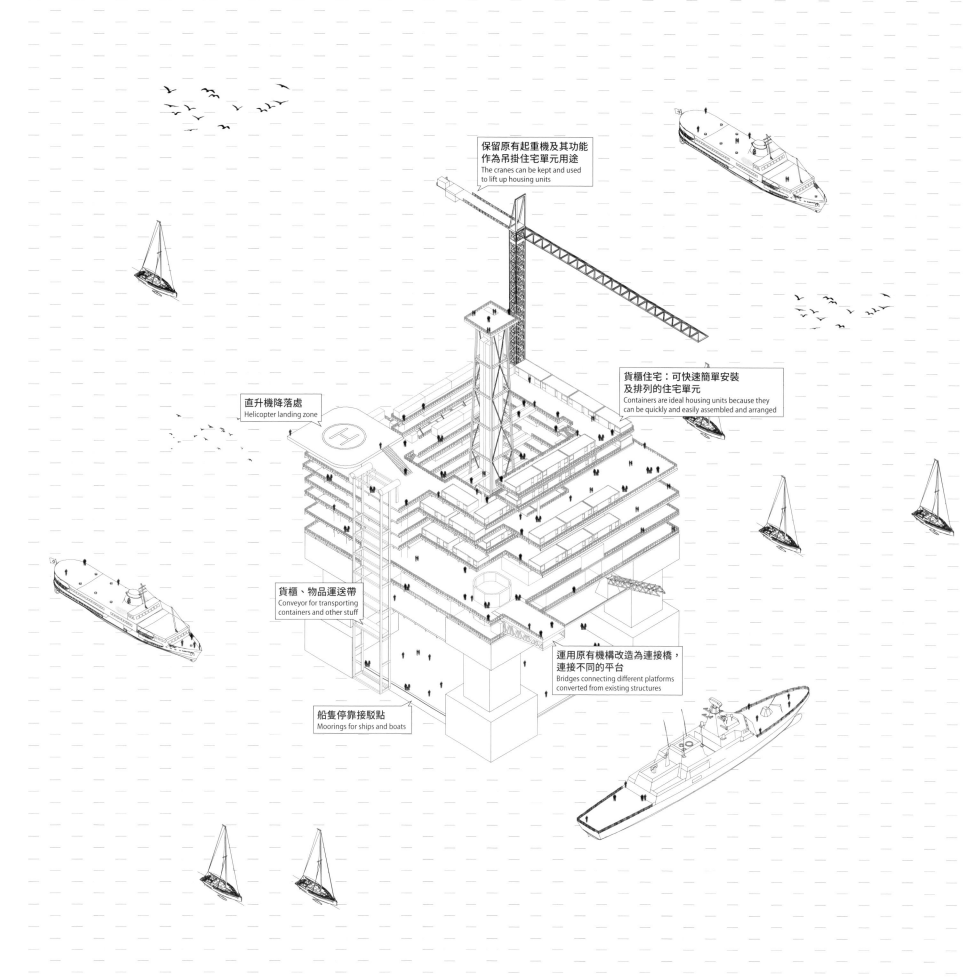

保留原有起重機及其功能
作為吊掛住宅單元用途
The cranes can be kept and used
to lift up housing units

貨櫃住宅：可快速簡單安裝
及排列的住宅單元
Containers are ideal housing units because they
can be quickly and easily assembled and arranged

直升機降落處
Helicopter landing zone

貨櫃、物品運送帶
Conveyor for transporting
containers and other stuff

運用原有機構改造為連接橋，
連接不同的平台
Bridges connecting different platforms
converted from existing structures

船隻停靠接駁點
Moorings for ships and boats

D
供電 Power Supply

D1
德國卡爾卡爾核電廠
仙境主題樂園
Wunderland Nuclear Reactor
Amusement Park, Kalkar, Germany

D2
英國倫敦泰特現代美術館
Tate Modern, London,
United Kingdom

D3
台灣高雄青埔
垃圾發電景觀公園
Kaohsiung Metropolitan Park,
Kaohsiung, Taiwan

D4
台灣桃園榮華壩（設計提案）
Ronghua Dam, Taoyuan,
Taiwan (Proposal)

W

供水 Water Towers

W1
荷蘭羊角村觀景塔
Watch Tower Sint Jansklooster,
Jansklooster, The Netherlands

W2
中國瀋陽水塔展廊
Water Tower Pavilion (Renovation),
Shenyang, China

W3
比利時安特衛普
森林水塔住宅
Water Tower Housing Brasschaat,
Brasschaat, Belgium

W4
德國布蘭登堡水塔住宅
BIORAMA-Projekt,
Brandenburg, Germany

W5
英國薩福克郡雲中之家
House in the Clouds,
Suffolk, England

W6
匈牙利德布勒森水塔
咖啡廳／畫廊／
攀岩場／觀景台
Nagyerdei Víztorony,
Debreczin, Hungary

D1

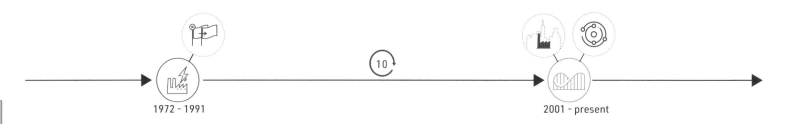

德國卡爾卡爾
核電廠
仙境主題樂園

**Wunderland
Nuclear Reactor
Amusement Park,
Kalkar, Germany**

DESIGNER Hennie van der Most

這座位於德國、耗資53億美元的核電廠,如今化身成為一座每年吸引數十萬遊客的遊樂園。

它建立在 SNR-300 核電廠舊址上,這是德國1972年開始建造的第一座快速增殖核反應爐電廠,這種反應爐設計使用鈽作為燃料並用鈉冷卻,完工後將可輸出327兆瓦的電量,但與常見的反應爐設計相比下,它有更高的核災風險。

1986年車諾比爾核災之後,當地居民感到更加恐慌,高度關注核電安全的抗爭使得工程不斷推遲,最後演變為政治和環境的噩夢(死因)。當反應爐最終完工時已花了53億美元,由於運作成本太高與長期的安全抗爭,政府當局最後決定不投入運作。該核電廠因完全未被使用而成為了世界上最複雜、最昂貴的巨型垃圾之一:這筆錢足以建造2萬戶獨立住宅。整個建築群規模大約占地80個足球場,所用的混凝土之多,足以建造從阿姆斯特丹到馬斯垂克的高速公路,而其複雜的電線系統連接起來足以環繞地球兩次。

這座核電廠建築群與土地最終賣給了荷蘭投資者 Hennie van der Most,該公司傳聞以300萬美元的價格購買了核電廠,並決定將它改造成遊樂園,最後命名為「Wunderland Kalkar」。總面積為136英畝,共有40個景點,包括類似於迪士尼樂園流行的 Splash Mountain。建築師善用核電廠的空間特色,在冷卻塔內部建造了一個輻射椅(鞦韆),而把冷卻塔外面變成了一個高40公尺的攀岩牆。許多為核電廠建造的設施也被整合到公園及其景點中。遊樂園還設有4家餐廳,8家酒吧和6家酒店,其中一個擁有450間客房。參觀遊樂園的遊客完全不用擔心輻射汙染,因為核電廠從未真正啟用過。

核電廠展開了重生之後的新角色功能,吸引了來自世界各地的遊客。Wunderland Kalkar 每年接待約60萬名遊客。在2011年福島災難之後,德國政府決定在2022年前分階段地全面性關閉境內所有核電廠,邁向非核家園。

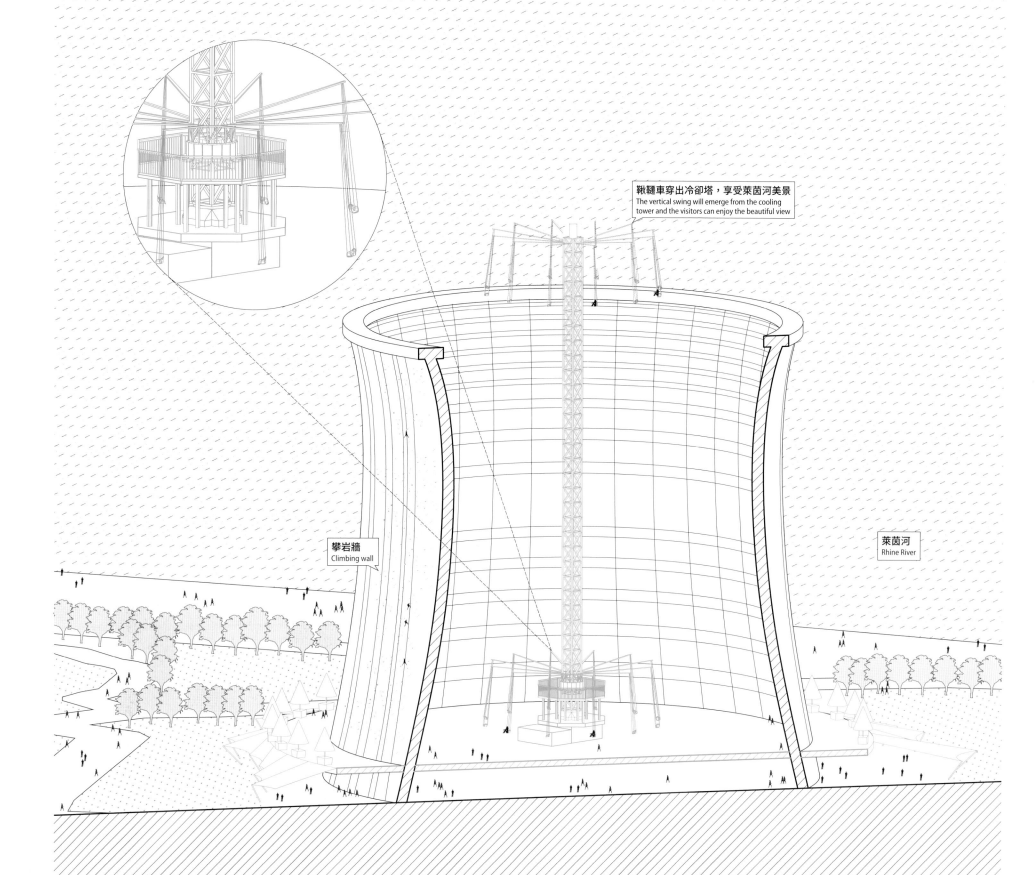

鞦韆車穿出冷卻塔，享受萊茵河美景
The vertical swing will emerge from the cooling tower and the visitors can enjoy the beautiful view

攀岩牆
Climbing wall

萊茵河
Rhine River

英國倫敦
泰特現代美術館
Tate Modern, London, United Kingdom

DESIGNER Herzog & de Meuron

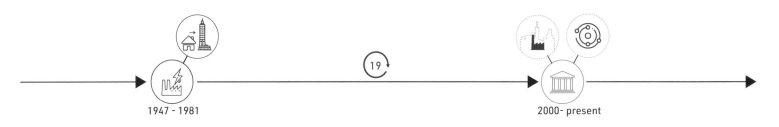

1947 - 1981 　　　 19 　　　 2000- present

泰特現代美術館是將基礎設施建築再利用、水岸開發都市設計、現代藝術展示與經營等三個維度結合，在三個維度都成功的案例。

最近 20 年裡，真正改變倫敦的動力來源竟然是一座古老的舊電廠，也就是如今的泰特現代美術館。建築物前身是坐落於泰晤士河畔，由建築師賈爾斯·吉爾伯特·史考特（Giles Gilbert Scott）設計，興建於二戰後期的發電廠，於 1981 年停止運作。由於商業開發的需求，發電廠面臨了拆除的命運，但透過許多人的努力，最後終於將其保留下來，以現代美術館的形式重生。

1995 年，瑞士建築事務所赫爾佐格和德梅隆的建築師，雅克·赫爾佐格（Jacques Herzog）和皮埃爾·德·梅隆（Pierre de Meuron），在國際比賽中從眾多知名競爭者中脫穎而出。他們意識到發電廠建築本身的潛力，提議的最小外部變化與館方對博物館的願景一致：設計上盡可能使用原有的空間架構去創造美術館的展覽空間。例如利用拆除了發電設備後的渦輪機大廳的宏偉空間，改造成為巨型作品的展演空間與寬敞的公共廣場，供遊客在此漫步或聚集。

赫爾佐格和德梅隆保留了電廠的大致布局、結構和房間名稱，主體分為三大部分，從北面沿河側到南面，分別是鍋爐房（Boiler House）、渦輪機廳（Turbine Hall）、開關塔（Switch House），在開關塔的地下則是油罐室（The Tank），而北側的大煙囪則改造成為觀光塔，鍋爐房區域的頂部加建了一條全長的玻璃光廊，由瑞士政府贊助，因此命名為「瑞士之光」。

美術館的增建工程從 2004 年開始，規劃在舊館的南邊，建造一座像旋轉金字塔的建築以容納更多訪客。增建耗資 2.6 億英鎊，樓高 10 層，增加了 2 萬平方公尺的展示空間。過去發電廠的三座巨型油罐展廳於 2012 年短暫開放，是該美術館獨有的另一種類型的展廳。2016 年油罐展廳重新開放，它將提供額外的展示空間，並提供教學與實作課程的公共區域。

除了身為標誌性建築外，泰特現代美術館一直以來都扮演整個泰晤士河濱水空間復興的重要角色，南岸的泰特現代美術館與北岸的聖保羅大教堂透過千禧橋連接，串聯成一條文化旅遊線路，每年吸引了成千上萬的遊客。泰特現代美術館可說是全球目前最成功、訪客流量最大的現代美術館，它改變了倫敦，對城市設計發展和藝術、文化、社會生活有著深遠的影響。

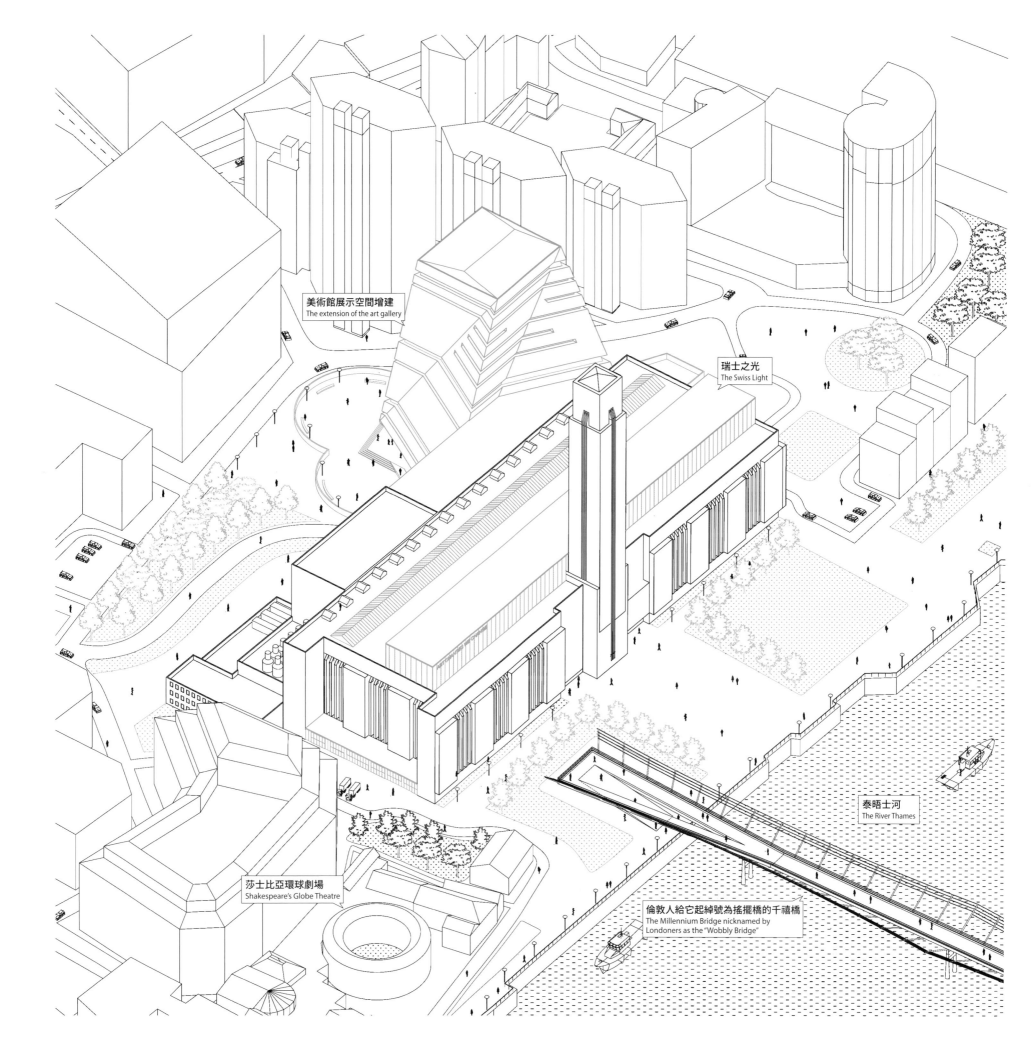

美術館展示空間增建
The extension of the art gallery

瑞士之光
The Swiss Light

泰晤士河
The River Thames

莎士比亞環球劇場
Shakespeare's Globe Theatre

倫敦人給它起綽號為搖擺橋的千禧橋
The Millennium Bridge nicknamed by
Londoners as the "Wobbly Bridge"

D3

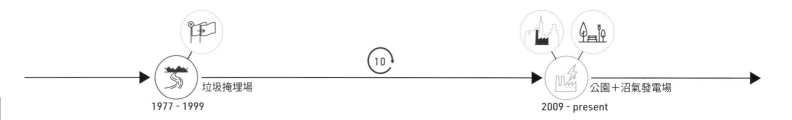

垃圾掩埋場
1977 - 1999

10

公園＋沼氣發電場
2009 - present

台灣高雄青埔
垃圾發電
景觀公園

**Kaohsiung
Metropolitan Park,
Kaohsiung, Taiwan**

DESIGNER 皓宇工程顧問股份有限公司
Cosmos International Inc.
Planning & Design
Consultants

大多的重生案例在重生後就不再是基礎設施，此案是少見重生為基礎設施的案例，可謂台灣之光了！將垃圾掩埋場轉變為公園與沼氣發電廠（基礎設施）。

高雄青埔垃圾發電景觀公園（高雄都會公園）綠地面積達95公頃，第二期區域原本是高雄市西青埔垃圾衛生掩埋場（48公頃）所在地，後因環保議題垃圾採焚化流程處理，於1999年停止掩埋封場（死因），是全台灣由掩埋場土地活化再利用的首例。1989年籌設都會公園時，主要目標除休閒遊憩外，希望增進環境景觀資源及改善地區環境品質，並分兩期進行建設。第一期及第二期工程於1996年及2009年完工開放後，近來已成為地方民眾休憩運動、騎單車的休閒新景點。二期公園為垃圾掩埋後土地再利用，連續壁及不透水層的鋪設隔絕了掩埋的垃圾及覆土，防止毀塌、滲水之情形，並將汙水引導至汙水處理區，避免汙染地下水。垃圾及覆土交疊堆置成了大面積的山丘，山勢平緩並呈東西兩座雙峰狀，最高海拔約43.5公尺。山丘上綠美化的植栽種類多樣，結合都市森林理念及生態綠化方式植栽，導入「成長型復育公園」理念。選取適合台灣南部氣候的原生種植物之地方樹種，現地育苗成長後依喬木與灌木組合栽植，搭配庭園景觀植物、動物蜜源與食草植物，營造自然化的景觀，配合

自行車道、步道環繞園區，以登山步道穿梭於小山坡之間。在園區山坡下方周邊，也設置多處景觀滯洪池，可收集雨季時由山坡上所流下的雨水，兼具水域生態及景觀功能。

過去為西青埔垃圾衛生掩埋場，最大的問題來自於垃圾在地底下分解所產生的沼氣，使掩埋場有自燃發生火災的危險及造成空氣汙染排放溫室氣體（甲烷）。故園區可見到掩埋區基地沼氣收集設施：加蓋不透水布再行覆土，阻絕沼氣自表層逸散（提升沼氣收集處理成果），並將集中的沼氣藉由沼氣收集井集中至收集站，再經由高密度聚乙烯材質製成的地下管線輸送至沼氣發電場發電。為達成環境保護及再生能源的多重目標，規劃燃燒（設置沼氣燃燒塔）及發電（沼氣發電場）兩種階段處理。西青埔處理沼氣發電設施被評估為台灣地區掩埋場中沼氣蘊藏量最豐富的一座，其防治汙染期程長達20年，沼氣蘊藏量可供20年以上的發電量。隨沼氣量逐年下降，每月沼氣處理量從高峰的200萬立方公尺降至約90~100萬立方公尺。高雄西青埔沼氣發電場是全台成功的首例，也是目前世界前十大，亞洲地區總設置規模最大的沼氣發電場。而在未來結束沼氣發電時，掩埋場上之沼氣抽取管路及沼氣收集站將保留於現場，見證這段歷史。

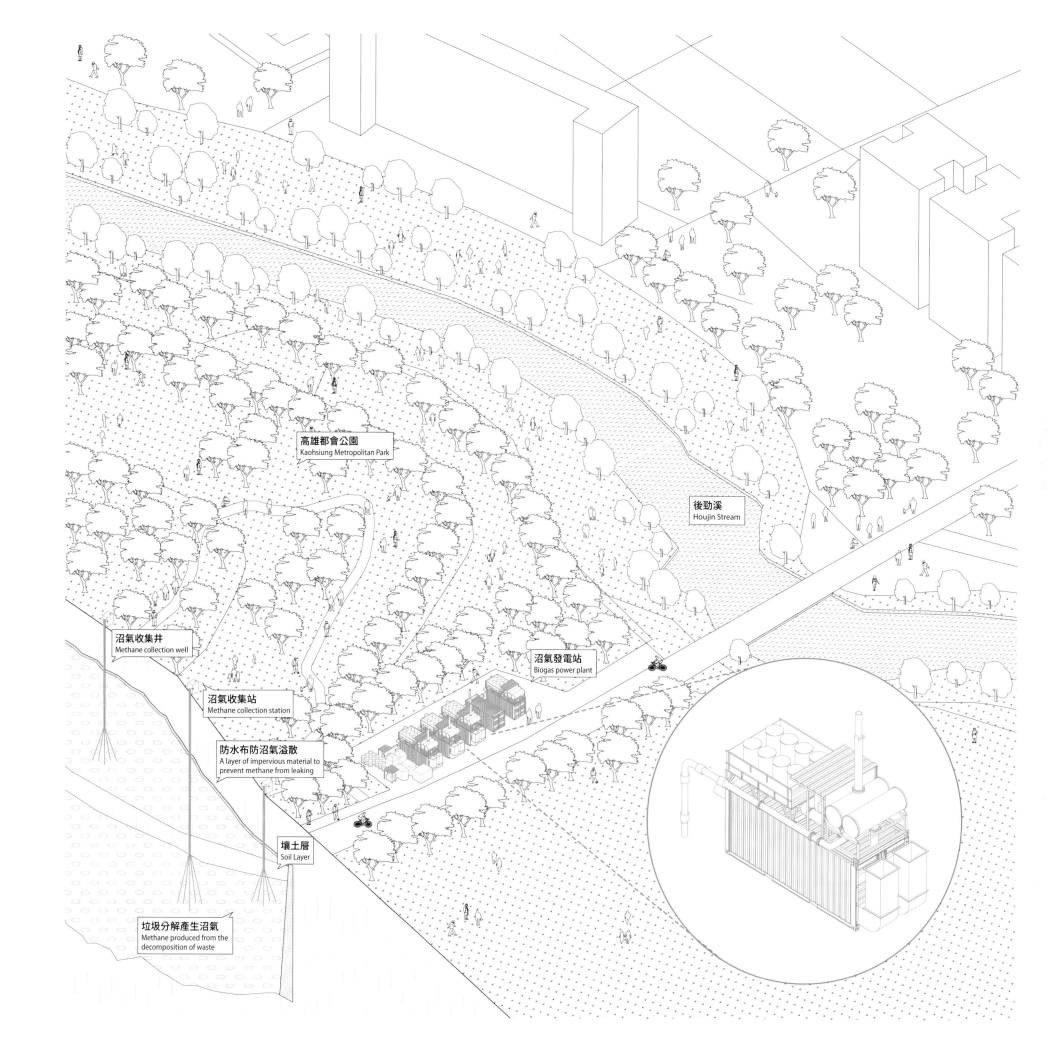

高雄都會公園
Kaohsiung Metropolitan Park

後勁溪
Houjin Stream

沼氣收集井
Methane collection well

沼氣發電站
Biogas power plant

沼氣收集站
Methane collection station

防水布防沼氣溢散
A layer of impervious material to
prevent methane from leaking

壤土層
Soil Layer

垃圾分解產生沼氣
Methane produced from the
decomposition of waste

D4

1983 - present　　　想像　　　proposal

台灣桃園榮華壩
（設計提案）

**Ronghua Dam,
Taoyuan, Taiwan**
(Proposal)

DESIGNER 陳天健

水是人類文明賴以維生的源頭，早在西元前4000年人類就開始建造水壩。現代的大型水壩多半是在20世紀上半葉建造，到了80年代漸漸開始有許多水壩達到其使用壽命，然而如此龐大的構造物要拆除，除了需要龐大的資金，工程也有一定的難度和危險性，因此大多數退役的水壩往往就閒置在原地。到了21世紀初，由於環保意識抬頭，歐美地區開始出現拆壩的聲浪，部分退役的水壩在民意的施壓下得以拆除。

然而，一座退役的水壩除了拆除之外，是否有可能作為其他的用途而獲得嶄新的生命呢？這是一個值得思考的命題。本案透過台灣桃園已逾使用年限的榮華壩作為設計的操作對象，試圖提供退役水壩除了拆除以外的其他可能。

榮華壩自1983年啟用，至2012年已淤積97%，然而由於地形關係，淤沙難以清除，攔砂及調洪功能盡失，僅剩發電功能還在運作。因此在本案中以保留發電功能並且保證汛期時河水能順利排至下游為原則，並參考200年迴歸期的洪水量為依據，將十座排洪閘門劃分為四座常態洩水、兩座輔助溢洪、兩座緊急排洪以及兩座可停用，將壩體作不同使用密度的劃分。

考量榮華壩地處深山，當地的活動以山訓、遊憩類為主，且榮華壩本身即為160公尺長、82公尺高的弧形牆體，適合發展垂直向度的活動，故期望藉由外掛於壩體上的增築，將榮華壩改建為山訓遊憩中心，提供遊客及山訓愛好者高空彈跳、山訓、攀岩、跳水、潛水、室內跳傘等垂直向度活動的體驗。

由於榮華壩位處峽谷，壩頂與聯外道路台七線（北橫公路）具有約100公尺的高差，因此挑戰之一是如何把遊客帶到新的山訓中心裡，同時也需要解決在水壩82公尺牆面上的垂直運載，因此設計中納入國外滑雪場常見的滑雪纜車系統，將原本位在台七線旁的管理站擴建為山頂纜車站，將遊客運送至壩頂的遊客中心，再轉乘類摩天輪系統，抵達各個位於不同高程的遊憩設施。

水壩是讓人類文明得以擴張的偉大工程，藉由這樣的增築帶入人的活動，讓人得以用自己的感官去體驗水壩的尺度，也讓水壩在原本的任務完成後，還能繼續參與文明活動的一部分。

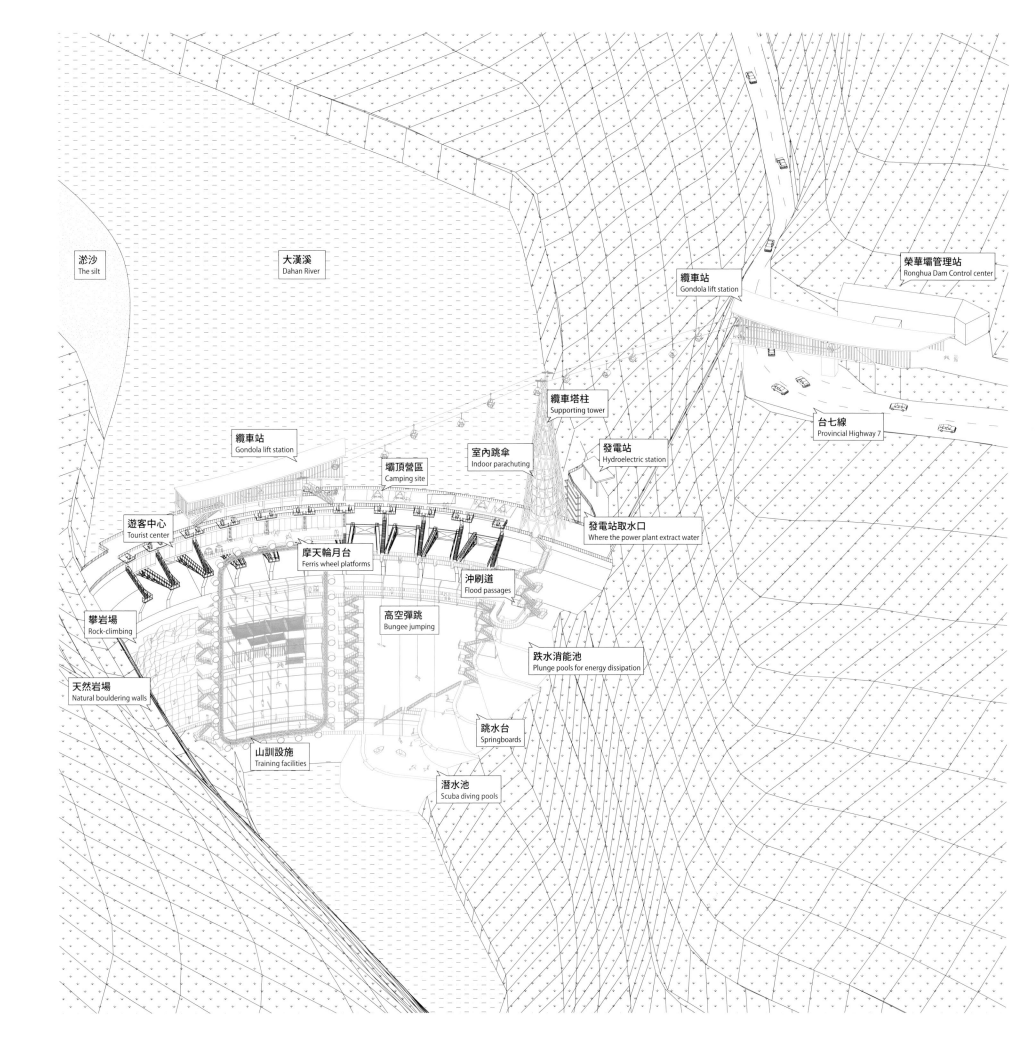

淤沙
The silt

大漢溪
Dahan River

榮華壩管理站
Ronghua Dam Control center

纜車站
Gondola lift station

纜車塔柱
Supporting tower

台七線
Provincial Highway 7

纜車站
Gondola lift station

室內跳傘
Indoor parachuting

發電站
Hydroelectric station

壩頂營區
Camping site

遊客中心
Tourist center

摩天輪月台
Ferris wheel platforms

發電站取水口
Where the power plant extract water

攀岩場
Rock-climbing

沖刷道
Flood passages

高空彈跳
Bungee jumping

跌水消能池
Plunge pools for energy dissipation

天然岩場
Natural bouldering walls

山訓設施
Training facilities

跳水台
Springboards

潛水池
Scuba diving pools

W1

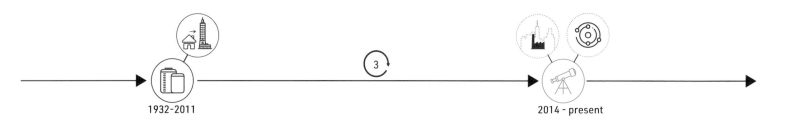

1932-2011　　　　　　　　　　3　　　　　　　　　　2014 - present

荷蘭羊角村
觀景塔

Watch Tower
Sint Jansklooster,
Jansklooster,
The Netherlands

DESIGNER Zecc Architecten

位於荷蘭 Overijssel 省的德維登（De Wieden）自然保護區內有一座古老的水塔，2014年由荷蘭 Zecc Architecten 公司進行改造，藉由在水塔內置入一系列扭曲的樓梯，使其轉變成為一座45公尺高的瞭望塔。

水塔的外觀如今被稱為「Viewpoint Sint Jansklooster」，除了在頂部增加四處大型窗戶外，基本維持舊貌，不過在內部新建了三段階梯，串聯水塔內部不同的部位。如果想自水塔頂端觀景平台眺望德維登美景，遊客得經過一系列的階梯。旅程的開始是一處封閉的樓梯，通往4公尺上方的一樓。接續路程的是由定向纖維板（OSB，由木料片交錯疊合、壓製的集成材）製成的方體階梯，溫暖的木材質感與水塔內壁生硬的混凝土形成強烈對比，一旁舊有的鋼製樓梯沿著圓柱狀牆壁向上，新設樓梯則是與舊有樓梯交錯折返於柱狀空間

之間，直達位於28公尺舊有混凝土水箱底部，鋼製樓梯由水塔的邊緣繞至水箱中間，新設樓梯則是由水箱底部中央貫穿直上。

最後一段路程則是走上新設的鋼製樓梯來到頂端觀景台，設計師移除部分水箱頂蓋增加可停留的空間，下方鋪設透明的地板，讓人在水箱中央上方，體驗暈眩的俯視感。塔頂除了原有的四個小型窗戶，另外增加四座大型觀景窗，8個框景創造出360度的視野。遊客可從登塔過程中獲得多種不同的驚喜和體驗，最後俯瞰德維登美麗的景色作為結尾。

水塔是一座國家紀念物並且被當地居民視為公共財產。這個設計方案極力克制對水塔外觀的更動，透過內部的巧妙改造，不止消除了居民對改造水塔的異議與疑慮，更加深了地方的連結與歸屬感。

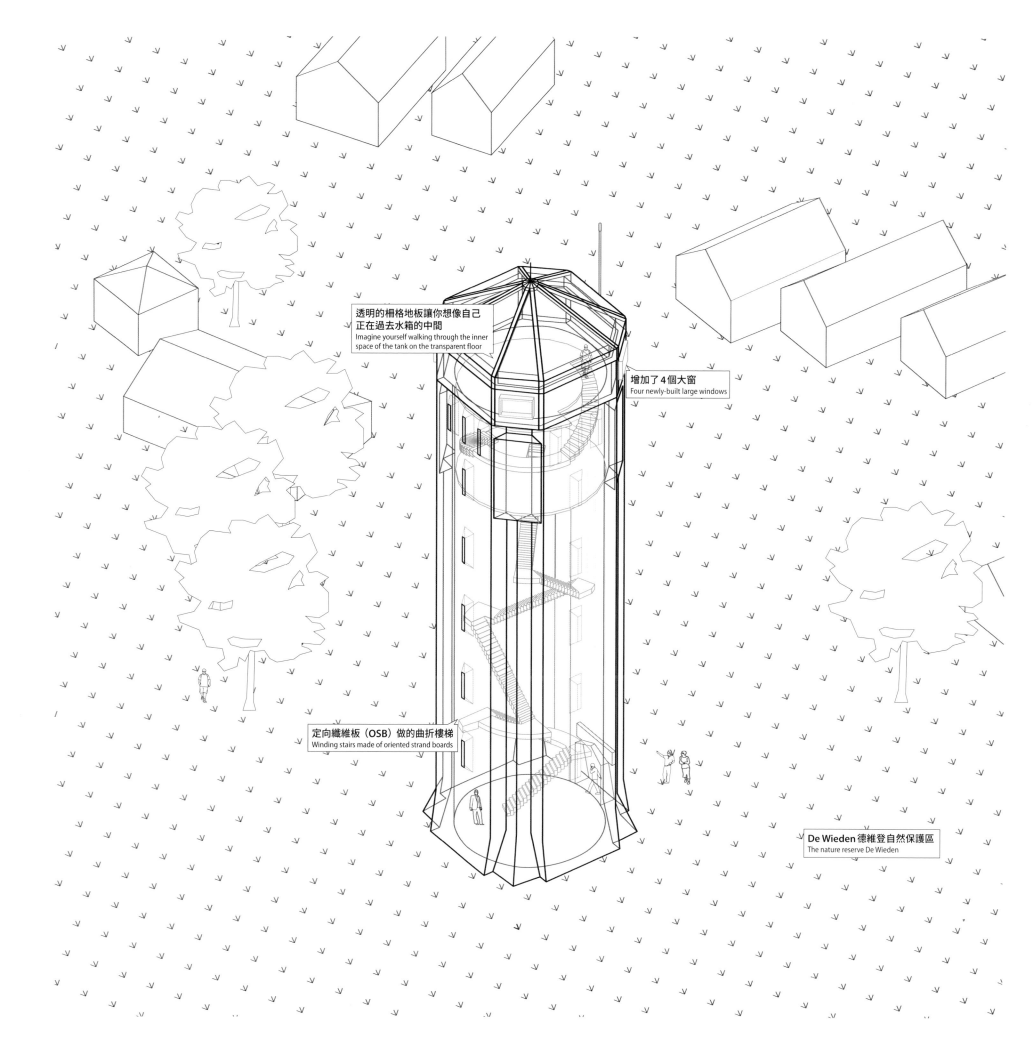

透明的柵格地板讓你想像自己
正在過去水箱的中間
Imagine yourself walking through the inner
space of the tank on the transparent floor

增加了 4 個大窗
Four newly-built large windows

定向纖維板（OSB）做的曲折樓梯
Winding stairs made of oriented strand boards

De Wieden 德維登自然保護區
The nature reserve De Wieden

W2

中國瀋陽
水塔展廊
Water Tower Pavilion (Renovation), Shenyang, China

DESIGNER META-Project

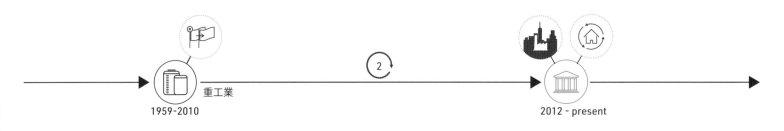

重工業
1959-2010

2012 - present

水塔展廊位於瀋陽鐵西區的一個老廠區內，其基址前身為中國人民解放軍第 1102 工廠，是成立於1959年大躍進時期的軍工廠。作為中國曾經最主要的重工業基地，瀋陽鐵西區充滿了從那一時期保留下來的大大小小的工業遺跡，隨處可見的水塔成了反映這一區域工業歷史的獨特印記。2010年萬科集團將其周邊的幾個工廠收購作為瀋陽藍山項目用地的同時，保留了這座水塔，並交給王碩的META團隊進行改造。建築團隊尊重它的歷史意義、保留其外觀，而在內部進行加固和改造，把它變成具有現實意義的水塔展廊——一個提供給周圍社區的公共活動空間和博物館。

META是在對待歷史與現實的審慎態度下，力圖將新的現實以一種複雜精巧的方式植入到舊的工業遺址內。使用雙層殼的手法：外殼是屬於過去的，是既存記憶的，而內殼是屬於現代的，是設計師想要的。內外殼用彩色開窗貫通連接。

META儘量不去碰觸完整保存的水塔本身，只進行必要的結構加固和局部處理了塔身上原有的窗洞口；另一方面，新加入的部分：一個複雜精巧的裝置被植入到水塔內部，中間的主體是兩個頭尾倒置的漏斗，較小的位於水塔頂部收集天光，較大的在塔身內部形成了一個拉長的縱深空間，並連接著多個類似「相機鏡頭」的採光窗——這些將環境光線經過間接反射導入水塔內部的採光窗，從塔身上每一個可能的開口「生長」出來。水塔的底部，連接入口與抬高的觀景平台之間，則用回收的紅磚砌成可坐的台階，這裡將成為一個「小劇場」，為周邊的公眾提供一個小型活動或集會放映的場所；由此處往上看，水塔成為一個間接感受外部世界的「感覺器官」，光線從頂部的光漏斗及每一個不同形狀及顏色的窗洞口進入到水塔中心的隧道內，在一天之中持續著微妙的變化。而從懸挑出塔身的觀景平台向周邊望出去，這一裝置則成為一個單純的觀察外部世界的「取景器」。

落成後的水塔展廊為萬科藍山小區提供一個新的公共活動空間——甚至在日常的使用功能之上，它可被視為一個城市的裝置藝術，一個具有精神性的場所。

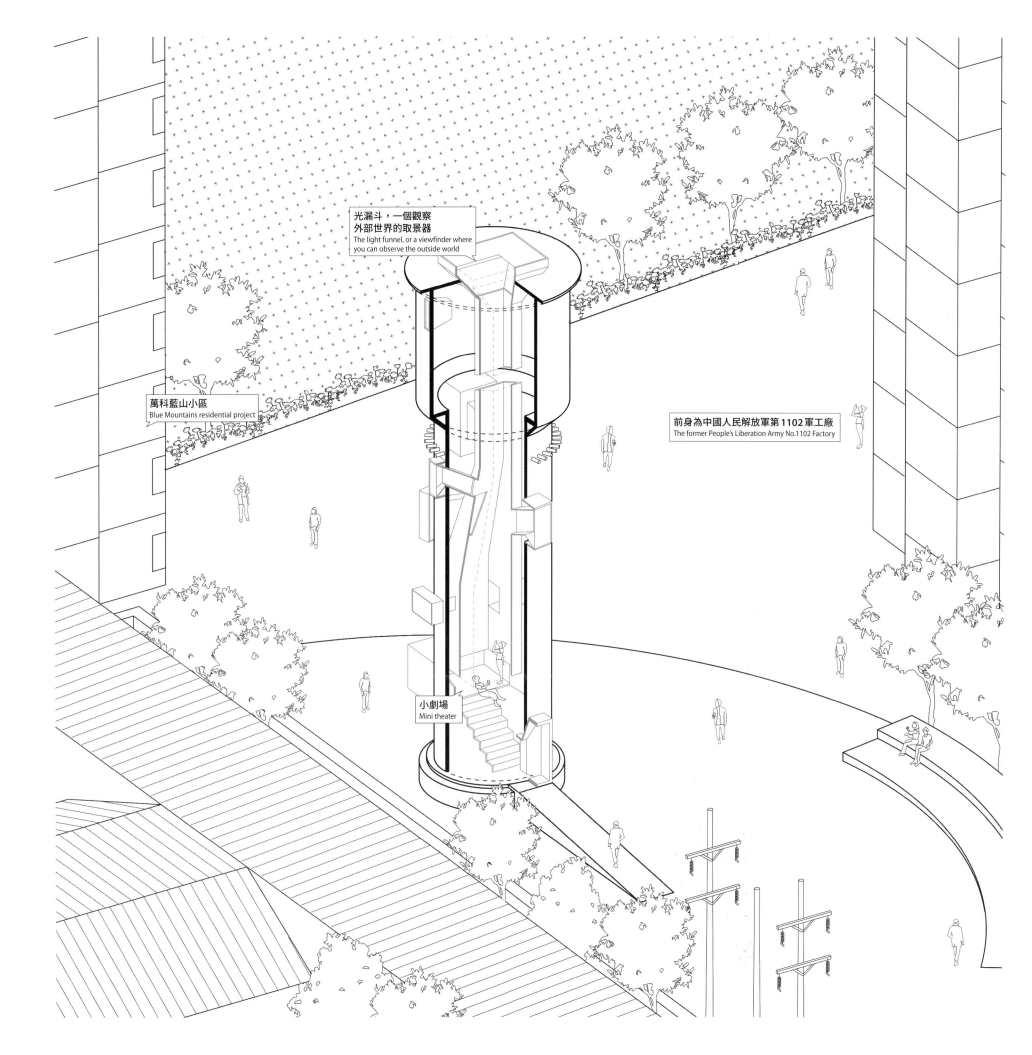

光漏斗，一個觀察
外部世界的取景器
The light funnel, or a viewfinder where
you can observe the outside world

萬科藍山小區
Blue Mountains residential project

前身為中國人民解放軍第1102軍工廠
The former People's Liberation Army No.1102 Factory

小劇場
Mini theater

W3

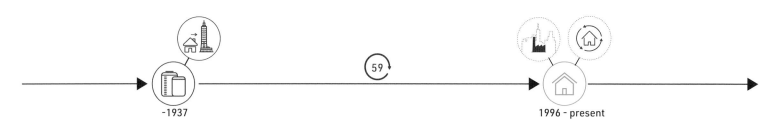

比利時安特衛普森林水塔住宅
Water Tower Housing Brasschaat, Brasschaat, Belgium

DESIGNER Binst Architects

位於比利時布拉斯哈特莊園（Brasschaat）的水塔，最早是為了向莊園主屋以及附屬建築提供自來水而興建設立。但在1937年水塔被布拉斯哈特（Brasschaat）社區興建的四座新水塔形成的全新供水系統取代而停用，水塔因其完全過時的技術和施工方法被遺棄（死因）。

1950年，市政府收購了包含水塔的莊園和周圍的土地。其中水塔的混凝土結構簡單，包括：由4個大型混凝土柱支撐的圓筒水塔，圓筒本身高約4公尺、總高度約為23公尺；放置在4根混凝土柱之間的方形平台，尺寸接近4公尺 x 4公尺；地下包含過濾池和水庫；周圍的土地與林地接壤，占地約680平方公尺；水塔下方一條蜿蜒的小溪將該區域與周圍的林地分開。

最終市政府理事會必須決定將此水塔拆除或出售賦予新使用。但在出售前水塔就已經空空如也。這樣一個貧乏不討喜的水塔，因緣際會遇到了一位伯樂，這位有特殊品味的人不僅欣賞它，並視之為「我的夢想之家」——一位當地的景觀設計師與Crepain合作，趁機入主夢想之家。

兩個人共同創造了一個簡單的計畫，打造一棟6層的住宅，將多天花園設置在舊混凝土水塔下，並將一樓包圍在一個堅實封閉的環境中。就像將豬腿肉放入半透明的腸衣似的，他們在中間四層的塔樓內裝設半透明玻璃幕牆（U型玻璃）。在這棟房屋位於6公尺高的起居空間內，可由陡峭的樓梯通向每個樓層，並在家中就可俯瞰毗鄰的Braaschaat森林的景色。

屋內照明使得塔樓在夜間發光，所以它被認為是鎮上的燈塔。玻璃幕牆在引入充足的自然採光的同時，半透明玻璃幕牆也保持了一定的隱私。舊的混凝土塔配上了新的玻璃幕牆強化了「新舊工業共生」的概念，在創造一個新的有用空間的同時提高了水塔的原始工業品質。

受限於經費，Crepain和他的客戶不得不精簡預算來完成這個項目。對此，建築師在2008年曾說過：「我們很幸運，我們沒有錢，所以我們現在擁有的是一個簡單俐落乾淨漂亮的建築。」有時候，做一些簡單的事情，比做一些複雜的事情更加困難。

要記住密斯凡德羅(Mies van der Rohe)的「少即是多」，而努力去做得「少」。

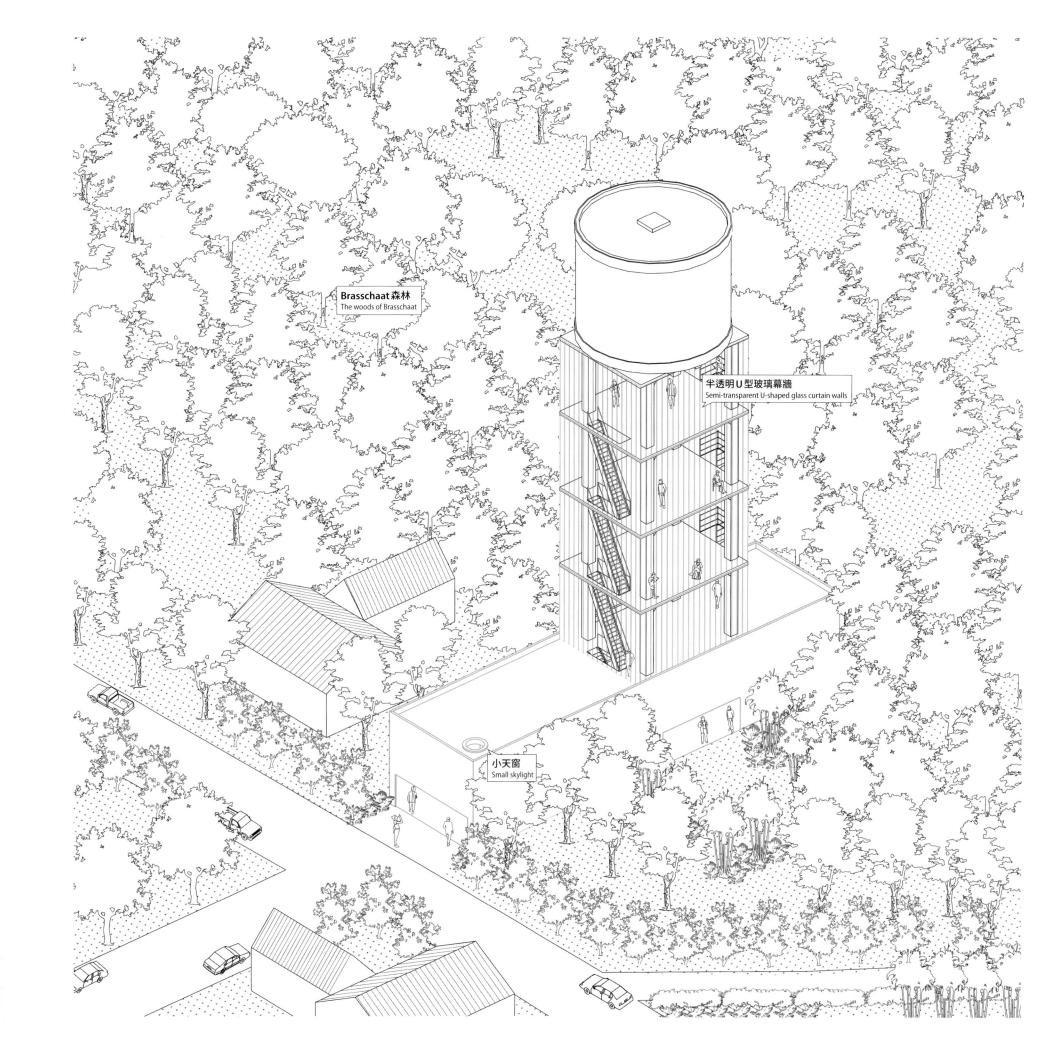

Brasschaat森林
The woods of Brasschaat

半透明 U 型玻璃幕牆
Semi-transparent U-shaped glass curtain walls

小天窗
Small skylight

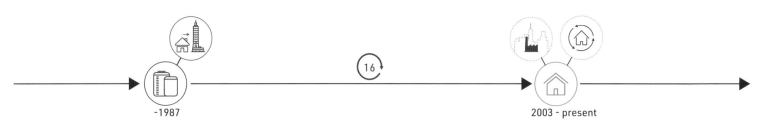

-1987　　　16　　　2003 - present

德國布蘭登堡
水塔住宅

BIORAMA-Projekt,
Brandenburg, Germany

DESIGNER Frank Meilchen

2002 年，時年 48 歲的產品設計師理查·賀爾汀（Richard Hurding）正在德國東北部的布蘭登堡州，參觀位於聯合國教科文組織自然公園的 Schorfheide-Chorin 生物圈保護區。在這裡 Hurding 發現了一座廢棄的水塔──後來被他變成了一個家，坐落在樹木包圍的山上。當時他和同為工業設計師的妻子，47 歲的莎拉·菲利普斯（Sarah Phillips），兩人在多國城市工作生活後正在尋找一個新的居住地點。

他們一直在尋求「沒有企業價值束縛的挑戰」和「具有環境價值和對社會更有意義的東西」。他們之前也在倫敦做過在工業建築中創造生活空間的想法。這座前東德水塔是一座芥末黃磚牆的歷史地標，一排水平窗戶繞過塔頂，有著橫跨森林的廣闊視野向他們召喚，住在塔樓的想法也非常有吸引力，總之，莎拉·菲利普斯看到了它的重生潛力。

為了將夢想變為現實，這對夫婦向 Barnim 縣的建築委員會申請將水塔改建為住宅。而為了使他們的申請案更具說服力，他們提出了將水塔變成旅遊景點的建議，在 21 公尺高的水塔上建造一個觀景台，遊客可以在那裡觀賞保護區及其混合生長的森林、湖泊、珍稀鳥類和兩棲動物。

2003 年他們的申請案被批准了，這對夫婦為塔周圍大約 3 英畝土地支付了 75,000 歐元。由於水塔是被指定的歷史性地標，政府決定不予出售，而是提供這對夫婦一年 500 歐元租金、為期 99 年的租約。

同年他們聘請了駐柏林的建築師梅林 Frank Meilchen 來幫助他們完成改造。對於梅林來說，最大的挑戰是移除水塔內巨大的混凝土水箱。工作人員用金剛石鋸將其切開，起重機操作員將每塊切片從窗戶中拉出放到地面上，單是這項工程就花費了一個月的時間。梅林在原來的建築物內建造了另一座承重塔，在結構上承載未來做為居住用途的各個新樓層。另外在原本的水塔旁邊再建造了一個紅色玻璃開窗的電梯塔，將保護區的遊客帶到觀景台。裝修費用約 600,000 歐元，花費大約 8 個月的時間完成。歐盟向電梯塔捐款 270,000 歐元使其成為輪椅者的無障礙空間。

2006 年，理查與莎拉夫婦搬進了水塔，現在這是一座 139 平方公尺的 6 層樓公寓。莎拉的辦公室位於一樓，而理查的辦公室在二樓，三樓則是他們的主臥室和懸掛在其上的主臥浴室。四樓為住家的第二個入口，加上客用洗手間和儲藏室；五樓是專用廚房和用餐區；客廳位於六樓頂層，提供 360 度全景。

夫婦倆將塔命名為 Biorama，將自然保護區之名與全景一詞作了完美的結合。

©Thomas Zimmermann (THWZ)　©Daniela Kloth

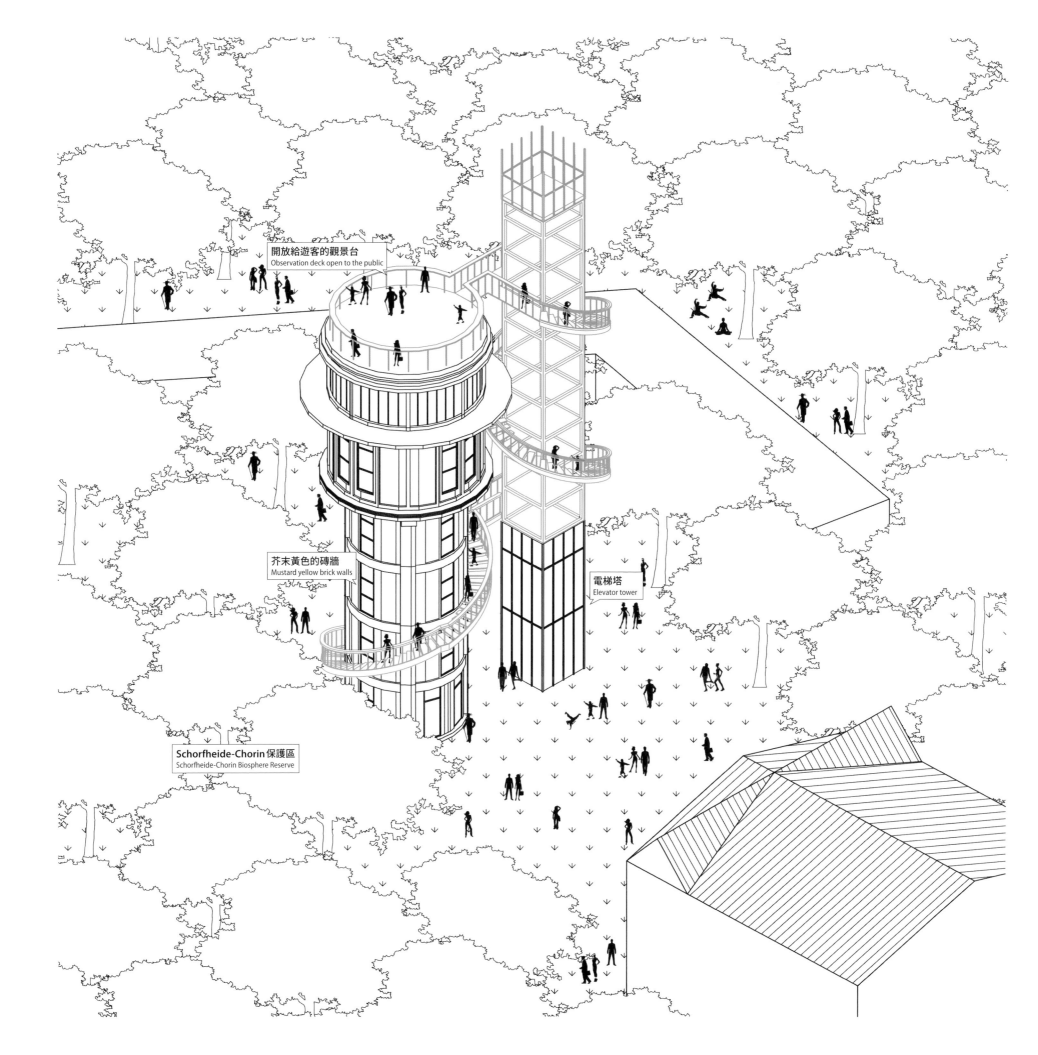

開放給遊客的觀景台
Observation deck open to the public

芥末黃色的磚牆
Mustard yellow brick walls

電梯塔
Elevator tower

Schorfheide-Chorin 保護區
Schorfheide-Chorin Biosphere Reserve

W5

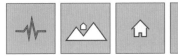

英國薩福克郡
雲中之家

House in the Clouds,
Suffolk, United Kingdom

DESIGNER Glencairne Stuart Ogilvie
& F. Forbes Glennie

英格蘭的一個老水塔，轉換成一間無與倫比的雲中民宿。

雲中之家的前身是英格蘭薩福克索普尼斯村的一座水塔，由倫敦的 Braithwaite 工程公司於 1923 年建造，其水箱容量為 50,000 加侖（230,000 升），用於接收並儲存由索普尼斯村風車抽取的水，由威廉騎士隊從 1923 年到 1938 年看守此區域。1943 年水塔在第二次世界大戰期間被空襲炸彈擊中，工程師利用水塔本身的鋼材進行水箱修復，這導致了 30,000 加侖（140,000 升）的儲水容量減少。

1963 年索普尼斯村引入了現代化的區域供水系統，該水塔的供水功能也因此停止，水塔僅用於儲水。1977 年雲中之家轉為 Ogilvie 家族私人所有，1979 年連水塔的儲水功能也顯得多餘，主水箱被拆除（死因），將水塔建築完全轉換成一般的普通房屋，同時也開始改造水塔內部，以創造更多的生活空間。

當第一次改造開始時，索普尼斯的居民擔心他們的景觀會受到破壞，因此為了改善水塔的外觀，Glencairne Stuart Ogilvie 與建築師 F. Forbes Glennie 在這巨大的結構上，透過巧妙的手法將它偽裝成一座房子，從幾公里遠的地方可以看到天際線上有一間小屋，彷彿是位於 21 公尺高的樹上。

1987 年 Ogilvie 再次為他的好友，兒童書籍作家 Malcolm Mason 女士進行建築內部修復，改造成她的住所及民宿。雲中之家為家庭度假提供寬敞的住宿，設有 5 間臥室，3 間浴室、客廳、餐廳和宏偉的「頂層遊戲空間」。它從上到下共有 68 個台階，高約 21 公尺，許多新增的小窗戶提供良好通風、採光，和美麗的景色。

1920 年由 Glencairne Stuart Ogilvie 創建的索普尼斯度假村與其他任何度假勝地都不同，他想為那些想要體驗真正英格蘭生活的人們創造一個理想的度假村。自 1995 年以來，雲中之家一直是二級保護建築，坐落在占地 1 英畝的私人土地上，俯瞰著索普尼斯高爾夫球場和薩福克海岸，享有薩福克最美的景色，成為全英國最著名的民宿之一。

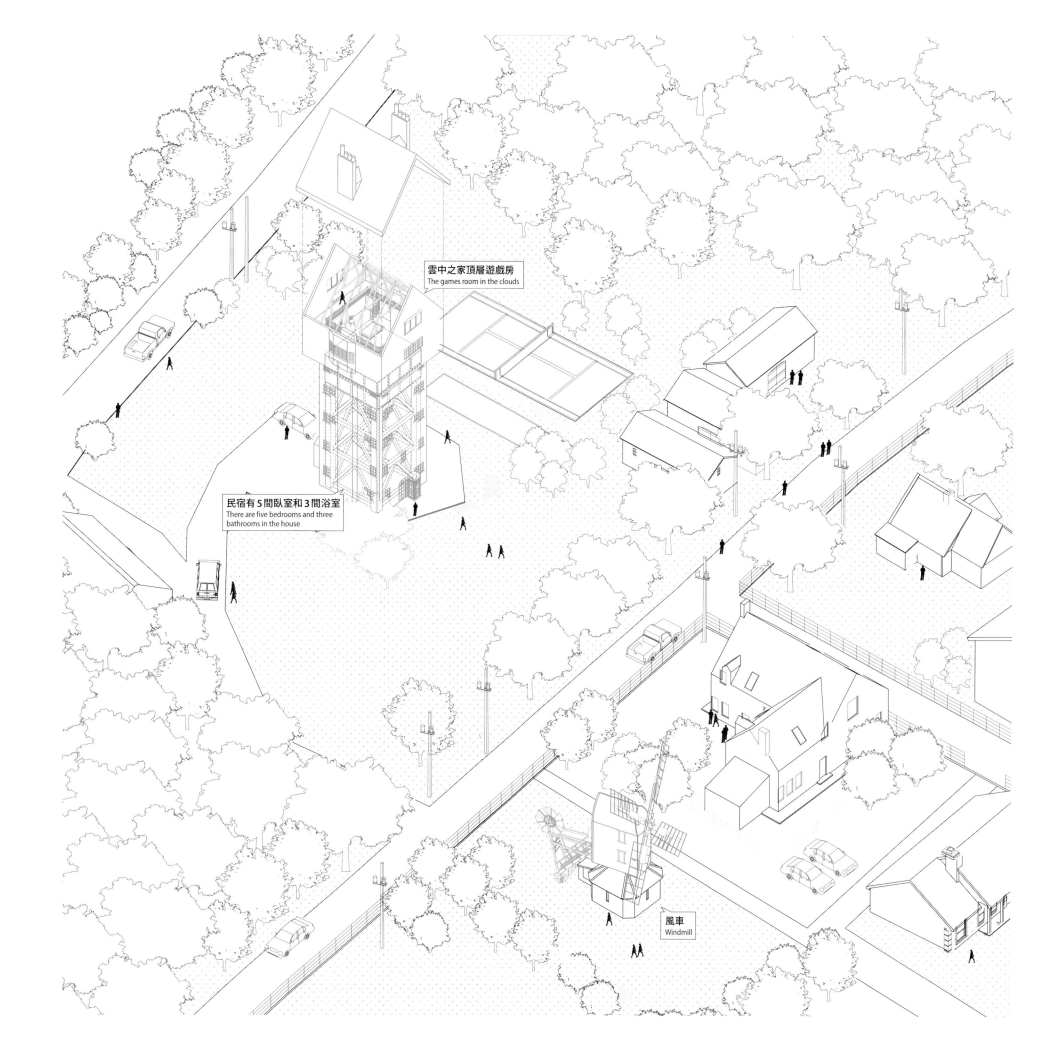

雲中之家頂層遊戲房
The games room in the clouds

民宿有 5 間臥室和 3 間浴室
There are five bedrooms and three
bathrooms in the house

風車
Windmill

W6

1913-1933 ―O→ 1933-1969 餐廳 ―O→ 1969-1994 公寓 ―O→ 1994 - present

匈牙利
德布勒森水塔
咖啡廳／畫廊／
攀岩場／觀景台

Nagyerdei Víztorony, Debreczin, Hungary

DESIGNER architects Zoltán Győrffy & Róbert Novák

1913 年興建的 Nagyerdei 水塔已經成為匈牙利德布勒森鎮的一個新的景點。

20 世紀初，水塔在工程建築中占有特殊的地位，這得益於鐵路網絡的建設，工業廠房的發展和城市的擴張。而如今它們大多數已經失去了原來的目的。由德布勒森大學所擁有的 Nagyerdei Víztorony 水塔卻依然在供水，為極少數的例外。最初由建築師 József Borsos 設計，它的重生是德布勒森大森林地區再生計畫最主要的元素之一。由建築師 Zoltán Győrffy 和 Róbert Novák 在仍運作的水塔上完成了 9 個月的翻新工程，在水塔混凝土框架內建造了一個餐廳、酒吧、咖啡館、商店、畫廊和觀景台，遊客還可以透過圍繞其中央支柱的攀岩柱來爬上建築物。

Nagyerdei Víztorony 水塔的重生非一時之作。1914，完工隔年就在一樓設計了咖啡館，並於 1933 年建成了一間自助餐廳。該建築於 1969 年轉為德布勒森大學所有，當時一樓被改成服務式公寓，此機能在水塔保護運動後全然消失。1994 年前，除了例行的維護，沒有對建築物進行大規模的改動。1994 年，在匈牙利的水塔保護運動下進行了全面重建，由當地政府和大學共同提交歐盟的聯合項目，設計的目標是保護現有結構與建築特徵和創造公共景點，因此，公寓移出了水塔，餐廳等公共景點又回到了水塔之下，甚至讓人們攀上了水塔。

一樓的拱圈改用玻璃包覆，因此一樓餐廳的內外空間顯得非常輕盈。一樓餐廳上的露台被保留下來，露台中心是攀岩柱。而畫廊占據了該水塔的地下室，有一個接待空間和展覽空間，之上的環形帶狀天窗提供地下空間良好的採光。為創造更多戶外活動空間，建築師在水塔周圍往下挖掘出一片人造山丘，也因此可以由外進入到一個被包圍的半圓形中庭，保留的擋土牆和堅固的大門，讓這個半圓形空間顯得受到保護。水塔頂部高 42 公尺，瞭望塔位於 34 公尺，周圍是覆蓋水箱的半球頂。過去的工程師塑造水箱周圍的拱廊令人驚豔，有了瞭望塔，環形的露台讓遊客欣賞到周圍森林和城市的全景。

Nagyerdei Víztorony 的重生創造了一個多層的、年輕的社區空間，更是一個有吸引力的旅遊景點。遊客在白日可以欣賞 Nagyerdei Víztorony 的工程魅力，晚上則可以搭配燈光藝術，參與各種節日及社區活動，如大學電影俱樂部、展覽、音樂會等。每個月都有為數眾多的訪客前來朝聖。

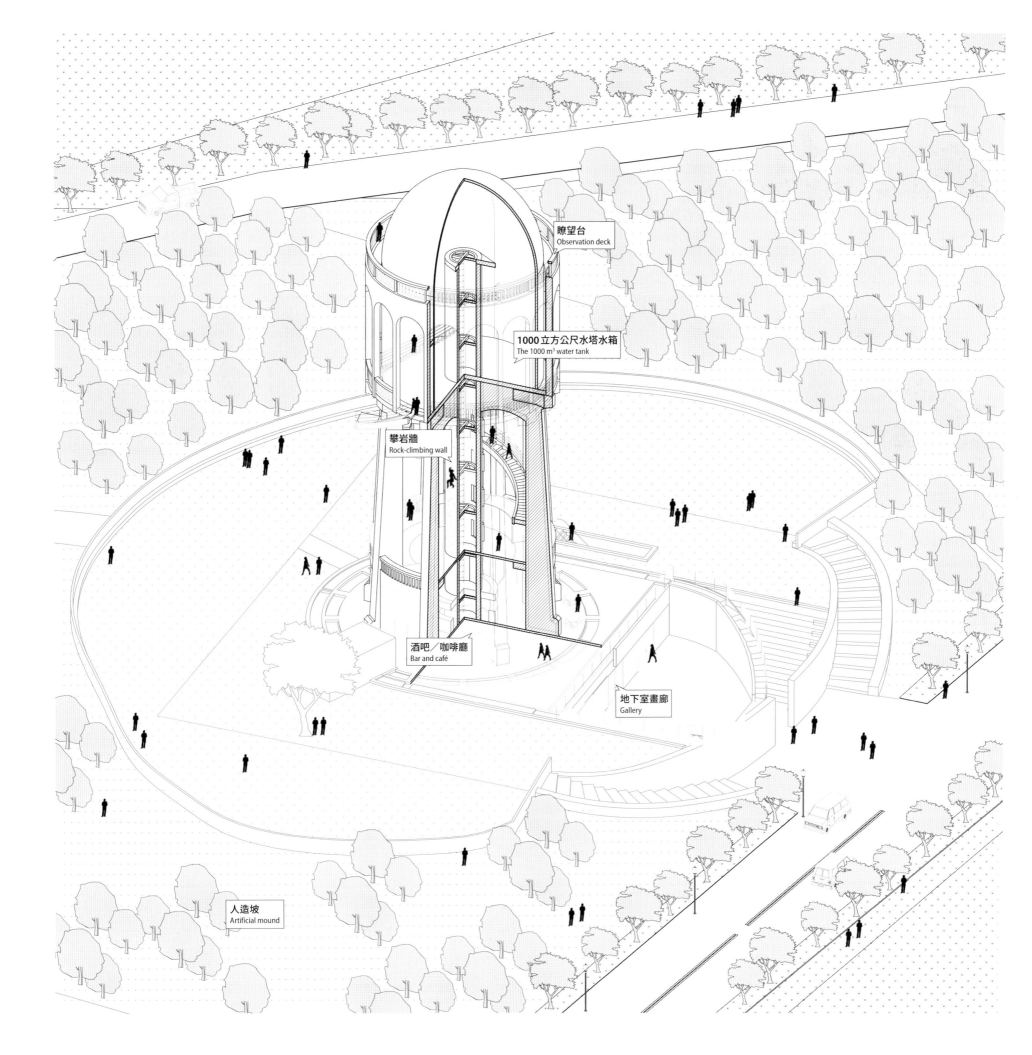

瞭望台
Observation deck

1000 立方公尺水塔水箱
The 1000 m³ water tank

攀岩牆
Rock-climbing wall

酒吧／咖啡廳
Bar and café

地下室畫廊
Gallery

人造坡
Artificial mound

S

倉儲 Storage

S1
丹麥哥本哈根
Frøsilos 集合住宅
Frøsilos SILO apartment, Copenhagen, Denmark

S2
芬蘭赫爾辛基
Silo 468 燈塔
Silo 468, Helsinki, Finland

S3
比利時韋訥海姆
酒廠筒倉住宅
Kanaal, Wijnegem, Belgium

S4
奧地利維也納瓦斯槽城市
Gasometer, Vienna, Austria

S5
英國倫敦國王十字
天然氣槽住宅
King's Cross Gasholders,
London, United Kingdom

S6
南非開普敦
非洲當代藝術博物館
Zeitz MOCAA,
Cape Town, South Africa

S7
加拿大蒙特婁
攀岩健身房 Allez-Up
Allez Up Rock Climbing Gym,
Montreal, Canada

S8
中國上海民生碼頭
八萬噸筒倉
80,000-tonne Silo Warehouse,
Shanghai, China

S1

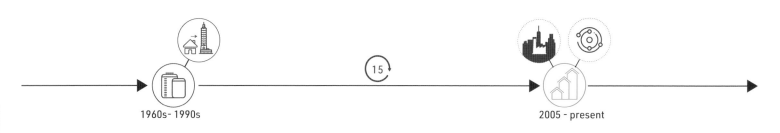
丹麥哥本哈根 Frøsilos 集合住宅

Frøsilos SILO apartment, Copenhagen, Denmark

DESIGNER MVRDV

在歐洲各地的老港口地區正在被逐步轉換成高品質的生活區。憑藉著絕妙的景色、海濱戲水區、靠近市區中心以及其「原有」的特色，這些都使老港口地區得到了快速的發展。

Frøsilos 雙子座住宅於 2005 年完工，是坐落在哥本哈根 Brygge 海濱地區的筒倉改造案例。MVRDV 的任務是將 1960 年代建造的雙筒倉改建翻修，轉變為公寓大樓，這些筒倉在 1990 年代因大豆加工廠產業的沒落而停用（死因）。

筒倉圓形結構的局限性，影響了解決方案的設計。在保持其內部特徵的前提下，要在環形混凝土承重牆鑿開大孔，在工程上十分費事。

因此，公寓的居住單元被設計成懸掛在筒倉的外部，而不是填充

在筒倉的內部，居住單元圍繞著兩個原始混凝土圓筒的周邊添加，外掛的居住單元可以使每個房間獲得最大化的城市美景。共有 8 層樓的公寓，每戶居住單元都有自己的私人陽台，而被釋放出的筒倉內部則作為公共空間與動線空間之用，以有兩個寬敞天井的大廳為中心。

地面層並沒有配置公寓的居住單元，而是直接裸露混凝土的核心筒。與該區域常見保留了歷史特徵的其它倉庫改造項目不同，MVRDV 將筒倉視為可以容納「未來主義」住宅的混凝土裸露結構。顯而易見，不同於世界上其他「向內填充」的筒倉改造案例，此案選擇「外掛」，讓整體設計可以更加靈活。

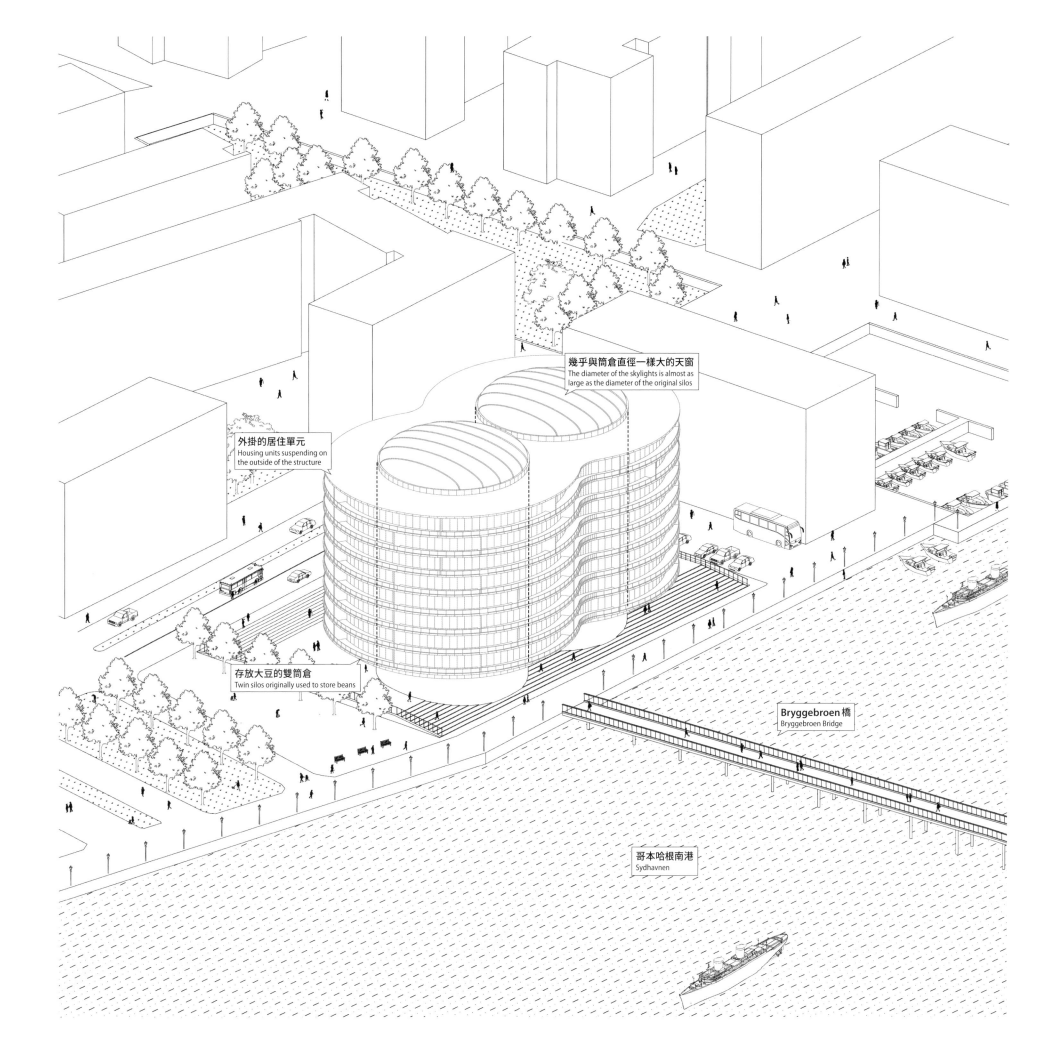

幾乎與筒倉直徑一樣大的天窗
The diameter of the skylights is almost as large as the diameter of the original silos

外掛的居住單元
Housing units suspending on the outside of the structure

存放大豆的雙筒倉
Twin silos originally used to store beans

Bryggebroen 橋
Bryggebroen Bridge

哥本哈根南港
Sydhavnen

S2

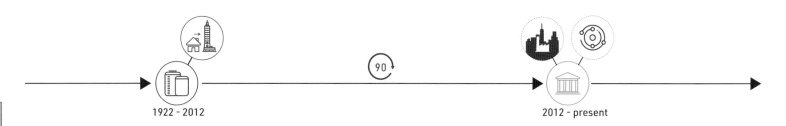

1922 - 2012   90   2012 - present

芬蘭赫爾辛基
Silo 468 燈塔

Silo 468,
Helsinki, Finland

DESIGNER Lighting Design Collective

冬天的芬蘭，黑夜可長達18個小時以上。當幾乎沒有日照時，公共空間的照明，成了城市中的庇護所，陪伴市民度過整個漫長的季節。

　　Silo 468是建築事務所Lighting Design Collective（LDC）設計的永久性公共空間。它原本是一個面向市中心、坐落海邊的圓筒狀石油倉庫。原本頹敗的外殼，經過重新油漆，穿鑿了代表2012年世界設計之都赫爾辛基的2,012個小圓孔，加上精心設計的照明系統，使得這個如燈塔般的裝置，標示著赫爾辛基市老舊區域重建的開始。

　　此地居民可感知的沿海風與海面反射的日光成為這個設計的出發點。LDC在450個圓孔後方，裝有可隨風擺動的鏡面鋼板，白天日光照射時，就如水面般的波光瀲瀲。而另外1,280個圓孔背後，則設置了色溫2,700K的白色LED燈。LDC開發了一組特殊的照明系統，使用群體智能（swarm intelligence）和自然模擬演算法（nature simulating algorithms），利用當地天氣的即時數據（例如：風速、風向、氣溫等）設定燈光模式，系統每5分鐘更新一次資料，產生大量且無法預測的排列組合，使觀眾能夠直觀地「看到」城市中的天氣變化。夜晚時，在好幾公里外都可以看見這些流暢自然且永不重複的圖案。

　　白天，穿透的小孔引進日光打在牆壁上，並隨時間位移。晚上，午夜時分，白色LED指示燈會變為紅色一小時，給內部空間帶來溫暖的紅色光芒，搭配漆成深紅色的筒倉內部，暗示曾經是密閉的石油儲存容器。筒倉內部重新施作地板，收納電線、水管與緊急照明等設備空間，使其成為更適合人潮聚集與活動的新場域。每天午夜02:30，當最後一班渡輪駛離岸邊，Silo 468就熄燈休息。等待下一次黑夜來臨，再次照亮並溫暖在它周邊生活的市民。

　　透過自然光和人造光共譜的巧妙設計，Silo 468為11,000人居住的郊區帶來光的希望，這個如公共藝術般獨特的市民空間，成為指引都市傾頹區域再生案例的先驅。

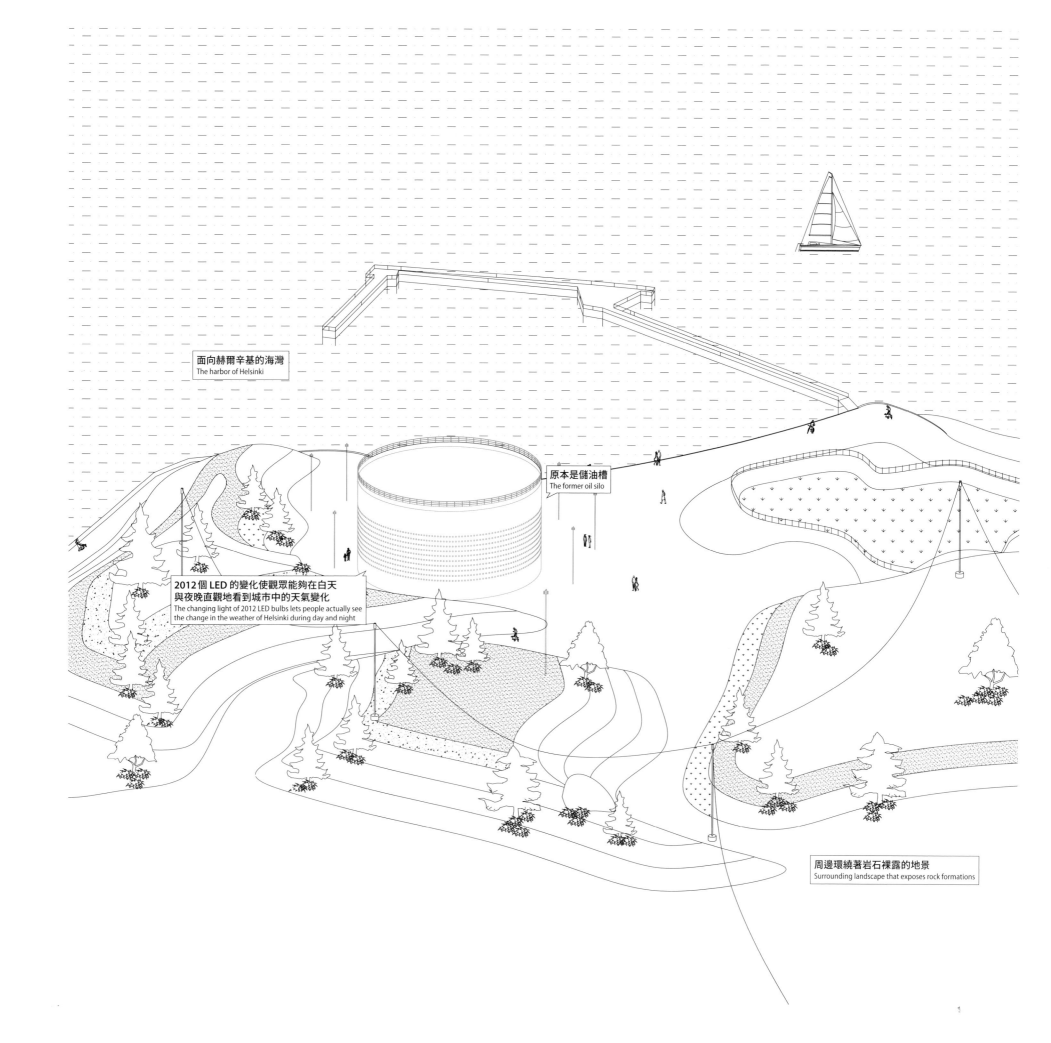

面向赫爾辛基的海灣
The harbor of Helsinki

原本是儲油槽
The former oil silo

**2012 個 LED 的變化使觀眾能夠在白天
與夜晚直觀地看到城市中的天氣變化**
The changing light of 2012 LED bulbs lets people actually see
the change in the weather of Helsinki during day and night

周邊環繞著岩石裸露的地景
Surrounding landscape that exposes rock formations

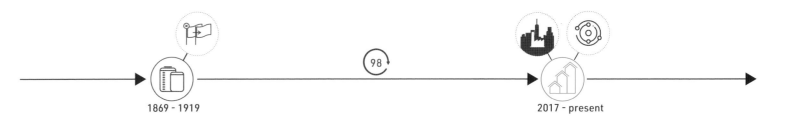

1869 - 1919

98

2017 - present

比利時韋訥海姆酒廠筒倉住宅
Kanaal, Wijnegem, Belgium

DESIGNER Stéphane Beel Architects

位於安特衛普郊外的 Wijnegem 的 Kanaal 住宅和商業開發項目，將坐落在阿爾伯特運河沿岸的一座已廢棄但有價值的 19 世紀工業基地，改造成一個混合使用的綜合體。除了工作室、博物館空間、辦公室和地下停車場外，大部分都被改建為住宅單元。

Kanaal 最初在 1869 年開業的 Jenever 酒廠的建築群中成形，酒廠成為法蘭德斯最大的酒廠之一，其出口遠至澳大利亞。1919 年因政府立法禁止在比利時銷售烈酒，導致了釀酒廠的停止運營（死因）。大約在第二次世界大戰結束後，Heineken（海尼根）公司收購了這個場地，並將它變成了一個啤酒廠。在此期間，又在現有的建築群中增加了 8 個筒倉。

Stéphane Beel Architects 負責改造舊啤酒廠的麥芽儲藏筒倉——這是整體開發項目中的一部分。2017 年春天工作完成後，該綜合體將包括一個專門用於藝術和骨董的基金會、高端購物設施和 100 間藝術導向的公寓。

灰色筒倉的升級：允許現有建築物生活運行，並且可以在不損害筒倉綜合體特色的情況下，實現新的住宅功能，並保證視野和自然光的宜居性。兩個灰色筒倉的高度分別為 31 和 28 公尺，並被一個新

的、細長且透明的矩形筒所取代。剩餘的 6 個筒倉被保留並開設立面上的矩形開窗。在平面上 8 個量體構成了兩戶的住宅單元，並呈現兩個 L 型的平面配置。每個住宅單元皆包含三個舊有的圓筒及一個新置入的透明矩形筒。所有舊有筒倉的底部都保持在一起，並且在升高的平台上提供博物館空間和進入住房單元的入口。灰色筒倉彼此分開，並由玻璃橋連接，串聯起四個原本獨立的空間，並構成一個完整的住宅單元。

新的住宅功能使筒倉綜合體能夠保持存在。反過來，舊有酒廠筒倉量體提供了現代住宅新的探索。在傳統現代住宅的平面設計上，多半是從一個完整的平面中區隔不同的機能空間。但在 Kanaal 的酒廠再利用案例中，每個既有的細長筒倉變成了獨立的機能空間。因此，必須串聯四個機能空間才能滿足完整居住單元的基本功能。對現代住宅而言，這不是一個經濟的平面規劃，但從實驗的角度來看，每個獨立的機能空間都具備了絕佳的通風及採光。更重要的是，Kanaal 成功地將新的建築機能放入了原本沒落的舊有空間，並重現了釀酒廠的歷史魅力。

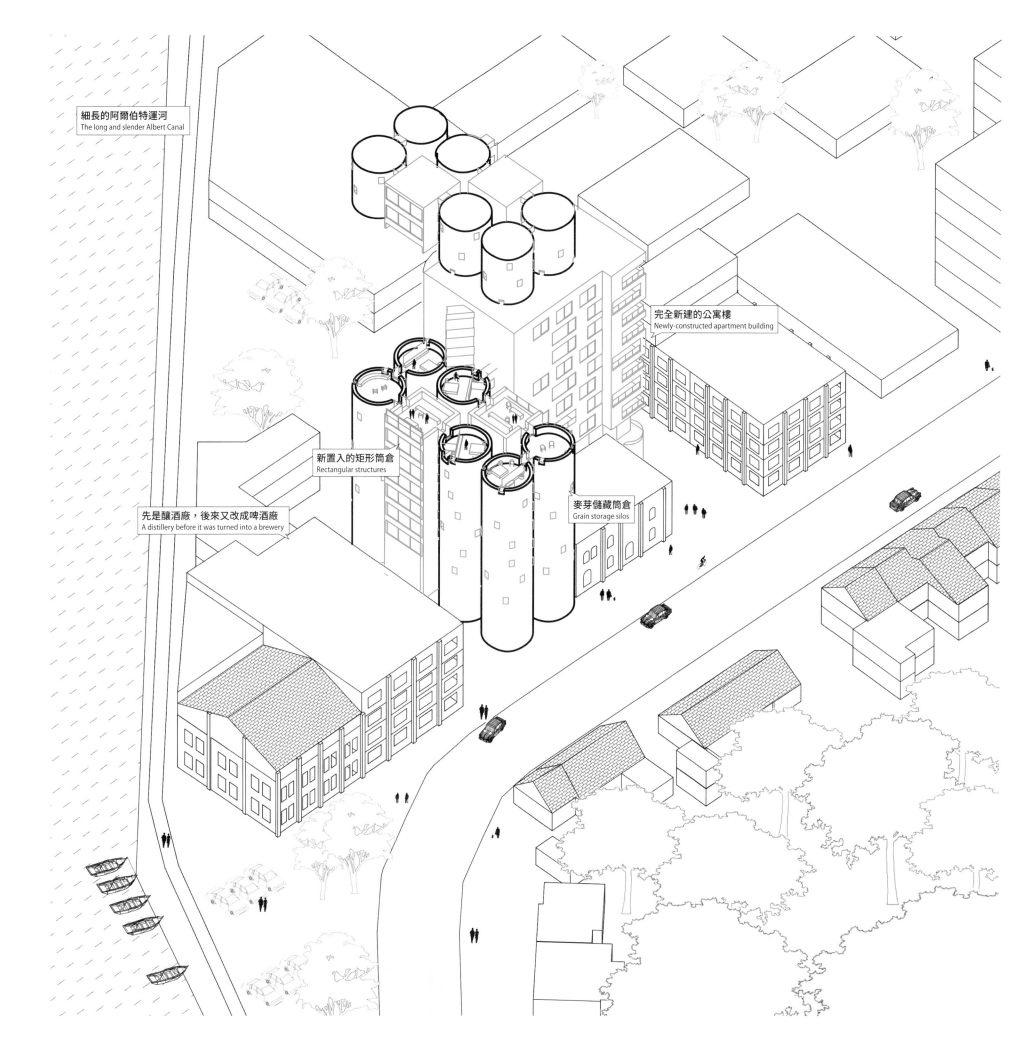

細長的阿爾伯特運河
The long and slender Albert Canal

完全新建的公寓樓
Newly-constructed apartment building

新置入的矩形筒倉
Rectangular structures

麥芽儲藏筒倉
Grain storage silos

先是釀酒廠，後來又改成啤酒廠
A distillery before it was turned into a brewery

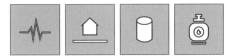

S4

奧地利維也納
瓦斯槽城市
Gasometer, Vienna, Austria

DESIGNER Jean Nouvel &
Coop Himmelblau &
Manfred Wehdorn &
Wilhelm Holzbauer

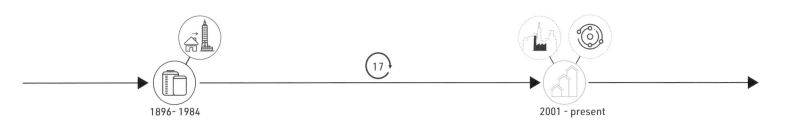

1896-1984 　　　 17 　　　 2001 - present

「瓦斯槽城（Gasometer City）」的英文字面意思，Gasometer 氣量計是儲氣槽外的圓形儀表（而不叫儲氣槽），但這個詞通常用來稱呼容器本身，到了今天仍然如此。

維也納的瓦斯槽城包含四個建於1896年至1899年的舊有儲氣槽，屬於位於維也納 Simmering 第11區附近的 Gaswerk Simmering 天然氣工廠，工廠內的設施原由英國公司 Inter Continental Gas Association（ICGA）建造，目的是幫助維也納供應城市燃氣。當時是全歐洲最大的儲氣槽設計，每座儲氣槽的容量為90,000立方公尺。即使 Simmering Works 每年可以提供超過1億立方公尺的天然氣，儲氣槽最終還是應付不了不斷增長的維也納人口，經過多次升級後，依然跟不上現代技術的迭代。1969年，城市開始使用天然氣取代煤氣，再加上儲氣槽建設的新技術，讓氣體可以更高的壓力轉變為更小的體積，儲存在地下或高壓儲氣球中（死因）。於是這四個儲氣槽最後只能於1984年退役。

自此之後它們被指定為受保護的歷史地標。維也納政府對受保護的古蹟進行了改造和重建，並於1995年鼓勵大眾對這些結構的新用途提出意見。儲氣槽是四個圓柱形伸縮式氣體容器，每個容器儲氣的容積約為90,000立方公尺。它們每個高70公尺，直徑60公尺。

在改造期間，儲氣槽被去除內臟，只保留了外殼磚牆和屋頂部分。這些建築物已經被用在新的住宅和商業用途。

2001年，由四位國際知名建築師分別獲得了一個儲氣槽，Gasometer A、B、C、D，進行改造設計。Jean Nouvel（Gasometer A座），Coop Himmelblau（Gasometer B座），Manfred Wehdorn（Gasometer C座），和 Wilhelm Holzbauer（Gasometer D座）。每個瓦斯槽分為幾個生活區（頂層的公寓）、工作（中間樓層的辦公室）、娛樂和購物（一樓的購物中心）。每個瓦斯槽的商場樓層都通過天橋連接到其他樓層。室內設施包括容量2,000-3,000人的音樂廳、電影院、學生宿舍、維也納縣市檔案館等。大約有800套公寓（歷史磚牆內有三分之二），有1,600名普通租戶，還有大約70套學生公寓，有250名學生居住。

瓦斯槽城營造出了一個獨一無二的城鄉風貌特色，它是一座城中城。真正意義上的社區已經形成，大型實體住房社區（租戶）以及活躍的虛擬互聯網社區（Gasometer 社區）。關於這一現象，不僅有很多新聞媒體爭相報導，也有許多心理學、建築學和城市規劃的專家學者為它提出相關論文。

頂層公寓生活區
Apartments at the top level

中間樓層辦公室工作區
Offices at the middle level

辦公室
Offices

購物商場空橋連接
Skywalks connecting the shopping mall levels

底層商場娛樂和購物區
Shopping mall at the bottom level

S5

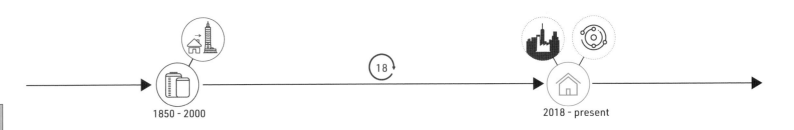
英國倫敦
國王十字
天然氣槽住宅

King's Cross
Gasholders, London,
United Kingdom

DESIGNER Wilkinson Eyre

Gasholder Park──以退役的儲氣槽為主角的都市空間再生與豪華住宅開發。位於倫敦 King's Cross 國王十字新的運河公園 Gasholder Park，雖然規模有限，但其獨特的工業個性卻彌補了尺度上的不足。巨大的鑄鐵框架，曾經站在運河的對岸，擁有 110 萬立方英尺煤氣的 Gasholder No. 8 是最大的標誌性儲氣槽，有著 82 英尺高的圓柱形金屬框架，曾經在 King's Cross 的天際線占據主導地位。而現在重生後的它，圍著一個有動態燈光裝置藝術空間般的圓形草坪，是一個都市的動態燈光景觀學校，也是當地鄰里的休閒場所。

King's Cross 的大型儲氣槽建於 1850 年代，為倫敦的大部分地區提供媒氣儲存，用於街道或家用照明，供暖氣和烹飪。儲氣槽一直使用到 2000 年終於退役。當 King's Cross 的重生剛開始時，Gasholders No. 8 以及 10、11 和 12 其複雜的鑄鐵結構曾經被拆解成一塊一塊運往約克郡，花了兩年時間修復，並在 2013 年重返 King's Cross，搬到了運河岸邊的新家，重建後的 No. 10、11 和 12 組成了 Gasholders Triplet，成為運河公寓 Gasholders London。

這個三連鎖儲氣槽是 King's Cross 工業遺產的特色，由 Wilkinson Eyre 將其融入到一個充滿現代設計的豪華住宅綜合體，該綜合體提供了豐富的開放空間及設施，如健身房、Spa、出租工作室與會議空間和宴會廳。該項目建在 8 層，9 層和 12 層的混凝土圓柱體上，由 145 個單元組成多種不同的戶型。現代化的玻璃和鋁製品設計，尊重被列入二級保護的維多利亞時代鑄鐵柱的特徵。室內風格將工業，手工藝和奢華元素融合在一個適合出現在倫敦市中心的家居和當代商業室內設計中。一系列的零售空間安置在 Gasholders 的地面層，讓公眾有機會進入該綜合體，最大程度地平衡工業與奢華，新與舊，公共與私人，外部與內部間的關係。

儲氣槽巨大的鑄鐵框架象徵著人類水封煤氣儲存技術的階段性成果，這樣的幾何鮮明的地標正逐漸從倫敦的天際線中消失，但要保留這些動人的工業遺產，必須投入大量資本，雖然我們都能理解這些遺產大大影響人們喜歡這個場所與否，但除了國王十字這樣地方以外的其他地區，這種程度的投資若要證明其合理性，勢必會面臨更大的挑戰。

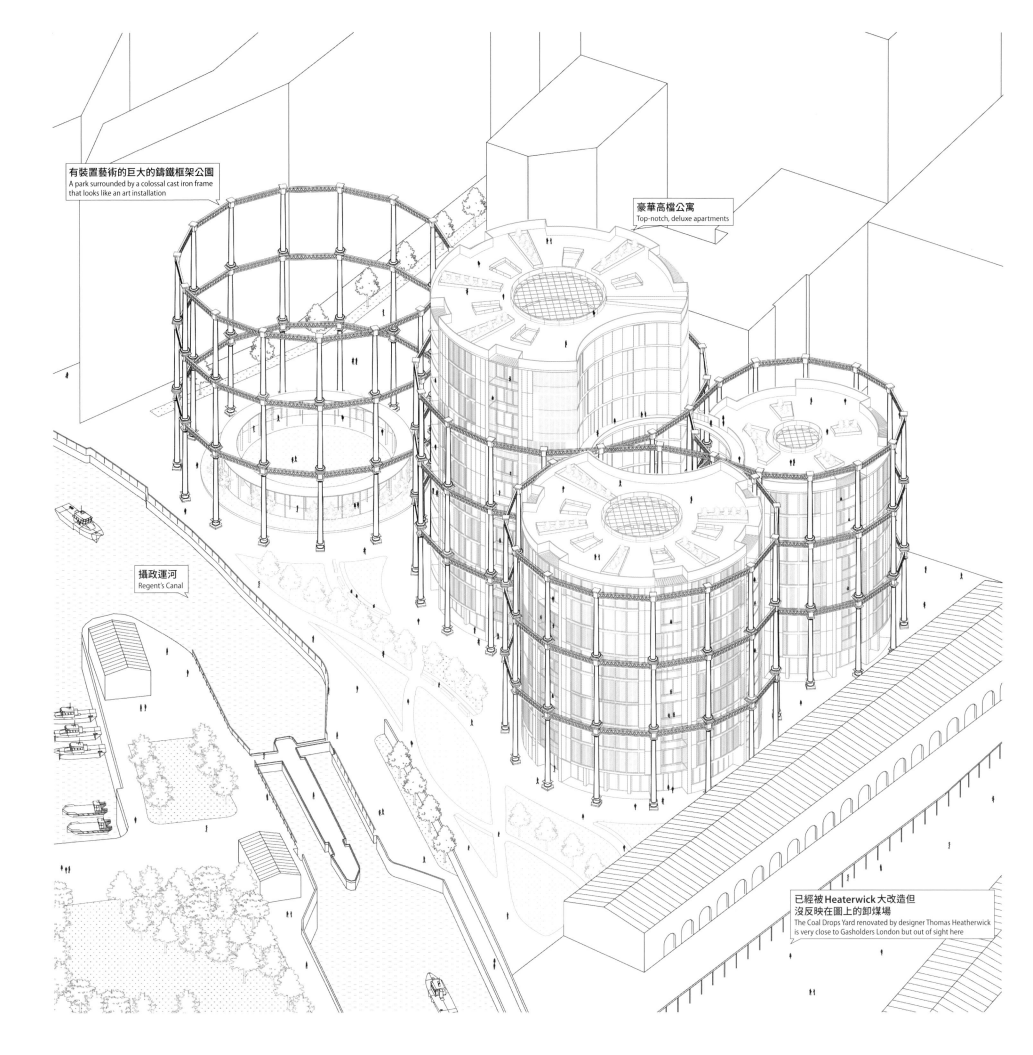

有裝置藝術的巨大的鑄鐵框架公園
A park surrounded by a colossal cast iron frame that looks like an art installation

豪華高檔公寓
Top-notch, deluxe apartments

攝政運河
Regent's Canal

已經被 Heaterwick 大改造但
沒反映在圖上的卸煤場
The Coal Drops Yard renovated by designer Thomas Heatherwick is very close to Gasholders London but out of sight here

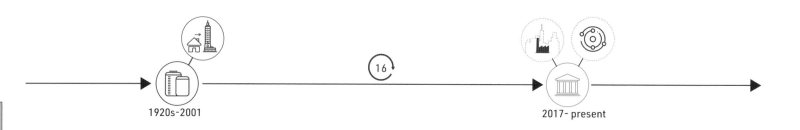

1920s-2001

2017- present

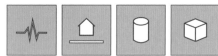

S6

南非開普敦非洲
當代藝術博物館
Zeitz MOCAA, Cape
Town, South Africa

DESIGNER Heatherwick Studio

南非最大的藝術博物館「非洲當代藝術博物館（Zeitz MOCAA）」，改建自挖空一座歷史悠久的糧食筒倉建築內部。英國設計師托馬斯・希瑟威克（Thomas Heatherwick）描述為「世界上最古老的建築」，Zeitz MOCAA 將成為世界上最重要的非洲藝術展覽空間。它位於 1920 年代開普敦海濱的糧食筒倉內，博物館坐落於 9,500 平方公尺的空間內，跨越 10 層樓的高度，依附著歷史悠久的 Grain Silo Complex 穀倉而建。該穀倉停用於 1990 年，算得上是開普敦工業歷史的一座里程碑。

如今 Heatherwick Studio 的改造為其賦予全新的面貌，在建築的管狀內部雕刻了巨大的部分，創造了一個由 80 個畫廊空間組成的複雜網絡。該建築物的機能包括了 Zeitz 博物館、V & A Waterfront 酒吧和餐館、酒店。博物館內包含總面積達 6,000 平方公尺的 80 個畫廊空間、屋頂雕塑花園、藝術品儲存和保護區、書店、餐廳、酒吧以及閱覽室。博物館還將設立多個以服裝、攝影、策展、動態影像、行為實踐和藝術教育等為中心的空間。這座型態看似單一的混凝土建築由兩個部分組成：一棟樓層分明的塔樓，以及 42 個蜂窩狀的高大筒倉。項目最大的難點在於將這些緊密結合的混凝土圓筒轉變為適用於藝術展示的空間，並保留建築既有的工業感。

其他的筒倉重生案例大多會避免對混凝土圓筒結構做大動作，但 Heatherwick 反其道而行，大膽地在建築內部開鑿出一個形似拱頂教堂的中庭，使之成為博物館的核心，並為環布於中庭的各層展覽區域提供了路徑。原本僅有 170 公釐厚的混凝土圓筒，加厚加強了鋼筋混凝土於其上，形成新的混凝土圓筒結構，擁有 420 公釐的厚度，並為舊筒倉的切割提供了依據，使 4,600 立方公尺大容積的中庭能呈現出彎曲的幾何形態。切割的邊緣被磨光並加上鏡面，與建築原有粗糙的混凝土形成對照。圓筒頂部皆裝配了直徑為 6 公尺的夾層玻璃，為中庭帶來自然光。位於地下的隧道則被用於一些需要在特殊場地進行的藝術創作。

在建築外部設有凸出的窗戶——由多面玻璃板組成。從外面看，玻璃板反射了桌山、羅本島、往來的人群還有天空中的雲彩。它們安裝於現有的混凝土框架內，將光線吸入中庭並提供萬花筒般的視覺效果，中庭的挖空打通，讓原本均值的圓筒狀量體分布，開始有了疏密，有了開放與收斂，進而產生流動、聚集。參觀者猶如身在一個大型的雕刻藝術品內，穿梭、流連、探險，並和建築師一起喚醒沉睡已久的糧食筒倉，見證它的重生。

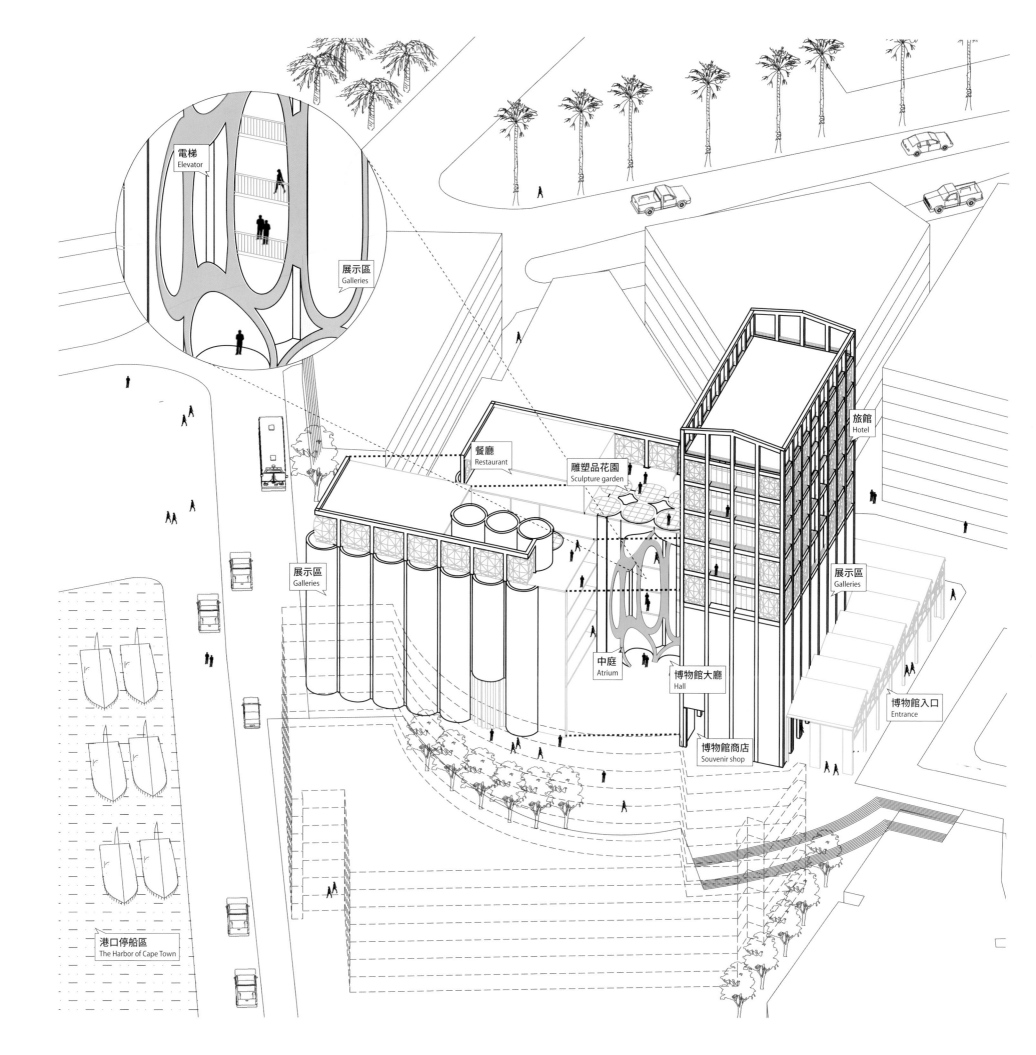

電梯
Elevator

展示區
Galleries

餐廳
Restaurant

雕塑品花園
Sculpture garden

旅館
Hotel

展示區
Galleries

中庭
Atrium

博物館大廳
Hall

展示區
Galleries

博物館入口
Entrance

博物館商店
Souvenir shop

港口停船區
The Harbor of Cape Town

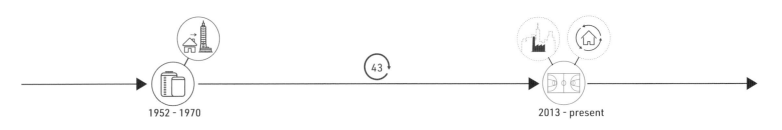

1952 - 1970　　　　　　　　　　　　　　　　　2013 - present

加拿大蒙特婁攀岩健身房 Allez-Up

Allez Up Rock Climbing Gym, Montreal, Canada

DESIGNER Smith Vigeant Architectes

攀岩健身房 Allez-Up 是加拿大蒙特婁西南部振興項目的核心。位於拉欽（Lachine）運河的側翼，舊雷德帕思（Redpath）煉糖廠的場地和筒倉，如今改造成獨一無二的室內攀岩設施，大大增加了運河娛樂與旅遊的吸引力。將廢棄的筒倉開發成一個攀岩健身房是一種獨特的方式，可以最大限度地發揮蒙特婁工業遺產的巨大潛力。

1825 年，拉欽運河開放通航，作為繞過聖勞倫斯河的近道，開啟了該片地區的工業化發展。到 1860 年代，這裡已成長為一個繁忙的工業區，有著全國最多樣化的工業集群。1952 年雷德帕思煉糖廠在此建造了 4 座筒倉，作為工廠的儲藏空間。不過，此時拉欽運河的發展已經進入晚期，因為它無法被拓寬，大型船隻難以通行。1959 年聖勞倫斯航道完工，意味著拉欽運河黃金時代的終結。1970 年運河正式關閉，很多企業搬遷（死因），最終這片社區陷入了蕭條。隨後，地區內其他工業建築開始被重新改造利用，但筒倉因為其特殊的型態較難改造，故在 40 年的時間裡完完全全遭到了遺棄。

Allez-Up 原先的健身房也改建自附近一個工業設施，其業主希望可以將其設備容納量擴大為原本的 3 倍，並且看好筒倉和其周邊場地非常適合改造為攀岩健身館。改造工程是將原先筒倉的幾個柱狀空間透過一個長方形的量體相連，將兩組閒置的筒倉連接起來。整個改建呈現了一種非常獨特的開發方式，受室內攀岩牆動態的啟發，建築師希望為攀岩者最大化地利用豎向空間。原來筒倉的內壁覆蓋著兩層雪松木板，目的用於避免儲藏的糖受潮，改造工程移除了部分木板，將其再利用而製成家具或室內裝飾。

主樓內的攀岩牆實際上類似於糖懸崖，不時提醒來訪者雷德帕思筒倉曾經是煉糖廠的一部分。純白色的稜形攀岩牆為初學者和經驗豐富的攀爬者提供了許多不同的路線選擇。多色的攀岩架穿過牆壁，增添了這個獨特內部空間的動感魅力。建築的主要立面被幾個斜切開口隔斷，通高的玻璃為室內提供全天候的自然光照。最重要的是筒倉本身被仔細地融入了功能，接待室放置於西邊的一側筒倉中，而攀岩路線遍布筒倉混凝土結構內外，充分利用了它們的垂直感。筒倉全然成為了 Allez-Up 攀岩體驗的一部分，而它們自身作為歷史工業遺產，也在不斷向來訪者訴說著過往的功能。

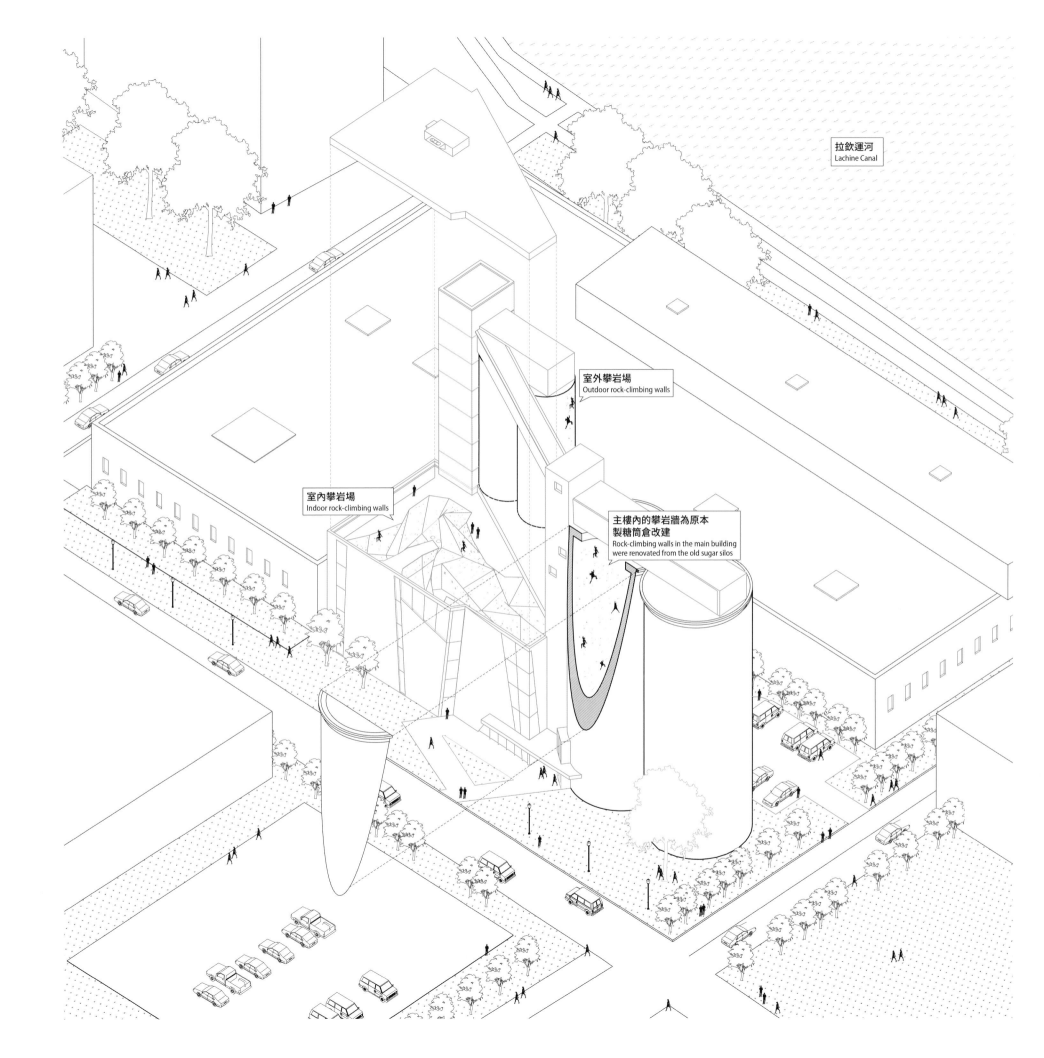

拉欽運河
Lachine Canal

室外攀岩場
Outdoor rock-climbing walls

室內攀岩場
Indoor rock-climbing walls

主樓內的攀岩牆為原本
製糖筒倉改建
Rock-climbing walls in the main building
were renovated from the old sugar silos

S8

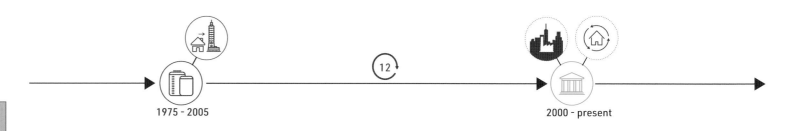

1975 - 2005　　　　　　12　　　　　　2000 - present

中國上海民生碼頭八萬噸筒倉

80,000-tonne Silo Warehouse, Shanghai, China

DESIGNER 大舍（柳亦春／陳屹峰） | Atelier Deshaus

上海城市公共空間設計促進中心作為「上海城市空間藝術季」品牌活動的運營機構，在全盤考慮區位選擇相對未成熟的地區、建築特色及環境空間適應性、方案可行性等要素，希望借助「空間藝術季」以啟動區域的活力，最終選擇了浦東新區民生碼頭以八萬噸筒倉為核心、與周邊257庫的室內外空間作為藝術季的主展館場地，結合民生碼頭的改造轉型形成新的文化焦點。

民生碼頭區域內既有建築15棟，其中四萬噸筒倉和八萬噸筒倉是其中的散糧筒倉，257庫是原來的灌包車間。民生碼頭的前身瑞記洋油棧碼頭建於1908年，由英國航行於歐亞之間最大的藍煙囪輪船公司（Blue Funnel Shipping Line）委託太古輪船公司興建。1924年，它成為遠東最先進的碼頭，此後名稱幾經更迭，於1956年更名為「民生碼頭」並沿用至今。其中的四萬噸筒倉在1975年建成，後隨著上海經濟的發展，碼頭裝卸輸送量需求逐漸增大，1991年開始建造八萬噸筒倉並於1995年完工使用，但由於之後糧食儲存方式的革命性改變（死因），筒倉在10年後於2005年停用。

筒倉經過10多年停用後破損嚴重，因此首期改造目標是要保證參訪者的安全、結構補強及消防疏散，以及創造良好的觀展體驗。展場筒倉與257庫的主要入口，被白色半透明板包覆成現代感十足且顯眼的入口，透明輕質與筒倉混凝土的厚實形成一種對比。筒倉特殊的空間型態和巨大的體量感，為初次進入的參訪者帶來極大震撼，

而懸在頭上的漏斗距地面近10公尺，更刺激了觀者一窺筒體內部的渴望，被圍繞每一個漏斗而置入的藝術品感動。筒倉頂層位於40公尺的高處，展廳規模超過3,000平方公尺。由於對垂直動線的需求非常高，因此大舍的建築師們首先利用了30個筒中的4個筒放置了僅供緊急情況下使用的垂直疏散樓梯和消防電梯。其他則主要由中庭中1樓到3樓的自動扶梯、外掛在筒倉外壁的3樓到7樓的扶梯、沿內筒壁而行的螺旋坡道和螺旋梯，皆是在滿足垂直動線的同時，為參訪者創造觀展時穿梭於筒倉建築的多重體驗。特別是到達3樓的平台後，通透的玻璃外的景色將人引到了筒倉北立面外側的通廊裡，參訪者被扶梯從3樓起分三段帶到了6樓約32公尺的高度。行進的過程中，一邊是近在咫尺的混凝土筒，參訪者可以近距離閱讀筒倉內壁所留下的歷史痕跡——過去糧食的碎屑遺留；另一邊，透過與藝術家丁乙合作設計立面——有漸變白色波點紋樣的玻璃，逐漸展開的黃浦江美景盡收眼底。

所有的改造動作都是為了建立更好的筒倉建築和濱江之間與參訪遊客的關係，使工業遺產和這座城市獲得更好的連接。隨著黃浦江兩岸的濱江貫通與公共空間的更新發展，靠著這樣一次在藝文活動所掀動的改造，使得黃浦濱江工業遺產獲得更多關注，也得到更新利用的機會。

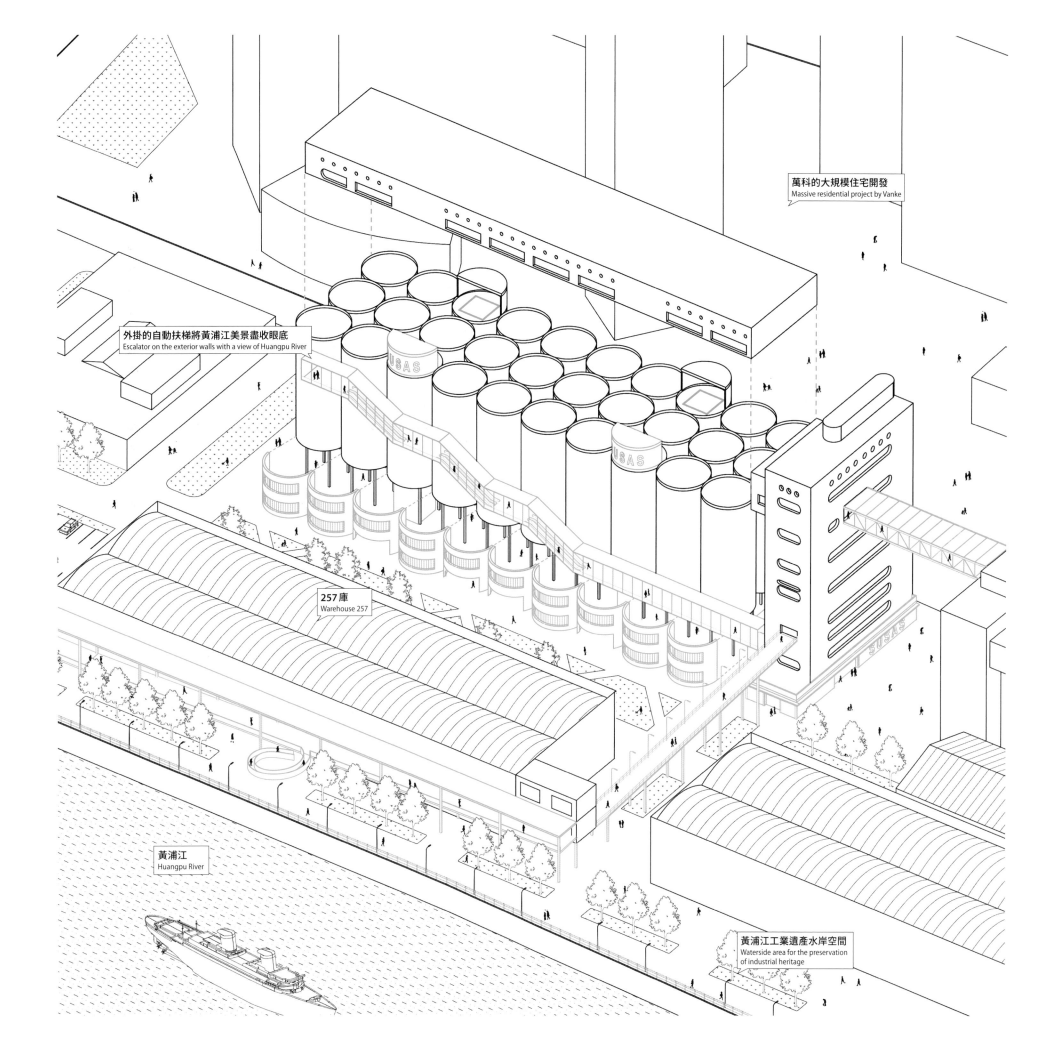

萬科的大規模住宅開發
Massive residential project by Vanke

外掛的自動扶梯將黃浦江美景盡收眼底
Escalator on the exterior walls with a view of Huangpu River

257庫
Warehouse 257

黃浦江
Huangpu River

黃浦江工業遺產水岸空間
Waterside area for the preservation of industrial heritage

M

軍事 Military

M1
德國柏林飛船機庫
室內熱帶度假村
Tropical Islands Dome Resort,
Krausnick, Germany

M2
德國柏林碉堡屋頂住宅
Fichte-Bunker, Berlin, Germany

M3
德國漢堡掩體再生能源中心
Renewable Wilhelmsburg
Energy Bunker, Hamburg, Germany

M4
荷蘭屈倫博赫
碉堡599水線紀念碑
Bunker 599, Zijderveld,
The Netherlands

M5
瑞典斯德哥爾摩核掩體
數據資訊中心
Data Center Pionen
White Mountain,
Stockholm, Sweden

M6
奧地利維也納
高射炮塔水族館
Haus des Meeres Flak Tower
aquarium, Vienna, Austria

M7
美國緬因空軍基地掩體
蝙蝠避難所
Air Force Base Bat Shelter Bunker,
Maine, USA

M8
英國薩福克海岸
拿破崙式海防塔樓住宅
Martello Tower Y, Suffolk coast,
United Kingdom

M9
英國樸茨茅斯港
海上堡壘旅館
Palmerston Sea Forts Hotel,
Solent, United Kingdom

M10
俄羅斯聖彼得堡亞歷山大堡
疫苗研究實驗室（廢墟）
Fort Alexander,
St Petersburg, Russia

M11
台灣新竹建功國小大桶教室
The Oil Storage Classroom
JianGong Primary School
Hsinchu, Taiwan

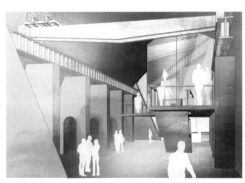

M12
台灣鳳山無線電信所
（設計提案）
Fongshan Wireless
Communications Station (Proposal)

M13
美國紐約無畏號航空母艦
海空暨太空博物館
The Intrepid Sea, Air & Space
Museum, New York, USA

M14
蘇聯颱風級核潛艇
（設計提案）
Typhoon class Nuclear Submarine
(Proposal)

M1

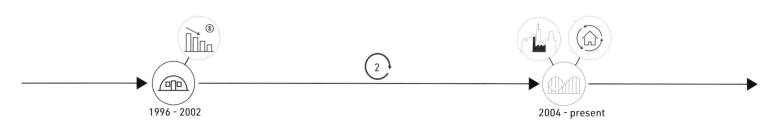

1996 - 2002   2   2004 - present

德國柏林
飛船機庫室內
熱帶度假村

**Tropical Islands
Dome Resort,
Krausnick, Germany**

DESIGNER Oscar Niemeyer

「我的熱帶島嶼（My Tropical Island）」度假村是一個熱帶主題公園，位於德國布蘭登堡前 Brand-Briesen 機場，距離柏林南部邊界50公里。它坐落在一個前飛艇機庫（被稱為 Aerium），這是世界上最大的獨立式機庫。飛艇機庫長360公尺，寬220公尺，高106公尺，一個獨立的鋼製圓頂「桶形碗」結構，大到足以容納艾菲爾鐵塔的側面，另外還配備了一個180公尺的切割台來製造飛艇的外殼。機庫建造工程於2000年11月完成，耗資7,800萬歐元，本屬於 Cargolifter 公司。CargoLifter 設法建造了飛艇 CL75 原型，但機庫原本打算安置的飛艇 CL160 卻從未建成，因為 CargoLifter 公司在 2002 年中破產（死因：資金缺乏）。2003年6月馬來西亞公司丹戎（Tanjong）以1,750萬歐元的價格買下機庫和500公頃土地，將機庫改造成一個名為「我

的熱帶島嶼」的休閒度假勝地，於2004年12月19日正式開放。

在機庫內，空氣溫度為26℃（78°F），空氣濕度約為64%，有海灘、雨林，約有50,000種植物和600種不同的物種，包括一些珍稀植物。設計看起來像珊瑚島的水域和潟湖，有噴泉、運河、漩渦，以及許多滑梯、游泳池、三溫暖浴室和 Spa 水療設施，這是歐洲最大的熱帶三溫暖浴場。此外還有酒吧和餐館、露營地、旅館、兒童遊樂區和一個瘋狂的高爾夫球場。全天候開放，全年無休。

「我的熱帶島嶼」每天最多可容納8,200名遊客，是全球最大的室內熱帶雨林與室內水上樂園。

機庫原本打算安置 CL160 飛艇
The airship hangar was originally intended to house the CL160 airship

熱帶雨林
The rainforest

以前是布蘭登堡 Brand-Briesen 機場
The former Brand-Briesen Airfield

1874 - 1922　　18　　1940 - 1945　防空避難所　戰爭結束　　1　1946 - 1963　戰爭難民收容所　安全衛生問題關閉　　0　1963 - 1990　參議院儲藏室　冷戰結束　　16　2006 - present

德國柏林
碉堡屋頂住宅

Fichte-Bunker,
Berlin, Germany

DESIGNER Paul Ingenbleek

德國柏林的費希特碉堡（Fichte-Bunker）可以說是隨著不同時期一變再變、多次重生的代表。由土木工程師 Johann Wilhelm Schwedler 設計，建於1874年位於德國柏林 Kreuzberg 區，是柏林現存的最後一座磚造煤氣槽，其教堂式磚牆外殼高21公尺，直徑56公尺。1920年代柏林的路燈開始改為電燈，使這個精心設計的煤氣容器逐漸顯得多餘（死因），終於在1922年停用。直到1940年納粹黨的工程師兼高級成員弗里茨‧托德（Fritz Todt）任命西門子—鮑恩恩（Siemens-Bauunion）將煤氣槽改造為6級防空避難所（第一次重生），用厚達3公尺的混凝土加固牆壁。最初的避難所計畫為6,000人，但在1945年2月3日的空襲期間，約有3萬人擠在大約750個5到7平方公尺大小的單獨房間裡避難，當代藝術家將其稱為豪華沙坑。儘管遭到猛烈的轟炸，但戰爭中的防空碉堡完好的保存下來。

戰爭結束後，Fichte-Bunker 最初是流離失所者的收容所，特別是那些從東柏林逃往西方世界的難民，又被用作少年拘留設施並作為老人的家。最後，它變成了一個無家可歸者收容所（第二次重生）。每晚房租為2.50馬克，由於內部條件非常糟糕，被稱為「無望的碉堡（Bunker der Hoffnungslosen）」。

該設施於1963年因衛生與安全問題關閉，從那時起直到東西德統一以前，該市將該建築用作參議院儲藏室的一部分（第三次重生），儲存緊急物資直到冷戰結束。1990年後費希特碉堡再次空無一人。

直到 SpeicherWerk Wohnbau GmbH 於2006年收購了它，並著手改造。儘管附近居民對於可能違反歷史建築規範提出異議，但此改造案仍獲得許可。工程師 Michael Ernst 和建築師 Paul Ingenbleek 得以發揮他們的魔力，將這個防空碉堡在2010年打造成為 Circlehouse（第四次重生）。該建築的頂部現在擁有13個2層樓高的豪華公寓及壯觀的私人屋頂花園，住戶可以通過外部電梯塔和橋梁進入大樓。而在 Circlehouse 下，為供大眾參觀的展示館，不僅展示了當時的建築技術，並以眾多展品記錄下戰爭轟炸期間，戰爭難民和無家可歸者的悲慘命運。過去黑暗的歷史在建築重生後，再度回到世人面前給與人們警惕。

從煤氣槽到豪華公寓一路走來錯綜複雜的歷史，意味著它占據了柏林時代與城市演化核心的位置。

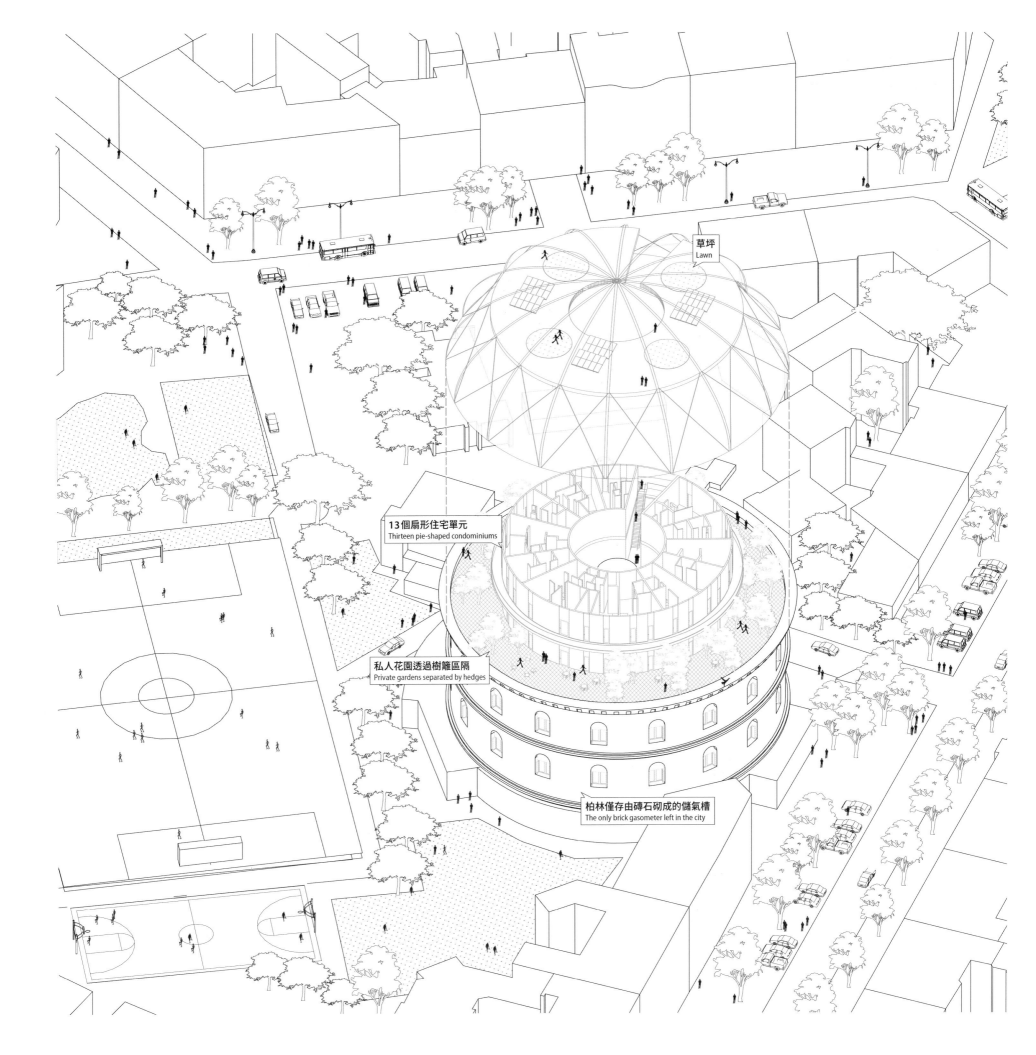

草坪
Lawn

13個扇形住宅單元
Thirteen pie-shaped condominiums

私人花園透過樹籬區隔
Private gardens separated by hedges

柏林僅存由磚石砌成的儲氣槽
The only brick gasometer left in the city

直到1947年，
內部結構體幾乎完全被英軍破壞

66

1940 - 1947

2013 - present

德國漢堡掩體
再生能源中心

**Renewable
Wilhelmsburg
Energy Bunker,
Hamburg, Germany**

DESIGNER HHS Planer +
Architekten AG

從戰爭第一線退役，變身為區域能源中樞，是重生案例中少見的再次變為基礎設施。

位於德國漢堡威廉斯堡（Wilhelmsburg）Reiherstieg地區的「再生能源中心（Energy Bunker）」，前身是德軍於1943年興建的巨大防空掩體（Air Raid Bunker），保護當地居民免於大量炸彈的轟炸，也曾作為支援前線的攻擊要塞。直到1947年，內部結構體幾乎完全被英軍破壞。之後長達60年一直處於荒廢狀態。

直到2006年，政府開始規劃將其改造。2010年進行了第一次靜態試驗，並在2011年開始針對改善安全性及空間機能轉換的工程。這個總成本為2,700萬歐元的重要設施再生計畫，由歐洲區域發展基金（ERDF）和漢堡能源公司（Hamburg Energie）出資，城市開發公司IBA Hamburg執行，目標是將它轉變為區域的永續能源中心。

原建築物雖然內部被嚴重破壞，但幸運的是，厚達3公尺的牆面與4公尺的屋頂讓骨架仍算穩固。為了使空間可以安全進駐使用，將混凝土噴灑在崩解風化的立面上使其穩定，並增加內部隔熱層。

室內拆除被炸毀的樓板並更換電梯及樓梯。室外則加裝大量太陽能板。

本案的核心是一個200萬公升的蓄水池，作為大型儲熱裝置。它儲存來自周圍生質能源電廠、木材燃燒、屋頂及南面太陽能系統產生的能源，另外還可引進附近工業設施的廢熱。透過有效地將不同來源的熱能匯集在一起，Energy Bunker可將能源重新分配到鄰近的社區，將來更計畫能為Reiherstieg的大部分地區提供熱能。同時，還可直接向公共電網輸入綠色電力。

閒置60多年之後，經過7年的項目開發和建設階段，這個再生能源中心如今成為全球第一個由戰爭遺留的龐大構造物所改造的複合式能源儲存與再分配設施。這座兼具紀念意義與未來環境友善指標的區域公共建築，不僅提供乾淨的再生能源，還展示如何利用當地資源生產和儲存熱能。同時也規劃遊客中心、永久展覽空間和30公尺高的咖啡廳與觀景平台，成為漢堡熱門的旅遊景點。

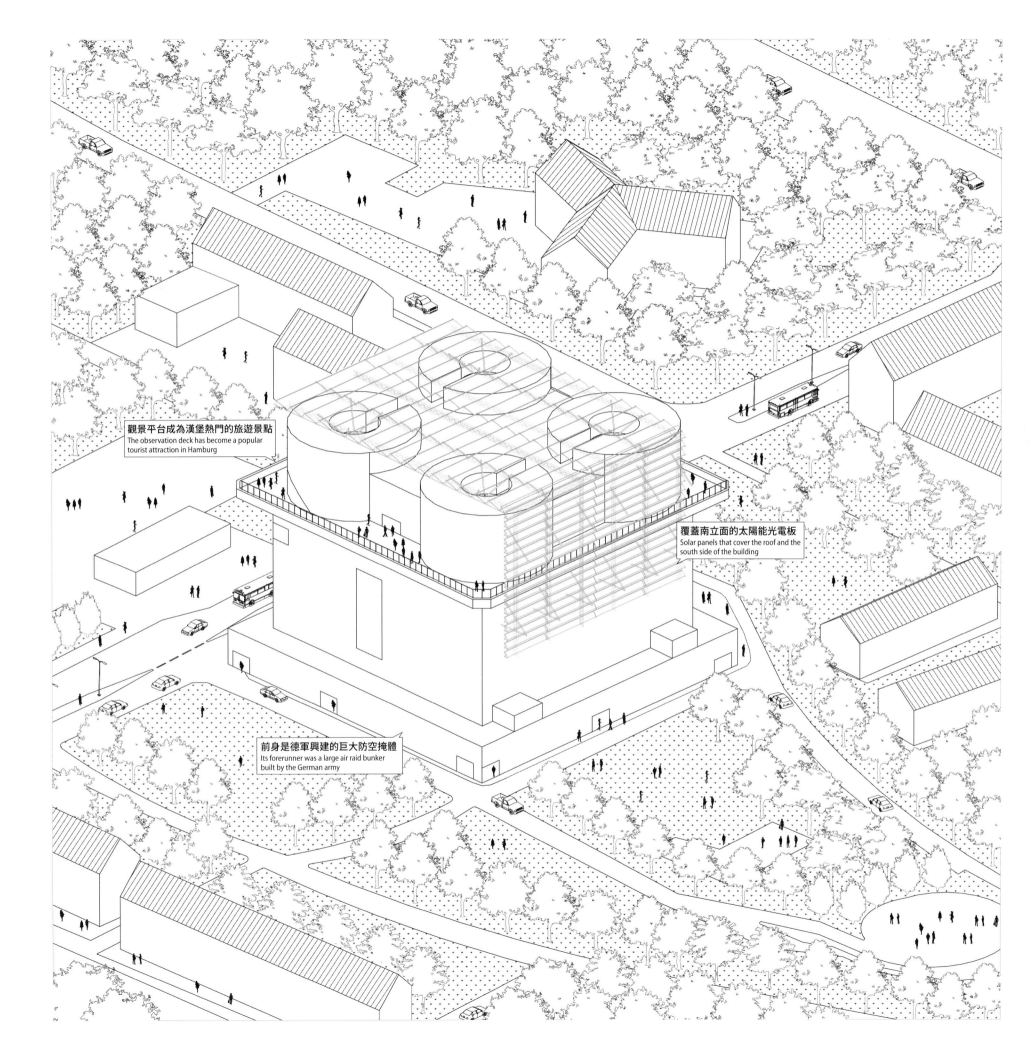

觀景平台成為漢堡熱門的旅遊景點
The observation deck has become a popular tourist attraction in Hamburg

覆蓋南立面的太陽能光電板
Solar panels that cover the roof and the south side of the building

前身是德軍興建的巨大防空掩體
Its forerunner was a large air raid bunker built by the German army

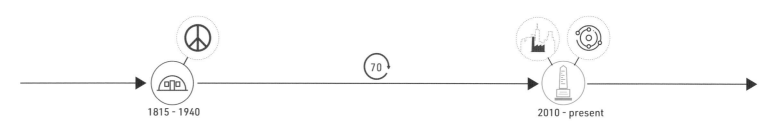
荷蘭屈倫博赫
碉堡599
水線紀念碑
Bunker 599, Zijderveld, The Netherlands

DESIGNER RAAAF + Atelier de Lyon

「碉堡599」是2000年開始的20年總體規劃的一部分，將荷蘭水線改造成國家公園，於2010年完工，這是對「有形性」地詮釋景觀和歷史做出了貢獻，展示了荷蘭多餘的二次大戰掩體，與荷蘭軍事防禦思想的歷史悲劇一起發揮：如何透過切割它來展現其內部而變成如雕塑、如遺跡般淒美的景點。

該項目揭示了新荷蘭水線NDW的祕密，這是從1815年到1940年的一段軍用防線，目的是透過洪水保護Muiden、Utrecht、Vreeswijk和Gorinchem等城市。制定這條防線的計畫雄心勃勃，它利用了荷蘭風景的地理條件和荷蘭獨立戰爭（1568-1648）中向水學習的經驗，以及拿破崙在俄羅斯（1812）的血腥失敗。另一方面，它在歷史上卻是一場無奈的錯誤：1940年，荷蘭軍隊準備在冰面上進行游擊戰，然而德軍帶著飛機轟炸鹿特丹，把傘兵放在前線後面。新荷蘭水線的防衛功能因此從未真正發揮作用。

Bunker 599建於1940年，在轟炸襲擊期間內部可容納多達13名士兵。荷蘭RAAAF和Atelier de Lyon工作室合作，為民眾揭開了碉堡內部的小型黑暗空間。他們將看似堅不可摧的NDW 700型掩體內部剖開：用一把金剛石線鋸切割整個碉堡的結構中心，前後花了40天的時間才切斷堅固的混凝土掩體。而後用一台起重機將其分開，在中間形成一條狹縫，一條長長的木板通道穿過了小卻極其沉重的建築，將訪客引向洪水區和相鄰自然保護區的人行道。

該碉堡隨後從市政紀念碑提升至國家古蹟，現在是荷蘭新水線的一部分。設計師另外建造了一套樓梯，將附近的道路連接到一條通過碉堡中心的通道上，到達洪水區上方的木製浮橋。碼頭和支撐浮橋的木樁提醒訪客，周圍的水域不來自於泥沙的移除，這裡原是在戰爭期間被淹沒的淺水平原。

這種激進的介入方式為荷蘭的文化遺產政策提供了詩意的可能。同樣，時間也讓人們以新的方式看待周圍的環境。

堤路擁有俯瞰碉堡 599 的完美視角
Street by the bank with a perfect perspective of Bunker 599

連接道路的入口階梯
Stairs that connect a nearby street

在敵軍轟炸時可容納 13 名士兵的內部空間
The interior can shelter 13 soldiers during bombing

被切分為二的碉堡本體
The bunker cut in half

浮橋為激烈的改造手法提供
另一種詩意的觀賞角度
Wooden boardwalk that provides a
poetic point of view

M5

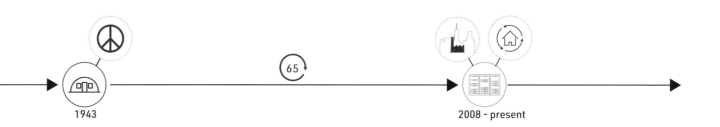

1943

65

2008 - present

瑞典斯德哥爾摩
核掩體
數據資訊中心

Data Center Pionen
White Mountain,
Stockholm, Sweden

DESIGNER Albert France-Lanord
Architects

瑞典最大的一家網路服務供應商 Bahnhof AB 於 2008 年落成一座安全性極高的數據中心，建在斯德哥爾摩市中心 Vita Berg 公園的地下岩層花崗岩下方 30 公尺處的一座核掩體裡。40 公分厚的入口大門使它與世隔絕，此地原本的用途是冷戰時期的軍用核掩體和指揮中心。如今這個數據中心沿用了軍用時代的代號「Pionen White Mountain」作為其名稱。Bahnhof AB 在瑞典有 5 個數據中心，Pionen White Mountain 是其中最大的一座，你也可以把自己的伺服器託管於此；例如「維基解密 WikiLeakes」就在此處放置電腦設備的機房。

這個地下的數據中心裡面除了機房外，還有溫室、瀑布，以及德國潛艇引擎和模擬陽光。由 Albert France-Lanord Architects 建築師事務所設計，整個數據中心空間面積為 1,200 平方公尺，建築師於 2007 至 2008 年間對 Pionen 進行了全面的重新設計，把它變成了現在的數據中心，Albert France-Lanord 建築事務所需要解決的最大的挑戰是與一個「沒有提供一個方形角度的空間」合作：岩石。為了擴大空間，他們花了兩年多時間，炸掉了 4,000 多立方公尺的堅硬岩石，以便取得 Bahnhof 所需的額外空間，將其備用發電機和伺服器安裝到洞穴中。

Pionen 共有 15 名資深員工，只有一名高級技術人員。他們使用德國潛艇引擎用作備份電源：備份電源由兩台柴油機充當，發電功率 1.5 兆瓦。這兩台柴油機原本是為潛艇設計的，Pionen 員工出於娛樂目的，還安裝了原來德國潛艇上的警報系統（汽笛）。冷卻系統依靠 Baltimore Aircoil 公司的風扇，能產生 1.5 兆瓦冷卻效果，足以冷卻幾百台機架式伺服器。三重 CRC 網路主幹接入：網絡完全 CRC，光纖和銅纜從三條不同路徑通向這裡，Pionen 是北歐網際網路連接最強的地方。此外，為了提高工作環境的舒適性，數據中心還配有模擬陽光、暖房、瀑布，以及一個能容納 2,600 公升海水的魚缸。

Bahnhof 首席執行官 Jon Karlung 解釋，數據中心的設計者常常忽視了和這些冷冰冰的東西打交道的人，我們覺得，這個奇異的地方本身就能讓設計師產生偉大的靈感。「既然我們得到了這個獨一無二的核掩體，我們就不能把它建成平庸的傳統託管中心。」他說：「我們希望與眾不同。這個地方本身需要一些出色的設計，而科幻小說是靈感的天然源泉。」

奇特的設計也為這家網路服務供應商傳播了名聲，使它特別引人注目。而且因為這裡提供主機託管服務，有些客戶經常要來這裡工作。這些人自然會和他人分享他們不同尋常的親身感受。

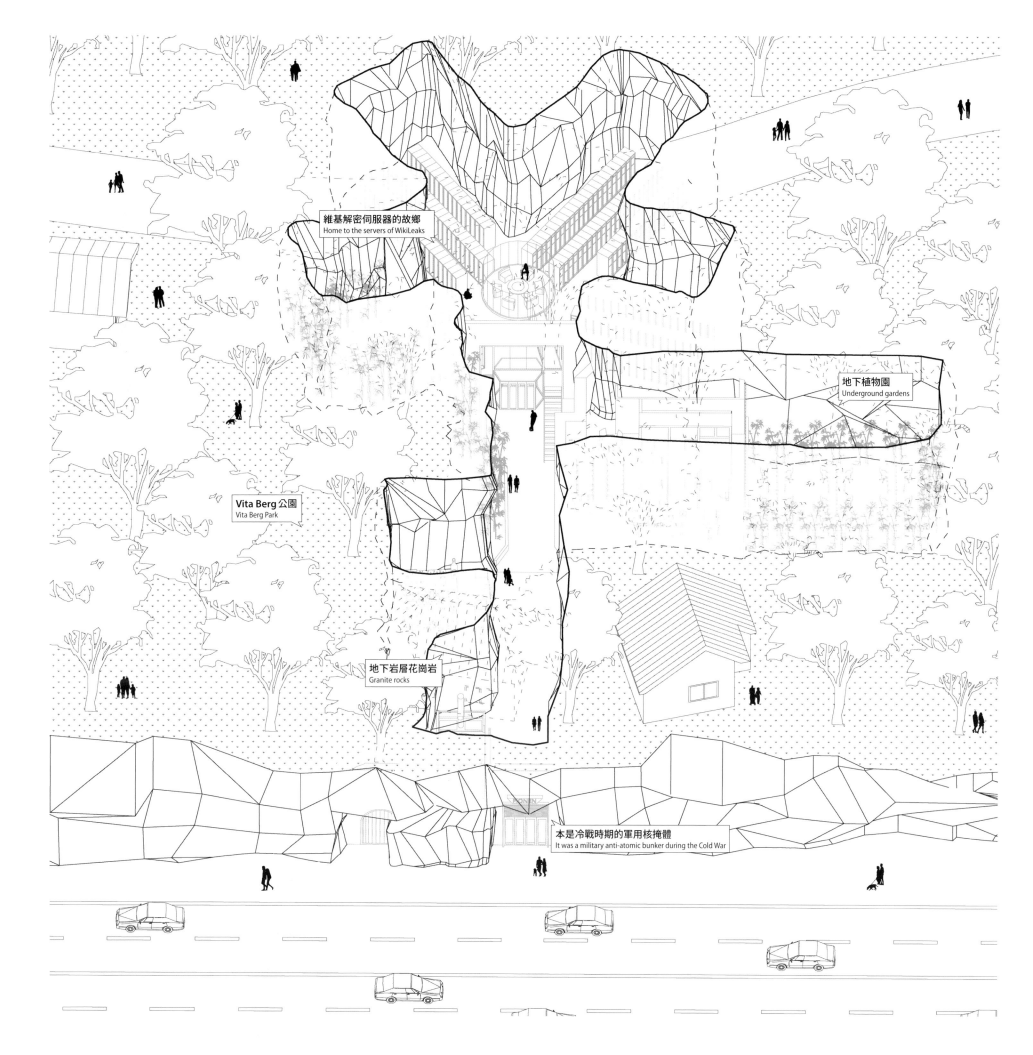

維基解密伺服器的故鄉
Home to the servers of WikiLeaks

地下植物園
Underground gardens

Vita Berg 公園
Vita Berg Park

地下岩層花崗岩
Granite rocks

本是冷戰時期的軍用核掩體
It was a military anti-atomic bunker during the Cold War

1943-1947 ② 1949 ⓪ 1949-1957 消防局? 34 1991 - present

奧地利維也納
高射炮塔水族館

Haus des Meeres
Flak Tower aquarium,
Vienna, Austria

DESIGNER Pesendorfer &
Machalek Architects

「海之家（Haus des Meeres）」可以說是不同時期一變再變、多次重生的代表，戰爭遺產特別會有這種現象。防空塔水族館「海之家」原本是二戰時期維也納高射炮塔防衛系統6座混凝土高塔中的一座，原先的設計是為了因應低空轟炸，但最後盟軍並沒有以這樣的方式攻擊，所以這套防空系統從未使用。除了防空功能外，戰時高射炮塔建築群還曾經提供了附近居民避難或醫院使用。如今全部6座建築都被列為紀念碑等級的保護：這些高射炮塔們部分歸國家所有。人們對它們的再利用計畫有多種版本的想像——從網路機房到咖啡館或圖書館。

戰爭結束後，埃斯特哈齊公園（Esterhazy park）的防空塔被暫時用作一間有38個房間的酒店（第一次重生）。然後轉換成消防局（第二次重生），到了60年代時，其地下室部分開始被用作青年旅館，名為 Stadtherberge Esterhazypark。消防局留下了一半可進駐樓層，而塔的其餘部分充滿了碎片與廢棄物。但同時，水族館開始一步一步在空的樓層擴張，形成多種功能複合使用於一座防空塔中，見證不同功能使用者彼此地盤的消長。最後，當水族館擴展到6層樓時，消防隊員離開了這座建築物，水族館工作人員終於可以清理戰爭遺跡般的地下室。

1991年這座建築被加上了一個輕巧的環繞式箱子，廣告牌標語是 Smashed to Pieces，在 The Still of the Night（英語和德語）中，由勞倫斯·韋納（Lawrence Weiner）設計的戰爭和法西斯主義紀念館（第三次重生）。市政當局不確定將 Weiner 的作品當作是一次性廣告牌還是藝術遺產，但到了2005年，後者的觀點占了上風。

在大規模翻新期間更換了電梯，讓它可以停在三個新樓層（7,8和9樓）。2000年和2007年增加了容納爬行動物和熱帶鳥類的玻璃溫室。至2010年，塔樓有10個可進駐樓層和一個開放的觀光屋頂甲板。第10層重建了一個高射炮控制拱頂，收藏著第二次世界大戰的展品，僅在周末開放，需要預約登記才能參觀。

防空塔的進一步擴建和所有權一直是有爭議的問題。該建築歸維也納市政府所有，故必須支付維修費用，但市政府決心擺脫這個財務負擔，於是在2008年向外部投資者出售該塔，並批准在高聳塔頂建造一座私人酒店的計畫。這些計畫遭到當地居民的強烈反對。海之家本身希望保留未來向上擴建的權力，因此為了獲得當地居民的信任票，只得選擇一個比較溫和的方案（第四次重生）。於2011年進行的600萬歐元擴建項目將增加一百萬升的魚缸和一個露天餐廳。市政府願意將建築物出售給海之家，條件是新主人保留 Weiner 的藝術品（標語 Smashed to Pieces），並由納稅人繼續支付防空塔的維護費。

目前海之家擁有超過一萬名水生生物，占地面積約4,000平方公尺，其中還設有一個私人 Aqua Terra 動物園，內部設有5層樓。在高聳的塔樓建造了一個多層溫室，裡面配有木走道和繩索橋梁，以及在小水景中繁衍生息的熱帶植被，展示著魚類和烏龜，而各種熱帶鳥類和猴在遊客中自由遊蕩。2007年，海之家安裝了最大的300,000升鯊魚水缸。在全球金融危機造成了維也納藝術博物館的遊客數量急遽下降的同時，海之家卻能逆勢成長，儘管票價大幅上漲，每年仍持續吸引更多遊客，在此防空塔欣賞城市美景。

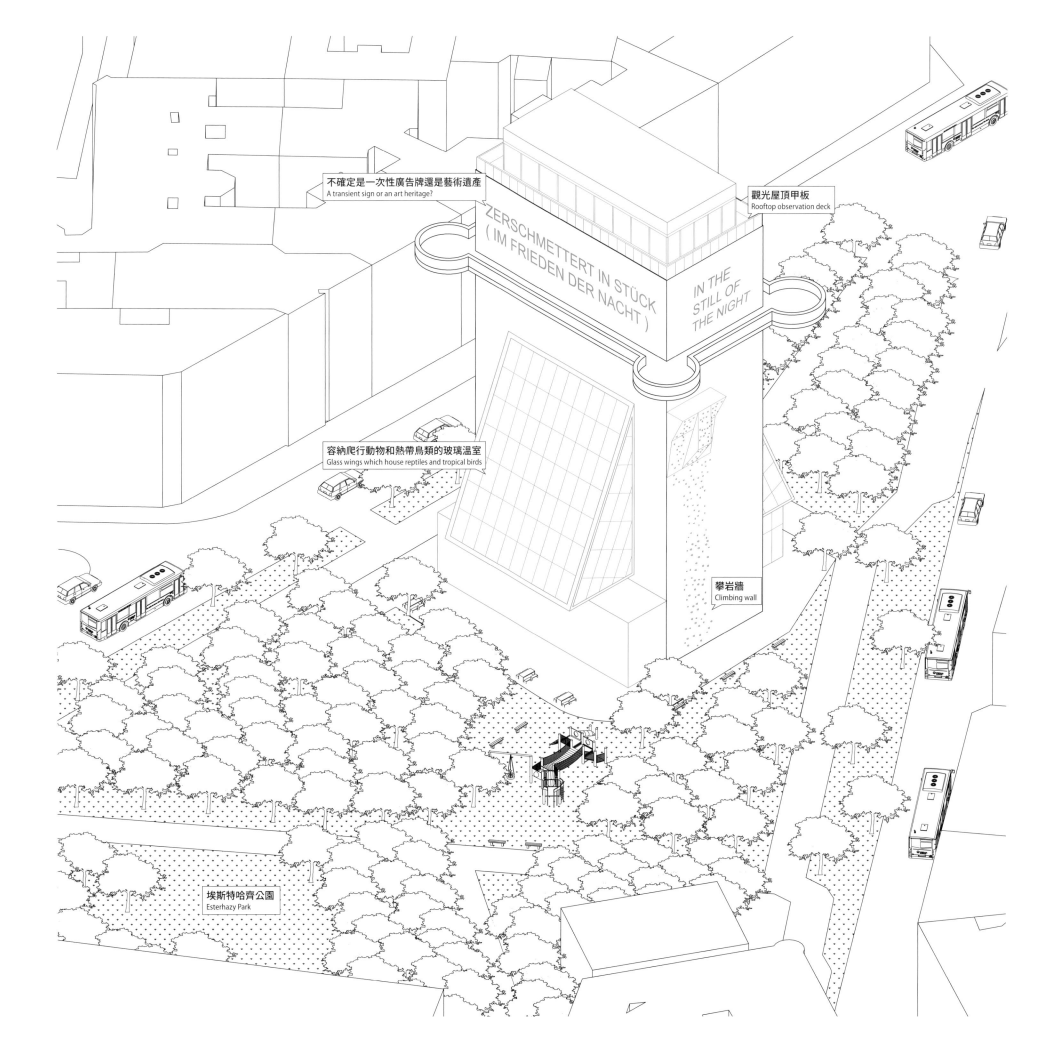

不確定是一次性廣告牌還是藝術遺產
A transient sign or an art heritage?

觀光屋頂甲板
Rooftop observation deck

ZERSCHMETTERT IN STÜCK
(IM FRIEDEN DER NACHT)

IN THE
STILL OF
THE NIGHT

容納爬行動物和熱帶鳥類的玻璃溫室
Glass wings which house reptiles and tropical birds

攀岩牆
Climbing wall

埃斯特哈齊公園
Esterhazy Park

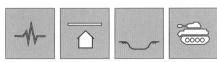
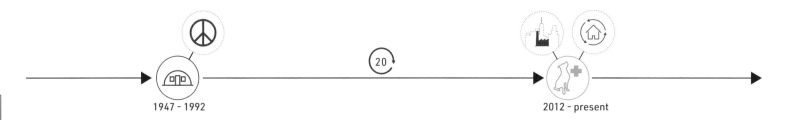

1947 - 1992

2012 - present

美國緬因
空軍基地掩體
蝙蝠避難所
Air Force Base Bat
Shelter Bunker,
Maine, USA

DESIGNER 不詳

緬因州北部的冷戰遺跡成為蝙蝠的冬眠場所。

美國魚類和野生動物管理局（United States Fish and Wildlife Service, USFWS）接收了43個位在緬因州的羅林空軍基地（Loring Air Force Base）的冷戰掩體。該基地在1994年關閉前，是核彈頭的空投和儲存基地，如今已轉變為Aroostook國家野生動物保護區。多年來USFWS一直研究如何讓這些早就被野草覆蓋，荒廢已久的冷戰遺跡有新的用途。2012年，他們決定將其中兩個轉換成用於安置生病蝙蝠的人造洞穴。

白鼻綜合症（WNS White-Nose Syndrome）是近代史上爆發最嚴重的野生動物疾病，已經在整個北美地區奪走多達670萬隻蝙蝠的生命，更使農業損失高達530億美元。2012年USFWS決定將掩體改造成蝙蝠的庇護所，除了提供無汙染的健康人造洞穴給冬眠的蝙蝠休息，也讓生病的蝙蝠可以獲得妥善照顧。

原本作為儲存武器的空間，即已針對隱密性及溫濕度做了初步設計考量，透過簡單的改造及儀器設置，就成為十分接近野外山洞環境的人造洞穴。

在監測了43個掩體中的溫度和相對濕度後，USFWS選擇先對其中兩個進行改造工程，以便為從佛蒙特州和紐約運來的30隻小棕蝙蝠創造一個宜居的冬季棲地。他們在掩體頂部放置加熱墊，並在掩體內部規劃小水池，以確保37-39°F恆溫和超過90%的相對濕度。接著，在室內添加基質（包括原木、電線、塑料網和蝙蝠窩），工程就完成了！

2012年12月，當雄性蝙蝠被安置到冬季的臨時家中後，科學家們利用洞穴內的熱像儀監測牠們的活動。發現蝙蝠往往聚集在人造洞穴後面的垂直牆上，而且只有一隻使用了原木，蝙蝠們偶爾醒來時會從建造的水池中飲水。

3月下旬牠們回到原來的冬眠環境，結果最後有9隻蝙蝠存活下來。Moosehorn和Aroostook國家野生動物保護區的避難所助理經理史蒂夫・阿吉烏斯（Steve Agius）在USFWS的客座文章中寫道，這次實驗證明蝙蝠確實能夠在人工的冬眠環境中存活下來，對抗白鼻綜合症，提高蝙蝠的生存率，這個結果激勵了科學家持續進行更深入的研究。

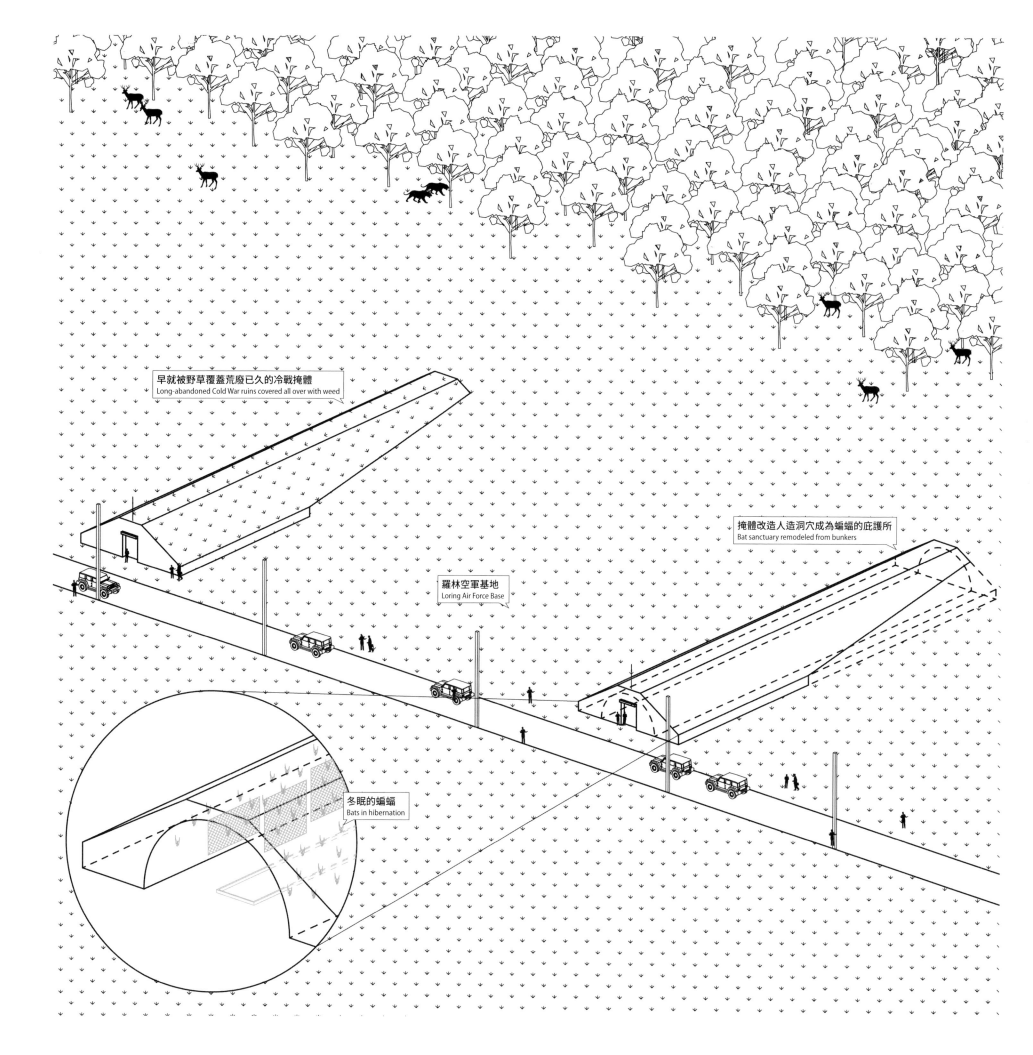

早就被野草覆蓋荒廢已久的冷戰掩體
Long-abandoned Cold War ruins covered all over with weed

掩體改造人造洞穴成為蝙蝠的庇護所
Bat sanctuary remodeled from bunkers

羅林空軍基地
Loring Air Force Base

冬眠的蝙蝠
Bats in hibernation

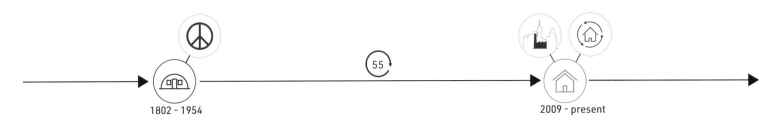

1802 - 1954　　　　55　　　　2009 - present

英國薩福克海岸
拿破崙式海防
塔樓住宅

**Martello Tower Y,
Suffolk coast,
United Kingdom**

DESIGNER Piercy & Company

將 1802 年所建、位於自然風景區，被英國歷史遺產單位指定為古蹟的拿破崙式防禦塔，轉變為 21 世紀的私人住宅，是一個苛刻的挑戰。這個有近 3 公尺厚的實心黏結磚牆的防禦塔靠近 Bawdsey 村，面向著大海和沙灘。由 Piercy & Company 的建築師 Stuart Piercy 與 Billings Jackson Design 的 Duncan Jackson 合作完成。英國歷史遺產單位將該案的成果視為重要歷史建築的重生典範。

重生後的新屋可由金屬樓梯進入，通往二樓，這裡有一個門廳，可以欣賞到兩層樓挑空的內部空間傘狀如雕塑般的結構。大廳裡有一個大衣櫥和一間廁所。在這個樓層還有一個燃木爐的接待室。樓梯環著傘狀結構的骨幹向下通往臥室和浴室，向上由兩個對稱、既有的樓梯神祕地穿出傘狀結構，通往包含廚房／客廳／用餐區的頂層空間。彷彿在輪船的甲板上，這個空間是由一個起伏不平的屋頂創建的，不僅為房間，美麗的天花板，而且還可以 360 度全方位地欣賞周圍的海洋和鄉村景色。

一樓（地下室）設有三間臥室，兩間浴室和兩個書房，原為儲水空間。空間策略上目的是明確區分舊的和新的，當代的介入盡可能輕微地觸及原始面料；紋理粗糙的磚石才是主角。該塔的面料包括 750,000 塊磚。據透露，它為設計設定了標準。新屋頂採用 3D 彎曲輕質結構，由鋼和層壓膠合板構成，由五對 Macalloy 鋼筋繫住。使用詳細的 3D 模型為非現場製造創建 2D 切割圖案。屋頂下方的無框弧形玻璃裙傳達了新舊之間的區別，並提供 360 度全景。屋頂朝向最小化視覺衝擊，採用單層薄膜覆蓋，帶有三個窗（採光井）。該系統是基本的，因為它必須從槍枝平台存儲和安裝。為了將光線帶入一樓（地下室），6 個 450 公分直徑的鑽孔（採光管）是通過厚磚砌牆鑽孔，從窗戶內部透到臥室和浴室。另外兩間臥室有 60 公分的洞，可以看到鄉村到西南和西北的「攝影暗房」。從護牆頂部鑽出 200 公分直徑的孔，以滿足通往底層和地下室的被動通風系統。這些磚管為廚房提供水、電、氣體和屋頂供暖。它們還適用於地下室和地面層的熱回收通風系統的供氣和排氣管道。

通過重新改造 1802 Martello Tower 拿破崙式防禦塔，將其轉變為住宅，歷史建物保存被推到了保存原則之外，而不是僅止於為「瀕危」的結構注入新的活力。

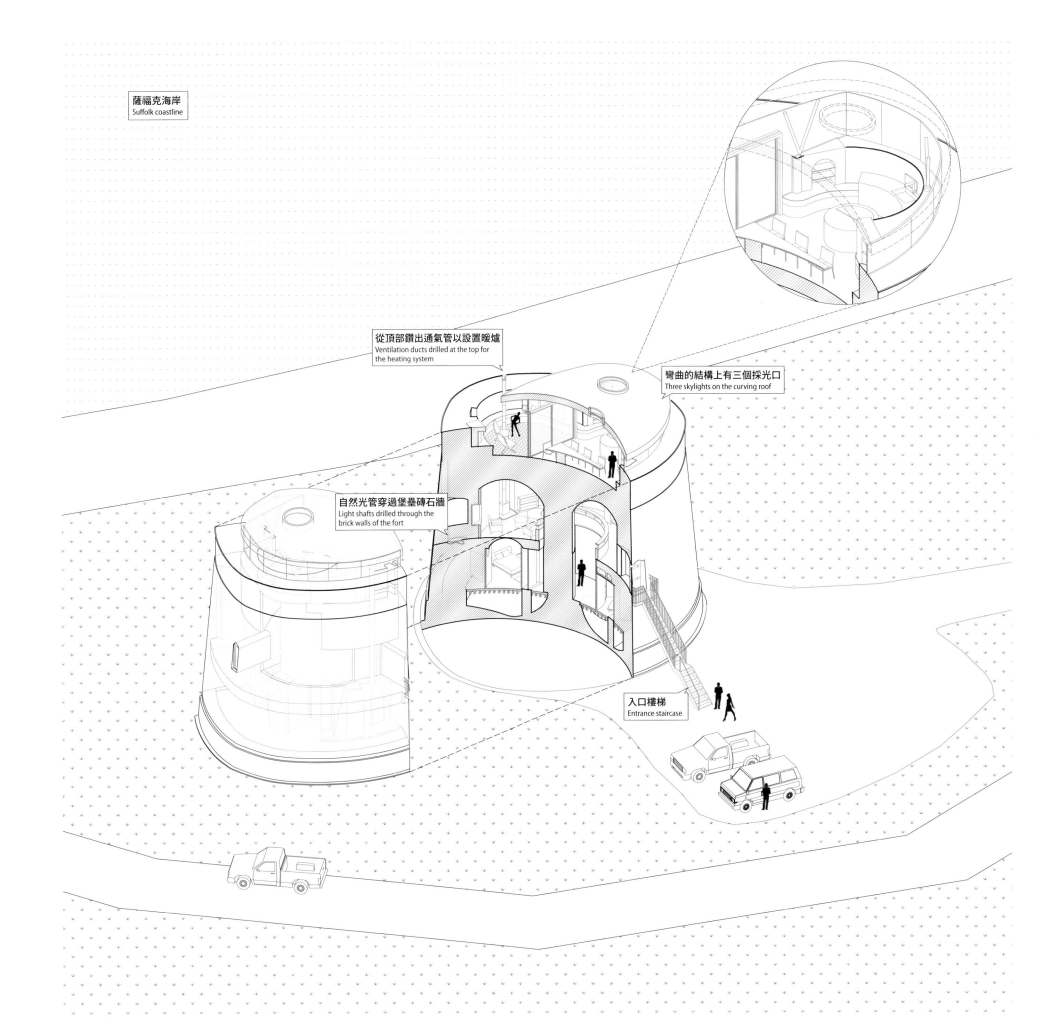

薩福克海岸
Suffolk coastline

從頂部鑽出通氣管以設置暖爐
Ventilation ducts drilled at the top for
the heating system

彎曲的結構上有三個採光口
Three skylights on the curving roof

自然光管穿過堡疊磚石牆
Light shafts drilled through the
brick walls of the fort

入口樓梯
Entrance staircase

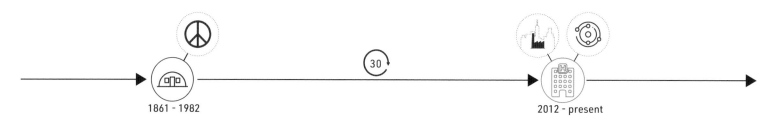
英國樸茨茅斯港海上堡壘旅館

Palmerston Sea Forts Hotel, Solent, United Kingdom

DESIGNER Amazing Venues

Spitbank Fort 或簡稱 Spit Fort，是 1859 年皇家委員會建造的維多利亞時代的海上堡壘，其直徑 50 公尺，最初設計用於保護樸茨茅斯港的入口。Spitbank Fort 位於英國樸茨茅斯 Solent 海峽，該海峽將懷特島與英國大陸分隔開來。Spitbank 比另外兩個主要的 Solent 堡壘——Horse Sand Fort 和 No Man's Land Fort——都要小。它的主要目的是作為進入這兩個主要堡壘的船隻的掩護防線。

1898 年，海上堡壘的角色被改為防禦輕型飛機，到了 1962 年被宣布為供過於求（死因），並於 1982 年由國防部收回處理。2009 年，它以 80 萬英鎊的價格出售，但在拍賣前就被超過 100 萬英鎊買下。該堡壘與其他兩個堡壘現在都由 Amazing Venues 經營，作為豪華酒店品牌以及博物館。三個堡壘總共有 50 個房間，其中 Spitbank Fort 設有 8 間套房，附設的舞廳、餐廳可用於私人饗宴或婚宴。

人們進入 Spitbank Fort 後可以體驗堡壘既有的特色。由加掛的起重機具將船隻從海上吊起，唯一的入口通道貫穿堡壘，人們進入後彷彿進入另個與世隔絕的聖地。Spitbank Fort 垂直分成三個不同層次：地下室式彈藥店，寬敞的炮台甲板和外部庭院，以及屋頂燈塔和前炮台。獨特的平面圖和原始結構的不可變，需要聰明的動線策略來形成連貫的布局，外部窗戶的數量有限且內部結構外露亦是一

種障礙，8 個炮室成為新臥室，剩下 7 個炮室則成為酒窖、桌球室等娛樂空間。然而，需要新的方法來解決現代機電系統在這座 19 世紀建築中的配管問題。外加的電纜和管道會對歷史建築造成難看的破壞，因此盡可能在新建的樓層空間內進行大量的布局，亦利用了舊有的通風管改善並延續其功能。

堡壘內部充滿強烈的機能感，裸露的結構明顯地揭示了其所在時代簡單有力的施工。成排的拱形紅磚結構從堡壘中心輻射出來，創造了迷人的用餐空間。原先通往炮台甲板的中央通道創造了一條路徑，引導遊客繞著磚砌走廊，通往 8 間臥室。屋頂區域的新設計採用更現代的做法，提供安靜的放鬆和沉思的庭院景色。在保有原堡壘特色的前提下，建築師對屋頂區域做了更積極的改造，不僅提供新的屋頂用餐空間，圍繞兩個前炮台的混凝土牆被重新定位為 SPA 區。

Spitbank Fort 自 1967 年以來一直是一個預定的紀念碑。顯而易見地目前重生使用的方式遠比單純的作為一個海上紀念碑要更為積極，也更具吸引力。該項目善用建築特色和有效空間管理，不混淆其歷史淵源，並讓它與強大的實用美學共存，透過一個成功的商業計畫將這些海上堡壘保存成一個獨特的印記。

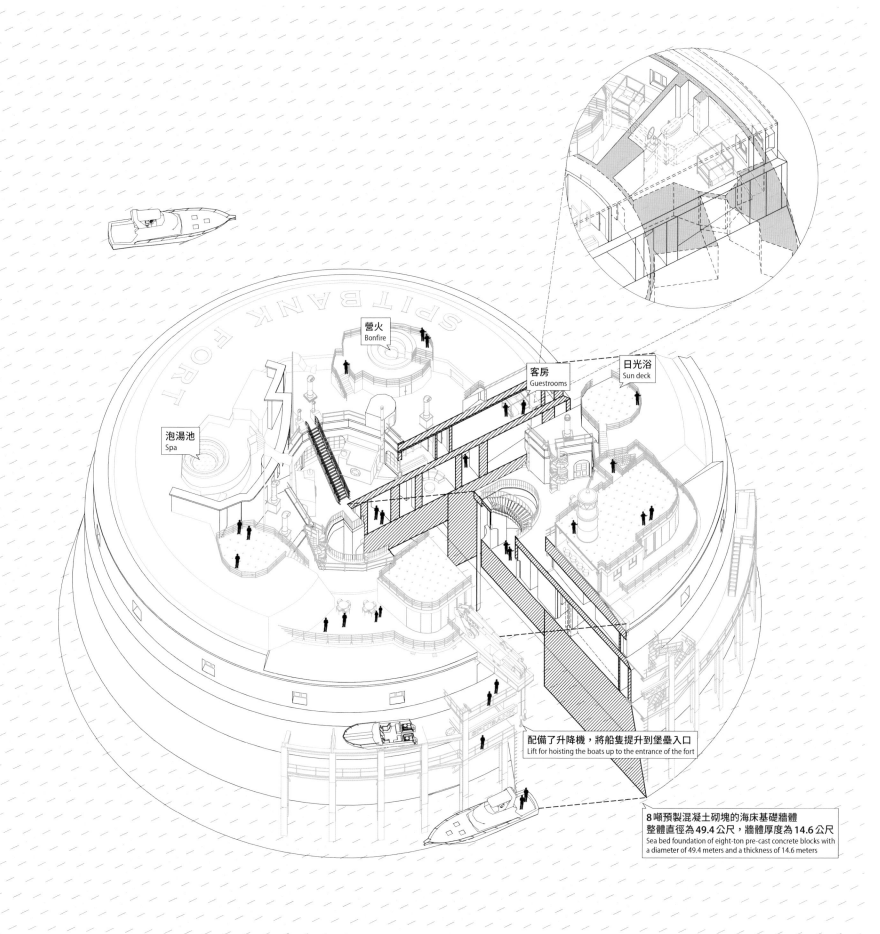

營火
Bonfire

客房
Guestrooms

日光浴
Sun deck

泡湯池
Spa

SPITBANK FORT

配備了升降機，將船隻提升到堡壘入口
Lift for hoisting the boats up to the entrance of the fort

8 噸預製混凝土砌塊的海床基礎牆體
整體直徑為 49.4 公尺，牆體厚度為 14.6 公尺
Sea bed foundation of eight-ton pre-cast concrete blocks with
a diameter of 49.4 meters and a thickness of 14.6 meters

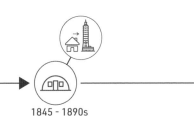
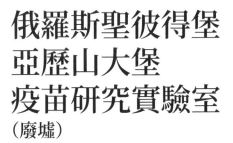

1845 - 1890s

7

疫苗研究室

1897 - 1917

廢棄

俄羅斯聖彼得堡
亞歷山大堡
疫苗研究實驗室
（廢墟）

Fort Alexander,
St Petersburg, Russia

DESIGNER Louis Barthelemy Carbonnier d'Arsit de Gragnac

俄羅斯瘟疫堡曾是黑死病醫學研究基地。

1703年聖彼得堡城市建立後，芬蘭灣的水道對俄羅斯具有極重要的戰略意義，為從波羅的海守衛這座城市，在接下來的兩個世紀中，俄羅斯持續加強芬蘭灣水道的防禦工事，在芬蘭灣南岸和北岸之間策略性部署了40多座堡壘，形成了抵抗來自海上敵人一股堅不可摧的防禦力量。其中建於1838年至1845年的亞歷山大堡是在人工島上的一個軍事基地，僅憑其雄偉的存在就足以威懾任何企圖入侵聖彼得堡的艦隊。原本由Louis Barthelemy Carbonnier d'Arsit de Gragnac起草了亞歷山大堡的初步藍圖，在他死後來自法國的俄羅斯軍事工程師Jean Antoine Maurice（又名Moris Gugovich Destrem）接手修改了堡壘的設計，興建工程由俄羅斯軍事工程師Mikhail von der Veide監造。尼古拉皇帝於1845年7月27日正式委任啟用，堡壘的命名是為了紀念他的兄弟亞歷山大一世。

這個橢圓形要塞的建築面積為90公尺乘60公尺，共三層樓，有一個中庭。總共55,355支長12公尺的椿被打入海床以加固地面，總建築面積超過5,000平方公尺，足以容納1,000名士兵駐軍。堡壘配備103門炮口，屋頂上有額外的空間可容納34門大型火炮，為其提供強大的軍事優勢。要塞本身在克里米亞戰爭中發揮了關鍵作用，阻止了英國和法國艦隊試圖進入克隆斯塔特的俄羅斯海軍基地。在那之後，亞歷山大堡只有兩次發揮其威懾作用：1863年與大英帝國的衝突，以及1877-1878俄土戰爭期間。但當時間走到19世紀末，亞歷山大堡的防禦功能在膛線炮和高爆炮彈的進步下，也漸漸走到了盡頭（死因）。此後亞歷山大堡開始主要用於彈藥儲存。

1894年亞歷山大‧耶爾辛（Alexandre Yersin）發現鼠疫病原體（耶爾辛氏菌），於是俄羅斯政府成立了一個預防鼠疫特別委員會，需要找一個適合的地點來加速研究。因為當時的亞歷山大堡已不再被用作軍事用途且與外隔絕，俄羅斯科學家便選擇在這裡研究各種致命病毒，比如霍亂、破傷風、斑疹傷寒、猩紅菌和鏈球菌感染，研究重點在於瘟疫本身以及血清和疫苗的製備。1897年帝國實驗醫學研究所委託這座堡壘作為新的研究實驗室，奧爾登的亞歷山大‧彼得羅維奇公爵（Duke Alexander Petrovich）捐贈了大量資金，對該基地進行了整修，以滿足其新的研究實驗室用途。

總體來說，亞歷山大堡的細菌性疾病實驗室是成功的，直到1917年共產黨接管後實驗室被關閉，堡壘被移交給俄羅斯海軍，細菌性疾病研究則轉移到莫斯科和聖彼得堡的研究所。亞歷山大堡於1983年終於正式廢棄，在1990年代末和21世紀初，它是狂歡派對的熱門地點，自2005年以來它由Strelna的君士坦丁宮總統會議中心管理。2007年堡壘管理部門宣布有意向投資者尋求擬議的翻修計畫，估計耗資4,300萬美元。今天，它被人稱為瘟疫堡，成為城市探險家和攝影師喜愛的去處。在海水結冰的冬天，遊人可以步行到達堡壘；在炎熱的夏季，訪客則可以乘船隻遊覽。

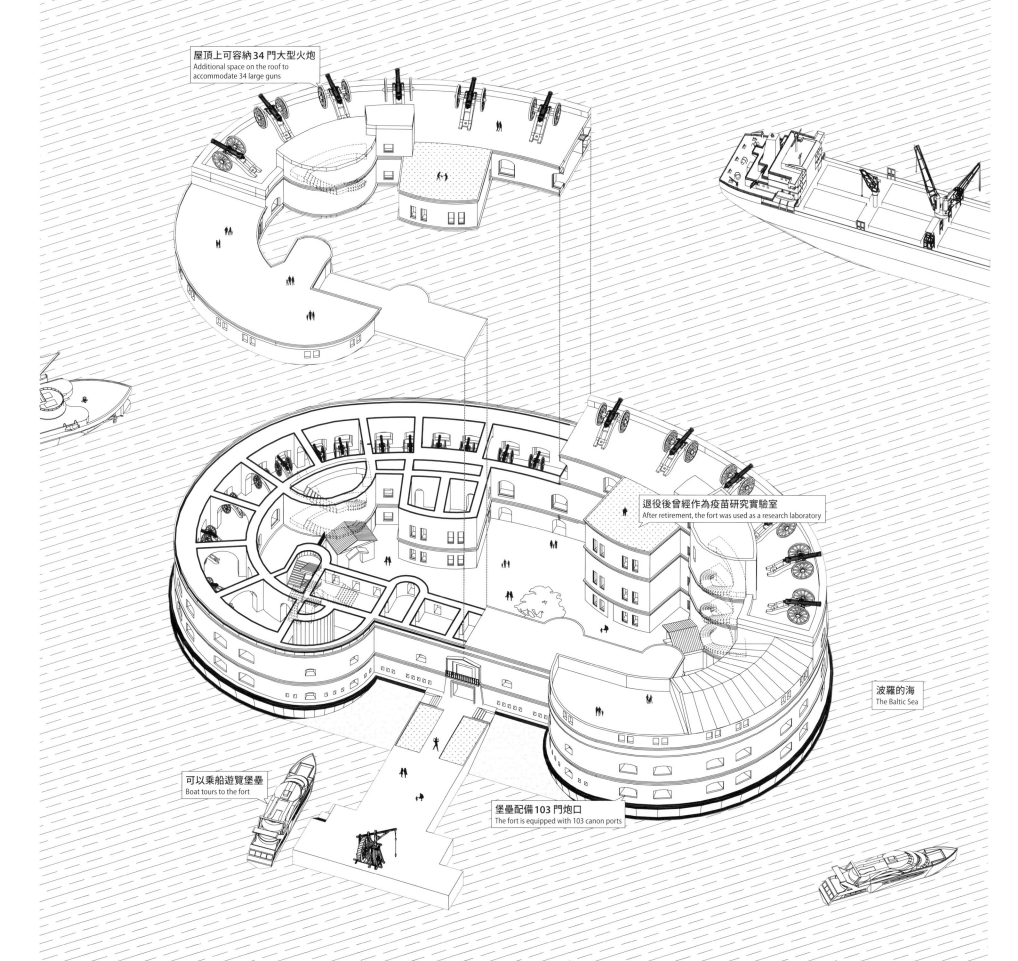

屋頂上可容納 **34** 門大型火炮
Additional space on the roof to accommodate 34 large guns

退役後曾經作為疫苗研究實驗室
After retirement, the fort was used as a research laboratory

波羅的海
The Baltic Sea

可以乘船遊覽堡壘
Boat tours to the fort

堡壘配備 **103** 門炮口
The fort is equipped with 103 canon ports

台灣新竹建功
國小大桶教室

**The Oil Storage
Classroom JianGong
Primary School
Hsinchu, Taiwan**

DESIGNER 九典聯合建築師事務所 |
Bio-Architecture
Formosana

整修日治時期儲油槽,改建為校園靜態展示與學習空間。

新竹市建功國民小學校園內西側小情人坡旁的「大圓桶」,是「日本海軍第六燃料廠新竹支廠」僅剩的三大圓桶形儲油槽之一,為台灣少數僅存二戰時期軍事工業遺址。九典聯合建築師事務所在建功國小的新建工程中,將當時廢棄被學校用來堆置資源回收的圓形倉庫,改造成多用途的教室。

日本海軍燃料廠是對海軍必要的燃料與潤滑油進行生產、加工、實驗、研究的機關。最初在日俄戰爭時設立的臨時煉碳製造所為其前身。大正10年改為海軍燃料廠,負責掌管有關燃料的事務。昭和16年,海軍燃料廠被改編設置第一至第六燃料廠,第一至第四燃料廠在日本本土,第五燃料廠在韓國平壤,第六燃料廠在台灣高雄左營。日本第六海軍燃料廠所屬的不僅是高雄左營煉油廠,事實上還包括新竹擁地四百多甲未建成的異辛烷廠(飛機汽油裡的一種主要成分),以及占地五百餘甲以後發展為新竹研究所的產業。

建功國小建校前,這個區域都屬於日本海軍第六燃料廠新竹支廠的廠區範圍。當時儲油槽剛建好還未啟用時美國就轟炸日本,日軍隨即撤離台灣,留下了燃料廠(死因),也就是現在建功國小的「大圓桶」。

雖然外表不起眼,只是拆模無粉刷的水泥表面,可是這直徑16公尺、高6公尺、防爆用外牆厚達60公分的圓形建築,即使不能算是古蹟,也必有過去可說的故事,且可塑性極高。

設計團隊首先掀開倉庫的鐵皮屋頂,讓日光進來,加蓋自然通風但避雨採光隔熱的中空板屋頂,讓內部空間起死回生。並以獨立的鋼管支柱支撐屋頂架構及夾層的展示空間,保留原有承重牆結構,打通原有窗戶,使空氣自然對流。中央6公尺高的圓錐空間為小型劇場使用,利用邊緣形成夾層上部展示空間及下部遊戲空間。將外牆清洗後,露出原有混凝土外牆痕跡,為多孔性外牆材料。設計團隊在周圍種植了攀爬綠藤,使外牆綠化,並新加屋頂的雨水回收系統至雨撲滿來負責澆灌。

新建的校園建築量體的西半部,向南偏離12.5度,以空出足夠的空間使「大圓桶」成為視覺焦點,讓軸線轉彎。在九典聯合建築師事務所的努力下,「大圓筒」搖身一變,成為了校園空間的主角。

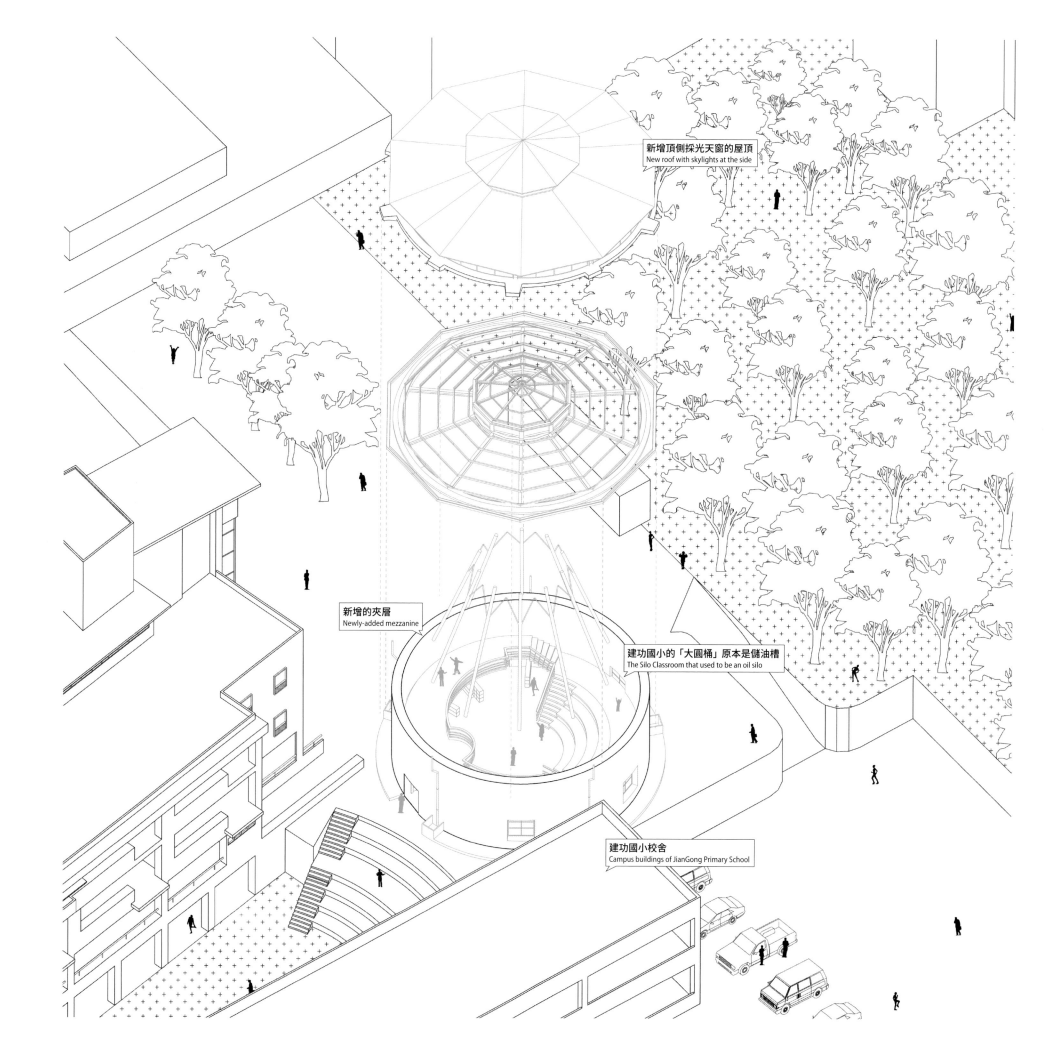

新增頂側採光天窗的屋頂
New roof with skylights at the side

新增的夾層
Newly-added mezzanine

建功國小的「大圓桶」原本是儲油槽
The Silo Classroom that used to be an oil silo

建功國小校舍
Campus buildings of JianGong Primary School

M12

1917-1949 → 0 → 政治犯招待所 1949-1962 → 0 → 海軍訓導中心 1962-2001 → 10 → 1991 - 2010 → 想像

台灣鳳山
無線電信所
（設計提案）

Fongshan Wireless
Communications
Station (Proposal)

DESIGNER 林靖涓

原日本海軍鳳山無線電信所於1917年開始建造，且於1919年完工，最初為了配合日軍南方作戰，對敵方進行電信干擾的戰術，以及監視遠東地區英美航艦和航空機之動態，作為日本殖民地最南端的重要無線電信所，與日本本地船橋無線電信所及針尾無線電信所構成海軍通信網的骨幹，其中鳳山無線電信所為目前保存最完整的電信所設施群。

無線電信所由第一送信所為主體，包含周圍軍官宿舍、辦公廳舍、小碉堡與後來增設的十字電台所等附屬設施，採用傘狀天線網以接收辨別軍艦方位及距離，進而使得電信所的外圍呈現獨特的同心圓紋理。最外圈半徑為400公尺，包含內外圈共有54座基座，用以固定通信電塔副塔；最內圈鳳山無線電信所房屋設施的外圍圍牆，所圍直徑為300公尺。原日本海軍鳳山無線電信所歷經了「日本無線電信所」、「鳳山招待所」、「海軍明德訓練班」、「文化資產保存」四個時期，由於其特殊的軍事設施地位，各個時期的機能轉換幾乎沒有重生期的等待。於1949年二戰後國民政府撤退來台接管後，除原有的無線電電台外，也曾作為情報隊的駐地，用來偵訊拘禁軍中政治犯與思想犯，為求隱匿稱為「鳳山招待所」。1962年作為海軍訓導中心，並於1976年成立海軍明德訓練班，負責管束軍中頑劣分子，同時海軍通訊大隊之分遣隊仍在使用十字電台。2001年因國軍組織調整裁撤，整體建築則被閒置，後於2010年被列為國定古蹟。

本設計面對都市擴張對於土地需求的問題，該如何重新將森嚴的軍事設施打開、模糊與都市和市民生活的邊界，同時也需要處理基地特殊的軍事歷史地位，如何在原有建物中置入現代機能，以現代的角度重回歷史的現場了解、感受與反思，重新塑造其價值。

設計由基地獨有的同心圓紋理下手，將形成都市斷裂的部分圍牆去除，以南北、東西兩個向度重新架構整體園區，將文化、歷史、教育以南北向梯度產生連結，使舊眷村、日治軍事設施與博物館形成完整的歷史體驗；將生活、休閒以東西向梯度產生連結，利用現有的通信電塔基座與綠地，串連鳳山共同市場創造在地文化生活圈。

第一送信所作為設計主體，以碉堡影院紀念館為機能，定義為提供在地居民的文化生活設施，以宣讀、反思、定位三個步驟重新審視基地歷史的更迭與空間重合。在原有的第一送信所中置入既公共、卻又封閉在黑盒子中的現代休閒機能「影院」，碎化原有空間序列，透過碎化原有的歷史空間架構，重新宣讀與自省。置入機能體以金屬反射面作為表面材質，強制疊合歷史空間與現代活動，在紀念館空間中創造疊合的時空、人與活動。以部分立面拆除形成入口，並重新整併成南北單一開口，以動線串聯活動大堂、物件展示、影像展示、資料收藏、反思小房，透過動線經歷重新宣讀歷史、重新反思歷史、重新定位自我。

拆去部分原有立面形成南面主入口
Part of the façade will be removed to create an entrance on the south side of the building

回復電信所園區內原有的主要幹道紋理
The main road of the communication center will be restored

所有展覽空間及動線均圍繞於盒狀影院之外圍
Galleries and paths around the cinema box

保留第一電信所原有樓梯作為動線之一，可直接由地面層上到二樓進入展覽動線
The existing stairs will be preserved, so visitors can enter the galleries on the second floor from the first floor

部分挑空的資料收藏區，作為收藏與展示鳳山無線電信所相關歷史資料的主要空間
The original elongated space will be another exhibition area to display historical artifacts

電信所原有長型的空間序列，作為展示相關歷史物件的展示區
The archives room where historical documents will be displayed

保留登上大碉堡頂層土丘的外梯
Stairs leading to the hilltop of the bunker

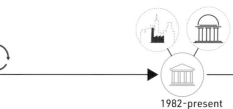

1941-1974　　　　　6　　　　　1982-present

美國紐約無畏號航空母艦海空暨太空博物館

The Intrepid Sea, Air & Space Museum, New York, USA

DESIGNER Perkins+Will

在紐約曼哈頓46街與12大道交匯處，有一艘航空母艦長期停泊於86號碼頭邊，上面還有各式戰機停泊於艦上甲板。這是讓你只需參觀一座博物館，就能看遍航空母艦、潛水艇、各種經典戰機、協和號客機與太空梭的「無畏號海空暨太空博物館」。

無畏號航空母艦（USS Intrepid CV-11）是美國第四艘以無畏命名的軍艦，屬於艾塞克斯級航空母艦（Essex-class aircraft carrier）。它於1941年開始建造，主要服役於二次世界大戰期間，參加過多次太平洋戰役，包括歷史上最大的列特海戰。它在數次日本神風特攻隊的自殺攻擊與魚雷襲擊中存活下來，繼續在1960年代的太空競賽中，為美國太空總署NASA擔任過雙子座和水星等太空任務的主要回收船艦，並參與越戰，一直到1974年於冷戰高潮時退役。無畏號曾兩次獲得海軍部隊嘉許勳表，並在二戰及越戰獲得五枚戰鬥之星。

無畏號退役後海軍原本打算將其出售拆解，但遭到民間反對。其中紐約房地產商兼慈善家扎卡里·費沙（Zachary Fisher）在1978年成立無畏號博物館組織（Intrepid Museum Foundation），並籌集資金，以購入無畏號作為博物館艦。在此民間組織努力下，海軍最後在1981年將無畏號轉交該組織管理。1982年無畏號除籍，並開始改建為博物館艦，同年無畏號海空暨太空博物館在紐約正式開放，自此成為曼哈頓的重要地標及旅遊點。而費沙在稍後購入更多海軍退役艦隻到博物館。

1986年，無畏號獲評為美國國家歷史地標。不過，無畏號只是整個博物館的其中一部分，博物館是占地18,000平方英尺的教育中心，涵蓋整個86號碼頭，哈德遜河公園信託基金的公共碼頭。其他還包括黑鱸號潛艦，這是唯一向公眾開放的美國柴油動力戰略導彈潛艦。同時展出的還有28架經過真實修復的飛機，包括間諜偵察機洛克希德A-12黑鳥，以及飛越大西洋最快的商用噴射客機協和號。2011年4月，太空總署宣布企業號太空梭將會轉移到海空暨太空博物館展覽，而企業號在2012年4月27日運抵紐約，並在6月6日運抵無畏號艦上，為無畏號博物館帶來了新高峰。

現在博物館每年迎接超過一百萬名遊客，每天都會主辦內容不同的主題活動，吸引絡繹不絕人潮前來參觀。退役太空梭企業號就停於該館中心舞台的位置，觀展的遊客們不僅可以漫步其下進行觀察，還能踏入一個平台以太空梭為背景拍一張完美角度的照片。該展廳還有一個太空艙，其間展出了任務控制對話的紀錄和音訊、視頻和照片等大量文物，讓大家領略太空艙內的情形，並瞭解到在推進太空旅行時航太飛船所起的作用。無畏號海空太空博物館的使命是透過其收藏、展覽和節目來提升人們對歷史、科學的認識和理解。

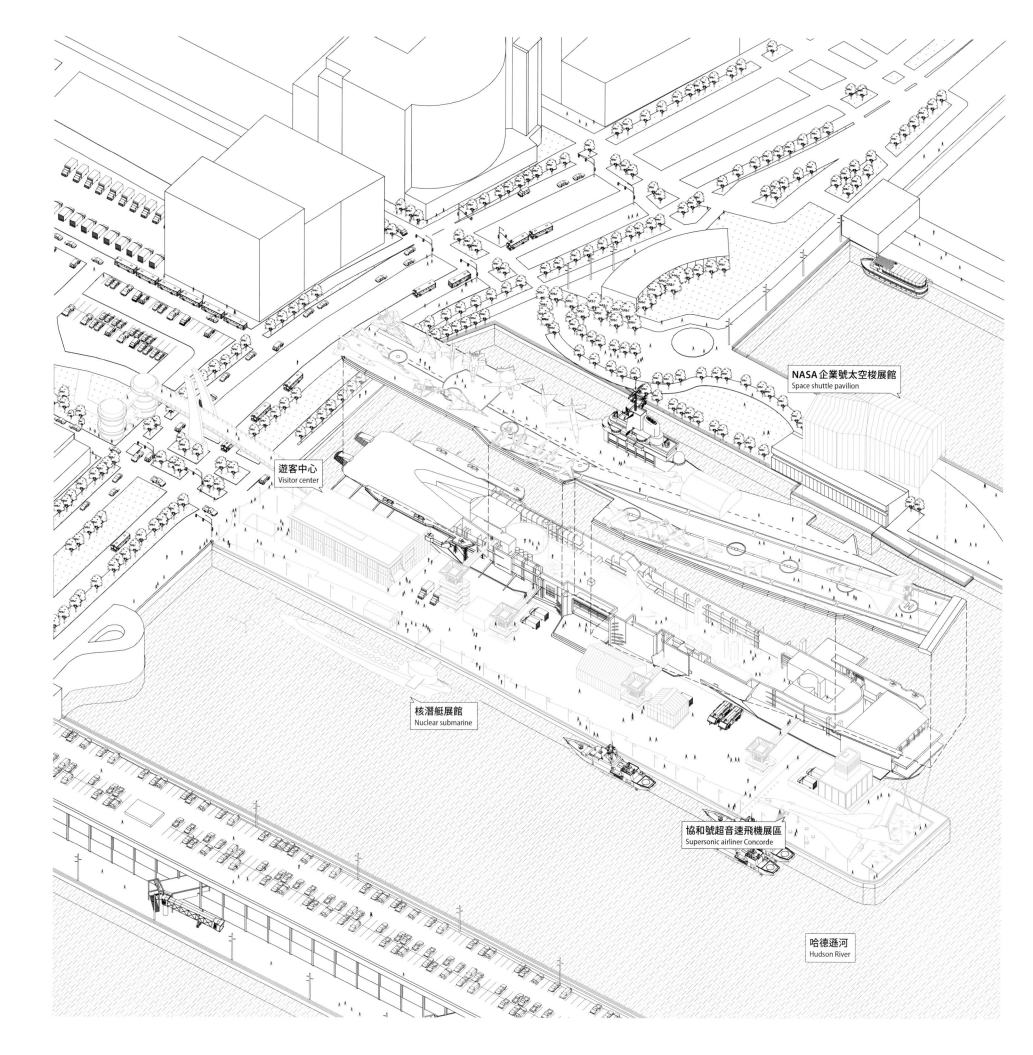

NASA 企業號太空梭展館
Space shuttle pavilion

遊客中心
Visitor center

核潛艇展館
Nuclear submarine

協和號超音速飛機展區
Supersonic airliner Concorde

哈德遜河
Hudson River

1989-2013　　　想像　　　proposal　庇護所

蘇聯
颱風級核潛艇
（設計提案）

**Typhoon class
Nuclear Submarine**
(Proposal)

DESIGNER 陳詠暄

人類所建造過的最大的戰略核彈核潛艇「颱風級」，是水下排水量可達48,000噸的龐然巨艦，全長175公尺；潛行航速可達27節，潛深500公尺，更可以在水下潛伏180天之久。颱風級建造計畫在1989年全部完成，共有6艘同級艦，是典型冷戰時期美蘇為了達到「相互保證毀滅原則」的產物。它擁有20枚彈道飛彈發射井，武器威力足以摧毀其所在的任何半球。冷戰結束蘇聯解體後，其中三艘潛艇已被拆解，剩下兩艘則於2013年底退役，只有一艘經現代化改造後仍處於服役狀態。目前聯合國正在尋求再利用方案，希冀三艘退役後的潛艦能重獲新生。

為此，設計師提出一個構想，盼能藉此解決一個長久以來的國際問題：

為了爭奪地中海和約旦河之間的領土，猶太人和阿拉伯人展開了將近一個世紀的戰爭，雙方都宣稱這片土地乘載著他們的民族歷史和宗教信仰。以色列西岸的加薩走廊地區，則是這場無止盡的爭奪下最悲慘的犧牲者。長年飽受戰火，土地空間不足，各種資源耗竭，人口超多卻不屬於任何國家的三不管地帶，在加薩走廊，大部分的家庭得要接受聯合國的援助才能勉強維持生活。2014年7、8月間，在以色列軍隊的攻擊中就有200多間學校被摧毀，光這一場軍事行動就產生了至少1,500位孤兒。以色列的空襲甚至破壞了加薩地區唯一的電廠，更因為缺電，連帶著中斷了淡水供給。

為解決加薩地區的慘況，退役後的颱風級核潛艇將被加以利用，原本足以毀滅地球的大規模殺傷性武器，將重生蛻變為加薩走廊維生系統的供給者，與孩童的守護者。

颱風級獨有的雙核反應爐足以提供一個海邊城市或一個小國足夠的發電量，且核潛艇遠離陸地可不受地震海嘯影響，並能在發生熔毀時立刻引入海水降溫。所以，可先派遣已隸屬聯合國的核潛艇航行至加薩地區外海，利用核潛艇上的核反應爐與海水淡化能力，為遭受以色列空襲後斷電停水的加薩走廊地區供電給水。

再以颱風級潛艇為計畫核心，環繞其周圍建造人工島，在人工島上收容孤兒並為其提供教育與庇護。這個希望和平島是一個類似海上避難所的城鎮設計，一個能讓孤兒們在島上成長的城鎮。加薩地區的許多孩童在兩軍交戰中造成終生殘疾，或成為孤兒。此由聯合國經營的和平島上提供孩童們所需的人道援助：從醫療、食品、收容、教育、職業訓練到心理輔導。

經過改造，颱風級核潛艇將從毀滅武器重生為救世裝置。而將資源投入照顧飽受民族衝突踐踏的孤兒，從長遠來看，受過教育的孩童長大後可成為穩定的力量，幫助重建已毀於戰爭的基礎設施，解決就業問題和減少貧困危機，無疑是未來建立和平社會最好的解方之一。

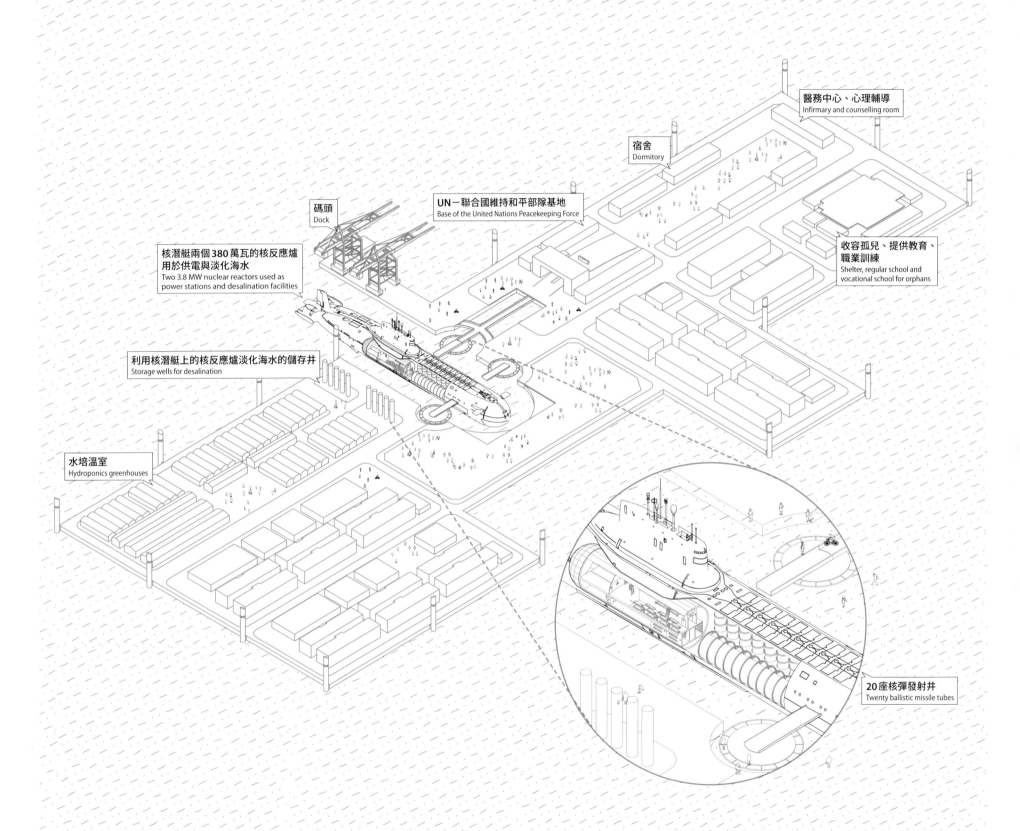

醫務中心、心理輔導
Infirmary and counselling room

宿舍
Dormitory

碼頭
Dock

UN－聯合國維持和平部隊基地
Base of the United Nations Peacekeeping Force

收容孤兒、提供教育、職業訓練
Shelter, regular school and vocational school for orphans

核潛艇兩個380萬瓦的核反應爐用於供電與淡化海水
Two 3.8 MW nuclear reactors used as power stations and desalination facilities

利用核潛艇上的核反應爐淡化海水的儲存井
Storage wells for desalination

水培溫室
Hydroponics greenhouses

20座核彈發射井
Twenty ballistic missile tubes

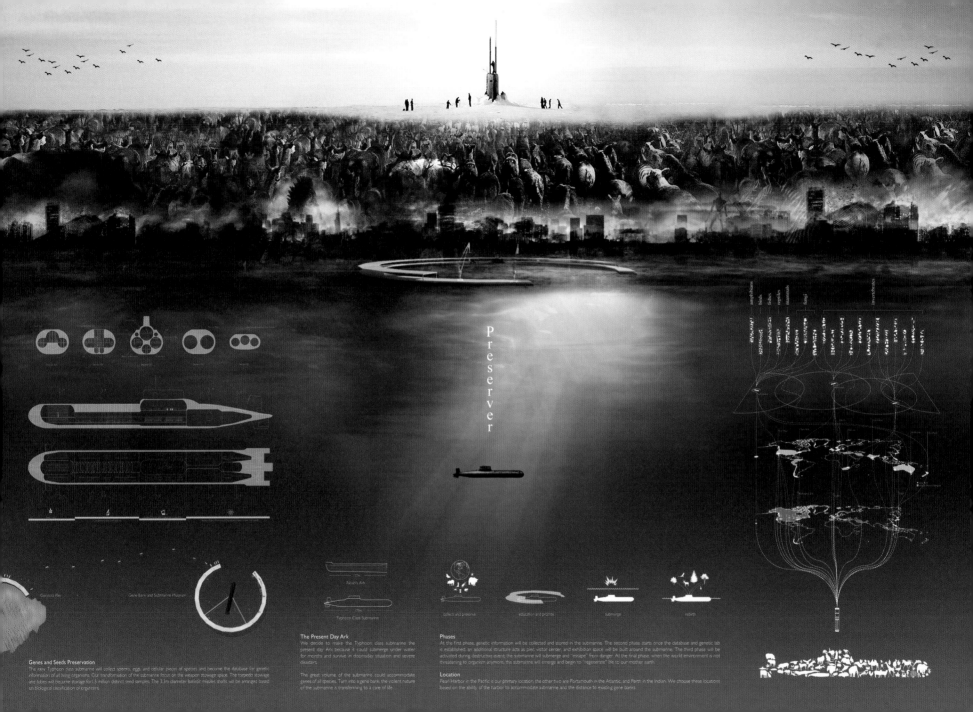

Preserver

Genes and Seeds Preservation

The new Typhoon class submarine will collect sperms, eggs, and cellular pieces of species and become the database for genetic information of all living organisms. Our transformation of the submarine focus on the weapon stowage space. The torpedo stowage and tubes will become storage for 1.5 million distinct seed samples. The 3.3m diameter ballistic missiles shafts will be arranged based on biological classification of organisms.

The Present Day Ark

We decide to make the Typhoon class submarine the present day Ark because it could submerge under water for months and survive in doomsday situation and severe disasters.

The great volume of the submarine could accommodate genes of all species. Turn into a gene bank, the violent nature of the submarine is transforming to a care of life.

Phases

At the first phase, genetic information will be collected and stored in the submarine. The second phase starts once the database and genetic lab is established; an additional structure acts as pier, visitor center, and exhibition space will be built around the submarine. The third phase will be activated during destructive event; the submarine will submerge and "escape" from danger. At the final phase, when the world environment is not threatening to organism anymore, the submarine will emerge and begin to "regenerate" life to our mother earth.

Location

Pearl Harbor in the Pacific is our primary location, the other two are Portsmouth in the Atlantic, and Perth in the Indian. We choose these locations based on the ability of the harbor to accommodate submarine, and the distance to existing gene banks.

「殺生」轉「護生」

簡介

冷戰期間，前蘇聯颱風級核潛艇是人類歷史上建造過最大的潛艇，美蘇相互保證毀滅原則下的產物，超級軍事強權和大規模毀滅的象徵。本項目的目標不單只是再利用颱風級核潛艇也重新賦與它別於昔日所象徵的意義。在此我們看到它具備拯救與保護地球物種的潛力與責任。

現代方舟

聖經裡的方舟不僅從天譴下拯救了諾亞和他的家人，以及世界上所有的物種，它同時也象徵著希望和一個新時代的起點。

縱使大洪水是過去出現在聖經上的故事，不變的是今日世界依然處於同樣足以滅絕文明與生物的威脅下——無論是天災，或是人禍。因此，我們決定讓退役的颱風級核潛艇重生為現代方舟，因其能力與容量足以面對末日災害的威脅：如核爆、地震、海嘯、隕石等嚴重的災難。核潛艇可以在海平面不上浮數個月以對抗海平面上最危險和惡劣的情況。颱風級核潛艇的尺寸為：艇長172.8公尺，寬23.3公尺，吃水12.5公尺。其大容量可容納地球上物種的基因使其變成一個基因庫，潛艇建造的殺生原意正在轉變為護生。

基因和種子保存

過去諾亞方舟進行對生物活體的收集，而改造後的颱風級潛艇將收集精子、卵子和細胞切片。它將成為所有生物體的遺傳資訊數據庫。

為了落實殺生轉變為護生的概念，過去的武器空間，魚雷室與飛彈井將被改造為存儲空間：53.3公分的魚雷發射管，和R-39彈道導彈井，每個空間將被改造以適應不同類型的品種，存儲單元，和實驗室設備。

魚雷室和魚雷管將成為儲存密封的種子。3,000立方公尺空間可容納約150萬不同的種子樣本。20個直徑3.3公尺的導彈井將根據生物的生物學分類排列；其中3個導彈井將被用來保存真菌，另外17個導彈井將被用於哺乳動物、爬行動物、魚類、鳥類、兩棲類和無脊椎動物。這些導彈井總共有2,800立方公尺（140立方公尺×20）來存儲約18,000液氮罐與遺傳細胞切片。

階段

計畫分為四個階段：基因和種子的收集和保存階段；教育及推廣階段；下潛和逃生階段；重啟與重生階段。

在第一階段中，遺傳資訊將被收集和儲存於潛艇。

第二階段，一旦數據庫和遺傳實驗室建立後，一個附屬的環形建築會被建造，平日充當碼頭，教育中心和展示空間並圍繞潛艇，邀請訪客來了解潛艇的使命與新角色。

第三階段，當末世災難發生時，颱風級潛艇會啟動緊急下潛，逃離危機，保護下一代生命的寶貴資源。

在最後階段，當世界的危機解除了，潛艇會重回到海平面上，開始「再生」和「復生」我們的地球。

地點

考慮這個項目的全面性，我們決定讓三艘僅存潛艇分布在不同的位置，選擇的這些地點的港口需要可容納潛艇，並且距離當今現有的基因庫不遠。我們選擇了這行星的三個主要海洋：太平洋的珍珠港、大西洋的樸茨茅斯和印度洋的珀斯。

From Killing to Preserving

Introduction

During the Cold War, the Soviet built the largest submarines there ever were: the Typhoons. These are the result of the doctrine of mutual assured destruction and can be seen as a symbol of military superpower and a typical example of massive destruction. The objective of this proposal is not merely to reuse the submarines; the designer also hopes to give them a new meaning essentially different from the old one. In this proposal, the Typhoons have the potential for and are given the purpose of saving and protecting the species on the planet.

Noah's Ark in the Modern Times

Noah's Ark in the Bible not only saves Noah, his family and all the species on earth from God's wrath, but also symbolizes hope and the beginning of a new era.

The flood is a biblical story that is supposed to have happened a long time ago. However, today the world is still under threat of disasters either natural or man-made that can wipe out all civilizations and creatures. Therefore, I decided to turn the retired Typhoon-class submarines into a Noah's Ark in the modern times. They are capable of and big enough to survive most catastrophic disasters, such as nuclear explosions, earthquakes, tsunamis and even the impact of meteorites. Moreover, the submarines can just submerge and stay underwater for several months, waiting for danger to pass of its own accord. In fact, the size of the Typhoons is similar to that of Noah's Ark which is 172.8 meters long, 23.3 meters wide and 12.5 meters draught. They are so huge that they can be a storehouse for genes of all species on earth. The submarines originally designed for killing are becoming life-preservers.

Storage of Genes and Seeds

Noah's Ark collects living animals; the repurposed Typhoons, on the other hand, will collect sperm and eggs. They will become a gene bank for all beings.

In practice, to turn a killing machine into a preservation facility, the space originally meant for torpedoes and missiles will be renovated as storeroom: the 53.3-centimeter long torpedo tubes and R-39 ballistic missile tubes will each be reused to store different classes of species and to accommodate laboratory equipment.

Specifically, the torpedo room and torpedo tubes will store seeds in airtight containers; the 3,000-cubic-meter space is big enough to house 1.5 million seed samples. Twenty missile tubes with a diameter of 3.3 meters will store species grouped according to their taxonomy: three of the tubes are for fungi while the other seventeen are for mammals, reptiles, fish, birds, amphibians and invertebrates. These tubes have a total of 2,800 cubic meters of storage space (140 cubic meters each) and can house about 18,000 liquid nitrogen containers with genetic samples and materials.

Phases

The project will be carried out in four phases: collection and storage of genes and seeds; education and advocacy; emergency submergence and escape; rebirth.

In the first phase, genetic samples will be collected and stored in the submarines. Once the gene bank and the laboratory are established, a circular auxiliary building that surrounds each of the submarines will be constructed (the second phase) to be used as a dock, education center and exhibition space where visitors can learn about the new role of the vessels.

In the third phase when a catastrophic disaster takes place and the end of the world seems to be at hand, the Typhoons will quickly submerge and escape from the crisis, protecting the precious cargo onboard and preserving life on earth. Finally, in the last phase, when the crisis is over, the submarines will return to the surface and a new cycle of life will begin on earth.

Location

Taking all aspects into consideration, we decided to deploy the three submarines at different locations. The candidates must have a port big enough to accommodate a submarine and should not be far from existing gene banks. In the end, we chose to put each of them on one of the three major oceans: Pearl Harbor on the Pacific, Portsmouth on the Atlantic and Perth on the Indian Oceans.

4

重生演化

Rebirth and
Evolution

形隨機能，是否存在？

鯊魚與鱷魚都是演化完美的生物
外型千萬年來沒有太大改變的生物
假如你要設計鯊魚，最後的結果就是鯊魚
假如你要設計鱷魚，最後的結果就是鱷魚

但是
鯊魚就是鯊魚
鱷魚就是鱷魚
鯊魚無法變成鱷魚
鱷魚也無法變成鯊魚

但是基礎設施與建築類型確實可以變成
變成另一種生物
其功能顯然沒有絕對性
所以建築上真有形隨機能這事嗎？

Is There Such a Thing as Form Follows Function?

Sharks and crocodiles are evolved to best suit their lifestyles.

Their appearance has not changed much for tens of thousands of years.

If you are going to design the most perfect shark, you will end up with a shark in the real world; if you are going to design the most perfect crocodile, you will end up with a crocodile in the real world.

The thing is,
A shark is a shark;
A crocodile is a crocodile.
A shark will never become a crocodile;
A crocodile will never become a shark.

However, infrastructure facilities or any other buildings can indeed (and they do) become another type of architecture.

Their functions do not necessarily conform with their forms.

重生前機能
Before Rebirth

維持人類生存所產生之活動 Activities that ensure the survival of human beings	生產 Manufacture	運輸 Transport	儲存 Storage	軍事 Military

因滿足從事這些活動而產生的設施
Infrastructure that is related to the above activities

發電廠
Power Station

高架橋／路
Elevated Highway/Roadway

儲存設施
Storage Facility

軍事設施
Military Facility

鑽油平台
Oil Rig

地下鐵
Subway

礦場
Mine

造船廠
Shipyard

起重機
Crane

重生後機能
After Rebirth

展覽空間／博物館
Gallery / Museum

動物保育所
Animal Shelter

運動設施
Sports Facility

資訊中心
Data Center

辦公室
Office Building

住宅
Private Home

開放空間
Public Open Space

溫室
Greenhouse

遊樂園
Theme Park

倉儲
Warehouse

學校
School

集合住宅
Multi-dwelling Unit

紀念碑
Monument

眺望台
Observation Deck

水族館
Aquarium

發電廠
Power Station

購物中心
Shopping Mall

旅館
Hotel

電影院
Cinema

表演廳
Theater / Concert Hall

疫苗研究室
Vaccine Laboratory

海上城市
Floating Town

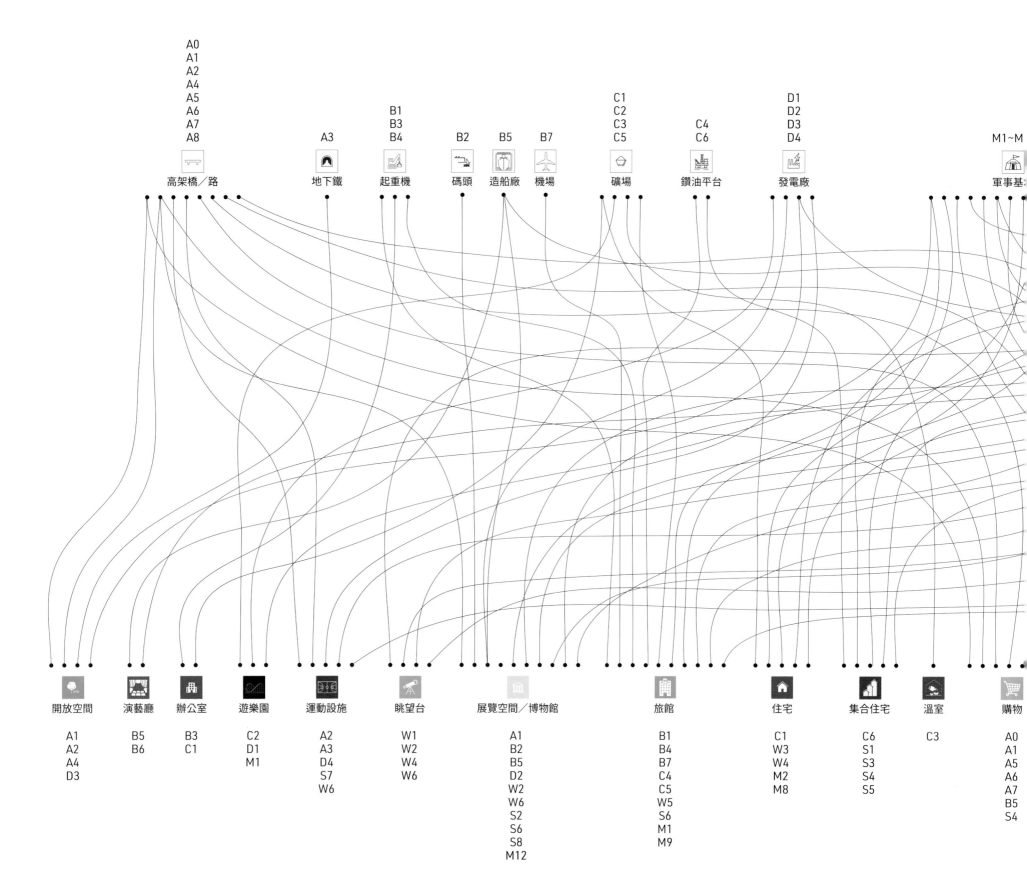

A0
A1
A2
A4
A5
A6
A7
A8

A3

B1
B3
B4

B2

B5

B7

C1
C2
C3
C5

C4
C6

D1
D2
D3
D4

M1~M

高架橋／路　　　　地下鐵　　起重機　　碼頭　造船廠　機場　　礦場　　鑽油平台　　發電廠　　　軍事基

開放空間　演藝廳　辦公室　遊樂園　運動設施　眺望台　展覽空間／博物館　　旅館　　住宅　集合住宅　溫室　購物

開放空間	演藝廳	辦公室	遊樂園	運動設施	眺望台	展覽空間／博物館	旅館	住宅	集合住宅	溫室	購物
A1	B5	B3	C2	A2	W1	A1	B1	C1	C6	C3	A0
A2	B6	C1	D1	A3	W2	B2	B4	W3	S1		A1
A4			M1	D4	W4	B5	B7	W4	S3		A5
D3				S7	W6	D2	C4	M2	S4		A6
				W6		W2	C5	M8	S5		A7
						W6	W5				B5
						S2	S6				S4
						S6	M1				
						S8	M9				
						M12					

B6+S1~S8

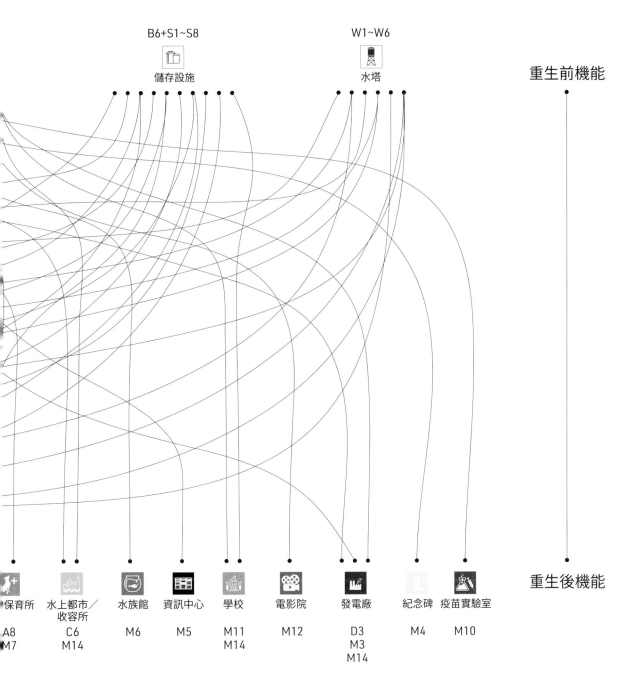

儲存設施

W1~W6

水塔

演化論的基本原理就是生殖成就最高者生存，基因的樣本數最多，以此點目前看來住宅或是飯店這種住宿類型的重生案例相對來講居主導地位。

雖然明顯有取樣不足的問題而不能下此結論，可是以人類目前建築產業與經濟活動的趨勢來看卻不難理解。

So, one cannot help but ask: when we are talking about architecture, is there such a thing as form follows function?

One of the basic principles of evolution is that the more offspring you produce, the more likely your species will survive because you have the largest sum of genes. Similarly, we can see that the dominant types of building reborn from an infrastructure facility are residential projects and hotels. Even though the number of samples in this book may not be great enough to make a definite conclusion, this liking for houses is not surprising when we take a look at the current trend in architectural application and the economic activity.

重生後機能

保育所	水上都市／收容所	水族館	資訊中心	學校	電影院	發電廠	紀念碑	疫苗實驗室
A8 M7	C6 M14	M6	M5	M11 M14	M12	D3 M3 M14	M4	M10

資訊疆界／研究上的取樣偏差

領土實質物理空間的疆界外，還有看不見的疆界——資訊向境外傳播的程度與影響力。因全球資訊版圖的極不對稱，各國經濟指標強弱的不同反映在資訊的映射製圖上，這真實反映了世界上的不平等，資訊發達程度與資訊版圖的不平均，導致網路搜尋與檢索程度的不平均。

書中選出的案例很多來自我們過去所認知的「已開發」國家。不過，世界銀行The World Bank目前已經廢除「已開發」、「開發中」的分類方式，自 2016 年，世界銀行已改用四個所得等級，將全世界經濟體劃入四個收入組別，以個人年均收入分出四個等級：

「第四級」高收入組：高於12,235 美元以上
「第三級」中等偏上收入組：介於12,235 - 3,956 美元之間
「第二級」中等偏下收入組：介於3,955 - 1,006 美元之間
「第一級」低收入組：低於 1,005 美元以下

就算不以世界銀行單以各國人均收入來區分，而是以考慮因子相形之下較為全面客觀，聯合國開發計畫署用來衡量各國社會經濟發展程度的人類發展指數 Human Development Index 來看，同樣是把世界各國分成四個等級：

極高：0.800 以上
高：0.799~0.700 之間
中：0.699~0.550 之間
低：0.549 以下

這四個組別是根據平均壽命、受教育年限、人均國民總收入計算出。只有被列入「極高」這一組的國家才能算是過去所認知的「已開發」國家。不過對我們的案例而言，人類發展指數的四個等級與全世界經濟體的四個所得等級大致上差不多，所以下面我們還是依世界銀行的分類討論。

案例多半出現在第四級與第三級所得國家。這些國家經過多次都市迭代，經濟發展的程度較高，退役的基礎設施眾多，加上無論母語為何，英文版本網路資訊為這些國家資訊載體的標準配備，而我們所找到的案例許多更是媒體寵兒明星建築師的作品，其中往往不乏他們一戰成名的代表作。

中國、印度，甚至從前的日本都有經濟發展急遽上升的過程，推測應有大量基礎設施迭代的現象，但可惜的是我們的取樣無法直接反映出這個事實。

回頭談一下已過時的分類：已開發國家大都處於後工業化時期，服務業與商業為主要產業；而開發中國家則大多處於工業化時期，製造業為主要產業；未開發國家則還停留在農業時代。已開發國家已經歷完產業轉型與升級的過程，所以這些國家境內確實淘汰了許多工業設施，但基礎設施涵蓋範圍遠不止工業設施。另外本書案例中占比例甚大的軍事設施重生案例，主要來源皆於二戰期間，更是與經濟發展程度無關的政治因素造成。

真正影響我們重生案例取樣的，是資訊取得的容易程度，這同時也包含了語言因素造成的資訊壁壘。撇開四個所得等級的區分不談，其中華語系國家的案例占有一定比例是因為情感與語言偏好，這些地區即使沒有英文的網路資訊對我們取樣沒影響，在這些地區我們也更容易找到、發掘那些非明星精英建築師的重生案例。

全球有超過12.4億個網站，非圖形的關鍵字檢索下我們的搜尋最多覆蓋六成，因為這些網站內所使用的語言英文占55%，華語占4%，總共59%。意思是該重生案例有沒有英文網路資訊的介紹與宣傳就已大致決定了我們能不能

Sampling Bias

In addition to the physical border between any two countries, there is also a kind of invisible border that marks the boundary and limit of the spread of information: the economic development of a country. Developed countries have more leverage when it comes to the creation, control and dissemination of information than developing and underdeveloped countries. Hence, strong bias can be discerned in the results of a search.

A lot of cases in this book are located in the so-called "developed" countries. However, it should be mentioned that The World Bank has already abandoned the classification of developed and developing countries. Since 2016, it divided all economies in the world into four groups by income level. Based on GNI per capita, the four groups are:

The fourth group (high income level): > US$12,235
The third group (upper-middle income level): US$ 3,956–12,235
The second group (lower-middle income level): US$1,006–3,955
The first group (low income level): < US$1.006

Even if we don't adopt The World Bank's classification which only takes GNI per capita into account and choose a more comprehensive and objective system instead, like the Human Development Index approved by the United Nations Development Programme to measure a country's social and economic development, the world is still divided into four levels:

Very high: 0.800 or higher
High: 0.700–0.799
Medium: 0.550–0.699
Low: 0.549 or lower

The four levels are determined by indices including life expectancy, education attainment and income, with the "very high" level corresponding to what we used to define as developed countries. However, since the result is more or less the same when we, using the above two classifications, compare the countries the cases in this book are located in, I will use the system of The World Bank.

As it turned out, the majority of cases are located in the third and fourth groups. The main reason for this is that cities in these countries have evolved for a long time, they are more affluent and they've got many disused infrastructure facilities. Moreover, despite their official languages, it's always easy to find an English version on their websites; English is a standard equipment for these countries to spread their messages to the world via the Internet. Last but not least, many of the cases (including those that made the architects famous overnight) are designed by star archetects in the architecture world.

Outside these well-off countries, there are also others, like China, India and Japan, that have experienced a rapid economic growth and much evolution in term of their infrastructure. Unfortunately, our samples do not show this.

Let's go back to the outdated country classification for a moment: developed countries are post-industrial societies where the service sector dominates; developing countries are societies in the middle of industrialization and their main industry comes from the manufacturing sector; while some of the lease developed countries are still farming societies. Developed countries have completed industrial transformation and upgrading, so there are many desolate industrial facilities in these places. However, industrial facilities are not the only type of infrastructure. In this book, there are quite a few renovated military structures, mostly from WWII. It's a political phenomenon and has nothing to do with a country's economy.

The only true factor that affects the sampling process is the accessibility of information, in which language plays an important part. That is why cases from Mandarin-speaking countries are not uncommon in this book, even though they may not belong to the same groups/levels as the other countries. Apart from the sentiment, the key reason is that it is perfectly OK if there is no English version for the information from these regions; the information is still accessible to me. Plus, I had the opportunity to discover some really good designs by those who would otherwise remain unknown to the world.

There are over 1.24 billion websites on the Internet. When I do the keyword search, no more than 60 percent of the results are accessible to me because 55 percent of them are in English and 4 percent are in Mandarin. What this means is that, whether the information regarding

夠找到它，沒有英文的資訊形式存在幾乎意味著它根本就不存在。第二級與第一級所得國家中沒有案例，不一定是因為這些國家可能連基礎設施建設都不完備，更不用談迭代重生，更直接的原因是如果只有母語而沒有英文版的資訊形式，那該案例資訊在世界其他地方被搜尋到的機會也就很小了。

在「有限」的局限下去討論「無限」

本書案例取樣並非完善完整，但依然出版並以此缺陷的說明作為結尾是要以它作為未來的基礎。引用下面這個比喻：過去只能根據片段的，局部的，非常枝微末節的觀察，人類就得開始試著從這些瑣碎微量的資訊去了解整個宇宙。並不是資訊不夠宏觀，就不能夠開始研究，就不能夠得出階段性結論，而非階段性的局部性結論事實上也從未局限人類對宇宙的想像。

人類並無因為觀察地球以外天體的「有限」能力
就停止過對「無限」宇宙的探索。
所有的進展

都是一次又一次，一代又一代建立在「有限」的基礎上往「無限」去推進的。

---

the reborn infrastructure is written or translated in English or not essentially determines whether I can include it in my collection. If there is no English version, it may as well be nonexistent. The fact that no cases are from the first and second groups doesn't mean there is no evolution of infrastructure or even no infrastructure at all in those countries. It just means that relevant information may be written down in languages other than English, so the chance of it being accessed is slim.

Finding Endless Possibilities with Limited Samples

Although the samples are nowhere near perfect or exhaustive, I still decided to publish this book and end it with an explanation so that future researchers can base their studies on it. It is the same in cosmology. When this field of study first emerged, people could only count on partial, fragmental and trivial observations and, starting from there, tried to understand the universe. We don't have to wait until enough information is gathered to begin a study or to draw a tentative conclusion, just like we never let a tentative conclusion limit our imagination about the universe.

Humans did not stop their exploration of the endless outer space
Because their means of observing celestial bodies were limited.
Advances towards endless possibilities are always established on
Limited steps accrued time after time, generation after generation.

B_交通運輸設施

B1 丹麥哥本哈根 起重機飯店
B2 丹麥赫爾辛格 海事博物館
B3 荷蘭阿姆斯特丹 吊車空中公室
B4 荷蘭阿姆斯特丹 法拉達起重機豪華酒店
B5 中國上海 船廠1862藝術中心
B6 德國漢堡 易北愛樂廳
B7 英國紐約日酒店碼頭 TWARK行中心飯店

C_礦產

C1 西班牙甲羅降那 水泥廠頂端公共住宅
C2 羅馬尼亞國際達 地下鹽礦博物館
C3 英國腹沃斯德郡 伊甸園植物園
C4 奧地利亞西巴丹 新油平台潛水旅館
C5 中國上海 余山地下深坑酒店
C6 英國蘇格蘭 克賽馬帝潛鑽油平台

A_運輸面空間

A0 直立本日
A1 豆田圖尿 塔諾圖尿
A2 美國拉杉磯濟離圖尿 英國倫敦公園
A3 地方線道神斯濟斯之家 荷電房尿
A4 A8ernA高速新高架橋林 秋葉原2k540商業藝廊
A5 瑞士蘇黎世顯新高架拱廊 日本東京
A6 台灣台北德河商路五金街
A7 巴西里約瀉林 Rio-Santos高速公路末段站

W_水利設施

W1 荷蘭羊角村觀景塔
W2 中國瀋陽水塔展廊
W3 比利時安特衛普 森林水塔住宅
W4 德國布蘭登堡水塔住宅
W5 英國隨福克郡雪中之家
W6 匈牙利德布勒森森水塔 咖啡廳/畫廊/攀岩場/觀景台

M_軍事設施

M1 德國柏林 飛船機庫室內熱帶度假村
M2 德國柏林 碉堡屋頂住宅
M3 德國漢堡 掩體再生能源中心
M4 荷蘭屈倫博赫 碉堡599水線紀念碑
M5 瑞典斯德哥爾摩 核掩體數據資訊中心
M6 奧地利維也納 高射炮塔水族館
M7 英國糖因 空軍基地掩體隧道避難所
M8 英國薩福克海岸 斜坡漸式潛塔樓住宅
M9 英國棧茅茅斯港 海上堡壘旅館
M10 俄羅斯聖彼得堡 岩漿山大峽谷地研究實驗基地
M11 台灣新竹 運艦碼頭入大游救團
M12 台灣風山 美國防空洞所(設計指)

M13 新聞核掩體 防空氣象觀所(設計指)
M14 新聞核掩體

D_能源設施

D1 德國卡爾卡耳 核電廠仙境主題樂園
D2 英國倫敦 森特現代生術館
D3 台灣高雄 青埔垃圾發電景觀公園
D4 台灣桃園八架媽藝碑

S_倉儲設施

S1 丹麥哥本哈根 Frosilos集合住宅
S2 芬蘭赫爾辛基 Silo 468燈塔
S3 比利時彩訪海姆 酒廠簡食住宅
S4 奧地利維也納 瓦斯槽城市
S5 英國倫敦國王十字 天然氣槽住宅
S6 南非開普敦 非洲當代藝術博物館
S7 加拿大蒙特婁 攀岩健身房Allez-Up
S8 中國上海 民主糧倉八萬噸筒倉

160 | 重生之路 The Road to Rebirth

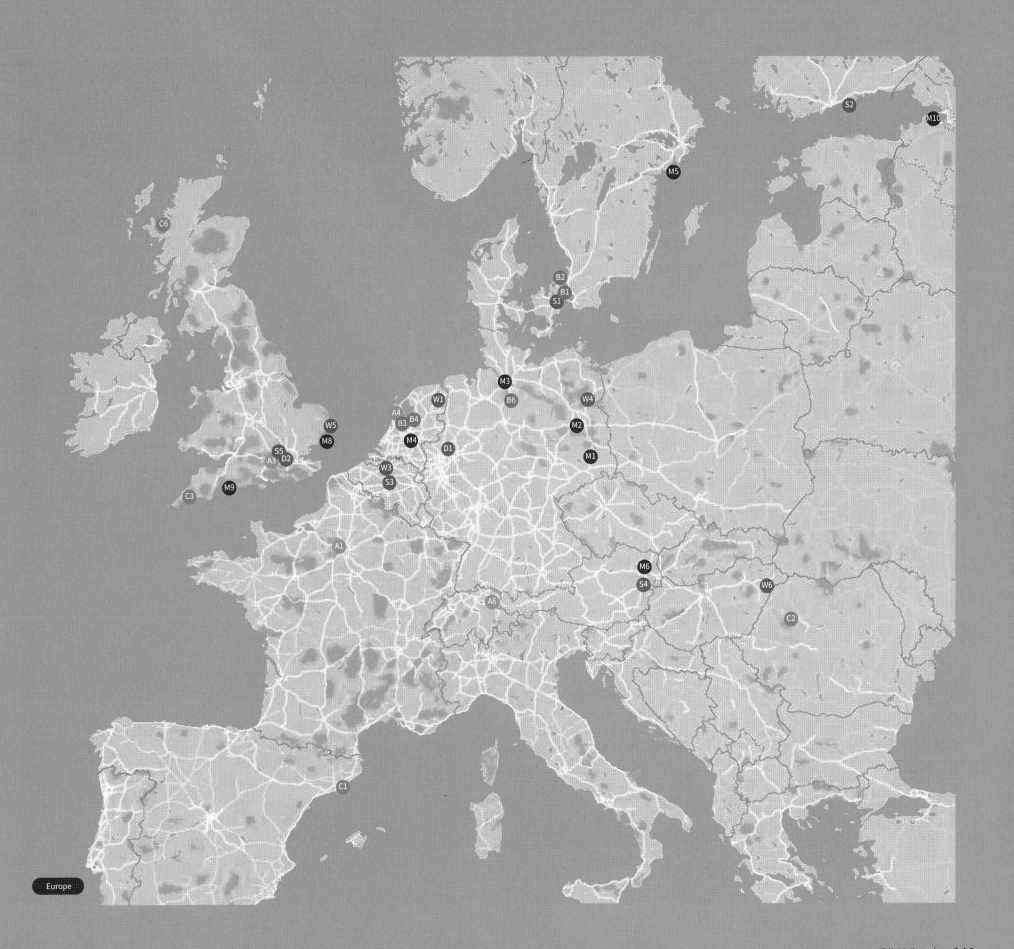

Europe

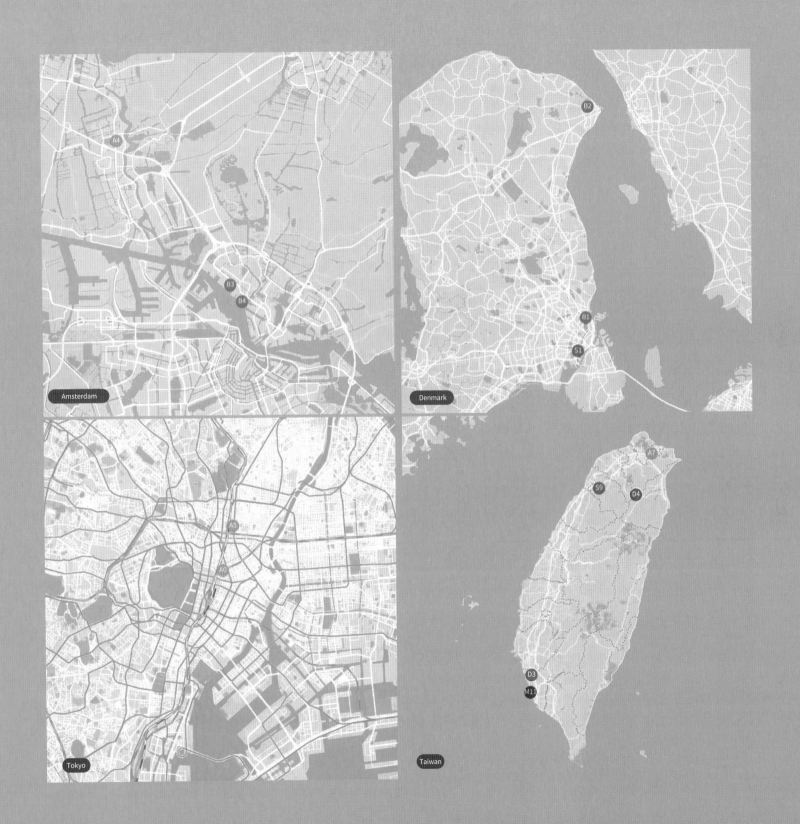

Amsterdam

Denmark

Tokyo

Taiwan

江戶時代的浮世繪常描繪日本橋及富士山的景色，日本橋這裡的老居民也期待能有類似波士頓永恆之掘 Big Dig 或首爾清溪川的計畫，拆除或改道從空中穿越日本橋的首都高速公路，重現往日風華。回應老居民的期待的解決之道是將首都高速公路經過日本橋週邊的這一段埋入日本橋川河岸的地底下，而埋入地下部分的高速公路也確實能避開複雜的的地下鐵與公共管道。然而此計畫雖然工程技術上可行，卻因其耗費鉅資而讓東京都廳一直沒通過這個計畫。

基地選擇讓我跨出了塚本由晴這門在哈佛的設計課：市場原本沒涵蓋的事──對廢棄基礎設施的再利用。

5

重生之路

The combination of Nihonbashi and Mount Fuji is not an uncommon theme in ukiyo-e prints from the Edo period. The elderly residents are hoping that, if similar projects like the Big Dig in Boston or the Cheonggyecheon in Seoul can be carried out and the Metropolitan Expressway that passes over Nihonbashi can be demolished or diverted, it's not impossible that the bridge restores its splendor of bygone days. To comply with the wish of the locals, the section of the Expressway over and near the bridge will have to be moved underground, beneath the Edo River. And it is plausible – the underground part can certainly make a detour to avoid the complex subway and pipeline systems already there. However, due to the enormous cost, the project hasn't been approved by the Tokyo Metropolitan Government, even though it's technically viable.

My choice of site for an assignment in a design studio taught by Yoshiharu Tsukamoto in Harvard gave me a chance to enter a whole new area that didn't appear in the traditional market: the reuse of abandoned infrastructure.

The Road to Rebirth

拆與不拆的辯證

目前高架橋這類的基礎設施，除了全部拆光，是否還有額外選項

第一種：存──大量保留不拆除

釋放出的土地沒有太高的利用價值；

土地政策或建管政策下，選擇保存較有利新的開發項目整體推進；

拆除工程難度高，曠日廢時；

具有歷史意義或工程價值，由文史單位找出理由；

新功能非常好整合入既有基礎設施，改造工程難度低，成本低；

反對拆除的抗爭強烈，保留的請願強烈；

政治力介入；

案例：法國巴黎藝術高架橋，美國紐約 Highline，韓國首爾路 7017。

第二種：不存──全部拆除

釋放出的土地有非常高的利用價值，或能帶動周邊發展增值；

土地政策或建管政策下，選擇拆除較有利新的開發項目整體推進；

拆除工程難度低；

影響公共安全已到非拆不可；

不具有歷史意義或工程價值；

無適合的新功能非可以整合入既有基礎設施，改造工程難度高，成本高；

拆除的請願強烈；

政治力介入；

案例：美國波士頓永恆之掘 Big dig，韓國首爾清溪川

在 1991 啟動的美國波士頓 Big Dig 大挖掘（永恆之掘），將市中心主要的高架道路系統地下化以釋放道路上空，把城市平面還給步行者，也還給波士頓更好的市容。後來受此影響的重大工程與開放空間案例著名的有韓國首爾的清溪川，2003 開始拆除其上於 1968 年所建造的高架道路。

第三種：介於存與不存，拆與不拆之間

有部分保留，也有部分拆除，但很可能都是拆除多於保留除了全部拆光，是否還有「存」與「不存」外的選項

在拆與不差的中間還是有一種可能性是存在局部的保留做再利用，這局部保留的本身戴上紀念性的色彩。我尋找東京的市場基地與設計高速公路市場碰巧的闖入這一個中間地帶：

我找到基地並不是一塊當時就存在的基地，而是在不確定的未來中，有被計畫拆除可能性的首都高速公路在日本橋周邊的那段。建議是此段可能被拆除的高速公路中，其中一小段拆除工程難度較高且離日本橋較遠，已對其周邊整體景觀影響度低的，保留一小段不拆除做為設計課要求的市場基地。塚本由晴老師十分喜歡此基地選擇的構想，且懸浮在河上的土地根本無法取得，除非使用高速公路本身。另外他表示　關於首都高速公路這段拆與不拆的辯證，日本橋此處有眾多的經濟力與政治力介入糾葛已久，至今未決，而這市場基地的提案正是介於兩者之間。

案例：柏林圍牆，東京高速公路市場

To Tear It Down or not to Tear It Down

Except for tearing down a retired elevated highway, are there any other options?

Option A: keep it, because…

The land has little potential even if the structure is torn down;

It's beneficial for a new development project to keep it;

It's difficult and time-consuming to tear it down;

The building has historical values or is a construction feat;

A new purpose is found for the building, the renovation is easy and the cost is low;

The public protests against its demolition and asks the government to preserve it;

Politics plays a role.

Examples include: Viaduc des Arts in Paris, France, High Line Park in NYC, USA, and Seoullo 7017 in Seoul, Korea.

Option B: tear it down, because…

The land has a lot of potential and can even add value to the surrounding area if the structure is torn down;

It's beneficial for a new development project to tear it down;

It's not difficult to tear it down;

It poses a threat to public safety;

The building does not have historical values and it's not a construction feat either;

A new purpose cannot be found for the building, or the renovation is difficult and costs too much;

The public protests against its preservation;

Politics plays a role.

Examples include: The Big Dig in Boston, USA and the Cheonggyecheon project in Seoul, Korea.

In 1991, Boston began the Big Dig that rerouted the chief elevated highway through the city into an underground structure to vacate the space above the ground and give the city back to the pedestrians; another famous megaproject that aimed to release urban space took place in Seoul in 2003, when the elevated highway built in 1968 over the river Cheonggyecheon was removed.

Option C: something in between

That is, a part of it is kept while the rest of it is removed. However, the portion removed is likely to be bigger than the portion kept.

There is a third option besides keeping it and tearing it down. It's an in-between zone where the structure is partially kept and reused. The part that is kept usually becomes something like a monument. When I was looking for a potential site in Tokyo to be re-designed, I happened to cross into this in-between zone.

What I had found was, as a matter of fact, not exactly an existing place, but a section of the Metropolitan Expressway near Nihonbashi that might be demolished in the future. I proposed to keep and refurbish a small part of the section. This part was more difficult to remove, relatively far away from Nihonbashi and did not obscure the entire view of the surrounding area.

My teacher Yoshiharu Tsukamoto was very pleased with this design because no land can be floating high above a river other than an elevated highway. He also mentioned that the debate over this section of the Metropolitan Expressway had been going on for a long time. Many factors, be they economic and political, caused the issue to remain unresolved. Now, this proposal can be the answer that brings about a win-win situation.

Examples include: The Berlin Wall in Germany and the Metropolitan Expressway Market in Tokyo, Japan.

第1階段：現在
首都高速公路，遮蔽日本橋川超過半個世紀的鋼鐵怪物

Phase 1: The present
Nihonbashi has been in the shadow of the steel monster,
the Metropolitan Expressway, for more than fifty years.

第2階段：拆除後
日本橋川將像過去一樣重見天日。唯一剩下的部分是懸於山手線之上的
（鐵路）的重疊部分這段，未來高速公路市場的虛擬基地。

Phase 2: After partially removed
Nihonbashi will regain its former glory. The only part of the Metropolitan Expressway that will be kept
is the section passing over the Yamanote railway line.
This is where the Metropolitan Expressway Market will be located.

第3階段：未來

前瞻的基礎設施可持續發展：日本橋與日本橋川過去的氛圍已恢復，而被留下的基礎設施殘骸重生再利用為市場。

這座高速公路市場將由拆除自高速公路的材料建造，構成了日本橋川上空的開放空間，

將與這條河及其曾經歷過的一切載沉載浮，見證首都高速公路服務東京超過半世紀的歷史。

Phase 3: The future

The view of Nihonbashi and the Edo River of bygone days being recovered, the part of the infrastructure that remains can be reused as a market.

The Metropolitan Expressway Market will be built with the material of the dismantled highway and form a public open space above Nihonbashi.

Together, the market and the river, along with everything it has experienced, will act as witnesses of the Metropolitan Expressway which has served Tokyo for over half a century.

拆除前
Before

拆除後
After

分段縫／伸縮縫
Joint

紅色十字為日本橋的位置，而白色發光段則為可能即將被拆除的部分。沿著拆除的部分由高速高路本身建造時的分段縫／伸縮縫，
就可以由此定義出有潛力作為高速高路市場的基地形狀。定義出的三塊基地都是較靠近交流道，這樣的的位置較容易將人從地面層帶到空中層。

The red cross indicates where Nihonbashi is while the white parts are sections likely to be torn down in the near future.
The dividing joints installed when the Metropolitan Expressway was first constructed define the shape and length of potential sites of the market.
The three candidates of sites, thus chosen are all close to an interchange, which is rather convenient when you want to bring visitors up there from ground level.

沿著高速公路的物件也成為被保留的一部分。例如長年都在維修狀態，沿著高速公路就一路可見的維修站。
高速公路上的標誌系統，路燈等等都會在拆除後成為大量的廢棄物，我們將回收這些物件使其成為新的高速公路市場上的元素。

Existing features of the elevated highway are also preserved – the maintenance platforms active all year round at various points along the Expressway for example.
More often than not, fixtures like road signs and lampposts go to waste after a removal, but actually they can be recycled and reused as elements of the new market.

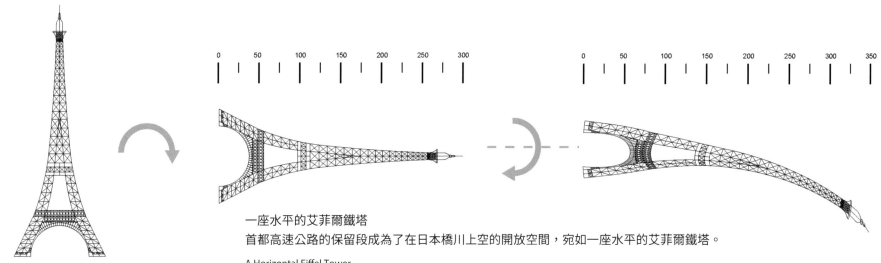

一座水平的艾菲爾鐵塔
首都高速公路的保留段成為了在日本橋川上空的開放空間，宛如一座水平的艾菲爾鐵塔。

A Horizontal Eiffel Tower
The section of the Metropolitan Expressway that will be turned into a public open space above the Edo River looks like the Eiffel Tower lying flat.

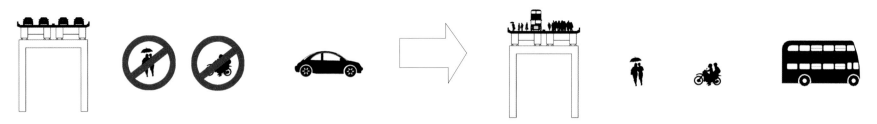

過去首都高速公路上為車輛所獨占的領域。

Right now, the Metropolitan Expressway is for cars only.

過去被獨占的領域未來將成為行人的開放空間，
但車輛將以另一種形式再回到首都高速公路上。

After renovation, it will be open to the pedestrians.
There will still be cars, but they are here for a different purpose.

高速公路市場裡需要商店，而車輛的尺度仍然是最適合在高速公路上運作。
各種不同類型不同尺度的車輛被改造為各種不同物業的商店。

Needless to say, shops are vital for the Metropolitan Expressway Market.
Since the size of a car is perfect in a former highway, vehicles of various types and sizes will be transformed into various kinds of shops.

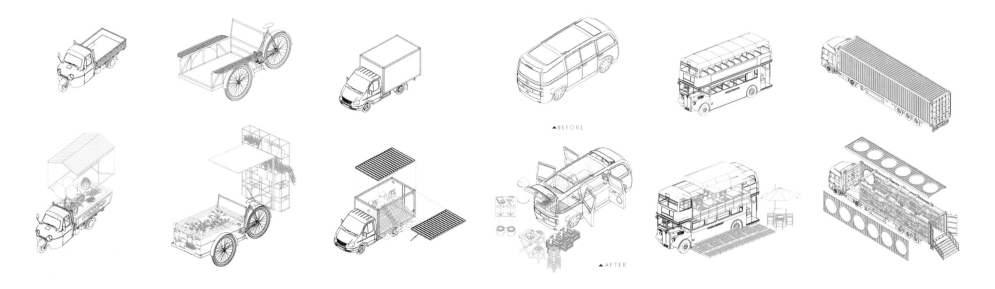

▲ BEFORE

▲ AFTER

首都高速公路保留段上的新增建部分，在入口與河道上構成了無形穹頂。在拆除了首都高速公路後，回復了昔日風光的日本橋川上，保留段重生為新的紀念物。

車輛被改造為商店，排列猶如車流仍舊運行在雖死猶生的首都高速公路市場上。

The newly-added structure over the section of the Metropolitan Expressway that will be preserved forms an invisible dome above the river. When the Metropolitan Expressway is torn down, Nihonbashi will regain its former glory and the section will be reborn as a new monument. Cars will be turned into shops and arranged in a way so it feels like there is still traffic on the highway that has died and then is born again.

一切
就像它的其它部分被拆除前，似乎它跟你一樣從未死過。
然而
每一天，當今日的你醒來時，昨日的你便已然死去。

今日即重生

Everything
Remains the same as the rest of the Metropolitan Expressway that has once been there, as if nothing ever dies, just like you never do.
However,
The fact is that you do die a lot. Every day when you wake up, the you from yesterday have already died.
We are always being reborn, day after day.

後記

《寄生之廟》,《重生之路》,《往生之間》三本書的計畫原本打算由《寄生之廟》的出版就直接畫上休止。但後來因為在交通大學建築研究所的教學,以及仍有出版社們願意繼續出版此系列作品,就順水推舟再次啟動,延續原計畫了。

2006-2007 在哈佛設計研究所 Havard GSD 的最後一個我在設計課的學生作品是《重生之路》這本書的起點,也是學生時代第一次在台灣以外的城市,尋找能適合自己設計構想的基地,(過去都是被指定好的基地):在東京城市漫遊 10 天內尋找幾個能為做設計課要求的市場基地,這次在東京的基地狩獵 Site Hunting 也讓我萌芽多年後試著到東京工作的念頭。

2007 我在哈佛研究所做此設計作業的同時,正在與一群韓國同學做韓國當時新首都的中央開放空間競圖,即使自己喜歡這個局部保留要拆除的基礎設施轉變市場的構想,也無法聚焦於此案上。沒想到畢業後拿它去交適合其主題的競圖反而拿到 Boston 建築師公會獎。從 2008 開始就不斷留意收集世界上基礎設施再利用這類案例,也是我任教

Epilogue

I was going to drop my plan to write a series of three architecture books after the publication of the first one, Parasitic Temples. However, after I taught in the Graduate Institute of Architecture at National Chiao Tung University and due to the fact that a couple of publishers had expressed their interest to publish the other two books, I changed my mind and re-started my writing.

The beginning of the second book, this book, is my last project as a graduate student. It was for a design studio at Harvard Graduate School of Design when I was studying there between 2006 and 2007. For the first time, I tried to find a site for my design project outside Taiwan; in the past, our teachers would assign a site for each of us. During my 10-day wandering in Tokyo, I looked for candidates that matched the requirements of the course. The idea of working in Tokyo in the future also germinated in this site hunting trip.

When I was working on the Harvard studio work in 2007, I also participated in a competition with a bunch of Korean fellow students to design a public open space for the new capital of their country. Although I liked the idea of turning a derelict infrastructure facility into a market, I couldn't have possibly focused on it myself. Little did I know that after graduating from Harvard, I would submit the design to a relevant competition The Future Design, hosted by Boston Society of Architects, and won an award. For all these years since 2008, I have been keeping an eye open for and collecting similar cases around the world that reuse an infrastructure facility. I also make it one of the topics for my students'

時，每年畢業設計會開給學生選的題目。從此之後的歲月我不斷關注世界上這類將廢棄基礎設施再利用視為重生的案例。

除了感謝眾多參與這本書的成員與我的家人外，也感謝 AECOM 公司曾經的同事與長官：林佑達，沈同生，林芳慧，劉泓志 (依評圖日期順序由先至後) 在 2018 年到交大建築研所「重生之路」這門設計課評圖，分享你們的寶貴意見。以及曾經在東京工作過的隈研吾都市建築研究所的同事，洪人傑，秋天。最後是目前我任職過最久，台北九典事務所的張清華建築師與郭英釗建築師兩位創辦人。

世界各地各種重生案例，能有幸被收錄應是當今成功知名度高的，可惜礙於有限人力物力，實際上有許多遺珠未收錄，然而更多得是有多少已死透的基礎設施沒有機會重生。此書亦為教學成果與教學生涯留下紀錄，書中的 54 案除了已經完成的真實重生案例外，也加入了我與學生們的想像與渴望，終究，Design is driven by desire。

《重生之路》是一本圖紙紀錄。

《寄生之廟》記錄的是那些沒有建築師的建築；相反的

《重生之路》收集而來的重生案例多是菁英建築師的作品，其中不乏他們一戰成名之作。歷史上，統計學走上獨立發展的道路從數學分離出來；建築學走上獨立發展的道路從藝術分離出來；都市學走上獨立發展的道路從景觀分離出來。圖學做為上述專業們的工具，早已經獨立發展分離出來。這些紀錄案例的等角圖與剖面透視圖可以視為一種建築都市與環境再現的形式。《寄生之廟》或《重生之路》這兩本書原本都可以開啟一個章節對建築與都市圖學本身的討論，然而這類偉大著作已眾，無須我輩多言。

水月・留痕

序中提過本書起源於我 2007 在哈佛研究所時的最後一個作品，此後的歲月我不斷關注世界上類似這樣的廢棄基礎設施「把再利用視為重生」的案例。畢業後十多年的職涯機緣，並無一絲一毫，一點一滴能有沾到這類形案子的機會，除了明白這類型項目是可遇不可求外，也接受了最後此案對我而言終究都是紙上建築，鏡花水月。

如夢幻泡影
如露亦如電

此書為此——留痕

graduation project every year.

Here, I would like to thank each and every member of the team that made this book happen and my family. Also, I would like to extend my thanks to my old colleagues and supervisors from AECOM, Youda Lin, Tongsheng Shen, Fanghui Lin and Hongzhi Liu (ordered by date of their evaluation), who helped evaluate the projects and gave valuable advice for my class at the Graduate Institute of Architecture at National Chiao Tung University in 2018. I want to thank my colleague, Renjie Hong, at Kengo Kuma and Associates from my days in Tokyo. Last but not least, I would like to thank the two architects who founded Bio-Architecture Formosana (a company for which is I have worked for the longest), Qinghua Zhang and Yingzhao Guo.

The reborn cases that are included in this book are among the most successful and famous in the world. Unfortunately, due to limited resources, there are a lot of hidden gems that are not mentioned; worse still, many infrastructure facilities didn't even have a chance to be reborn. This book is a record of my teaching job. The 54 cases comprise not only real projects but also the imaginations and desires of my students and me. After all, design is driven by desire.

The Path to Rebirth is a graphic record. While Parasitic Temples is all about buildings that do not have an architect, The Path to Rebirth is mostly about works by elite architects. Many of these works are what made them famous in the first place. Statistics branched out from mathematics and became a separate field of study; architecture branched out from art and became a separate field of study; urban planning branched out from landscaping and became a separate field of study. As for drafting, it had become a separate discipline long before the others, being the essential technique of all the above fields. The isometric drawings (sometimes with sectional perspectives) of the case studies in this book can be seen as a representation of the architecture, the city, as well as the surrounding environment. For Parasitic Temples and The Path to Rebirth, I could have written a whole chapter to discuss the art and science of drafting. In the end, I decided against this, since there were already numerous treatises on the subject.

A Moon in the Water Leaves nothing but a Trace

In Introduction, I mentioned that this book originated from my last project during my studio work at Harvard in 2007 and that ever since then, I have been dedicated a lot of time to the research of infrastructure facilities worldwide that were once neglected but were then reused and reborn. Now, more than a decade after my graduation, I have never had the opportunity to really carry out a project even slightly relevant. Finally, I come to the realization that such projects are not to be begged for and accept that they will always be a moon in the water for me.

Like dreams and visions in a bubble

Or a bead of dew, a flash of lightning.

And this book is the trace.

3

The Collection of Reborn Infrastructure

A Transportation (9)
B Port facilities (7)
C Mines (6)
D Power supply (6)
W Water towers (4)
S Storage (9)
M Military structures (13)

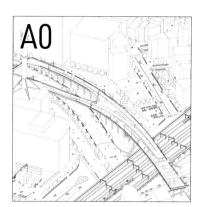

A0

Tokyo Highway Market, Tokyo, Japan
(Proposal)

賴伯威｜Po Wei Lai

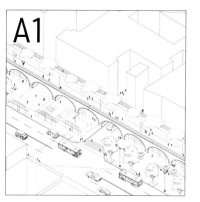

A1

Viaduc des Arts, Paris, France

Patrick Berger

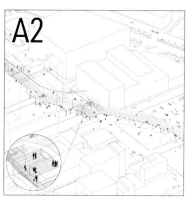

A2

New York High Line Park, New York, USA

Diller Scofidio + Renfro, James Corner

A3

House of Vans London, London,
United Kingdom

Tim Greatrex

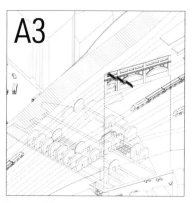

A4

A8ernA, Zaanstad, the Netherlands

NL Architects

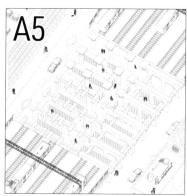

A5

2k540 Aki-Oka Artisan, Tokyo, Japan

東日本都市開發公司｜
JR East Urban Development Corporation

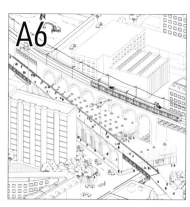

A6

Refurbishment Viaduct Arches,
Zurich, Switzerland

EM2N

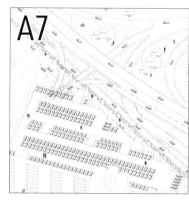

A7

Huanhe S. Rd Hardware street,
Taipei, Taiwan

Unknown

A8

Rio-Santos Highway Orangutan Jungle
Park, Rio, Brazil (Proposal)

侯雅齡｜Yaling Hou

B1

The Krane, Copenhagen, Denmark

Arcgency-Mads Møller

B2

Maritime Museum of Denmark, Helsingør, Denmark

BIG

B3

Kraanspoor, Amsterdam, The Netherlands

OTH Architecten

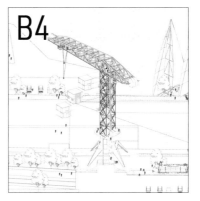

B4

Crane Hotel Faralda NDSM, Amsterdam, The Netherlands

IAA Architecten Amsterdam

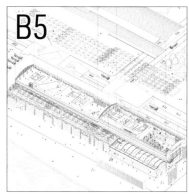

B5

Shipyard 1862, Shanghai, China

隈研吾建築都市設計事務所｜
Kengo Kuma and Associates

B6

Elbe Philharmonic Hall, Hamburg, Germany

Herzog & de Meuron

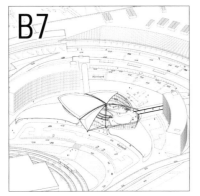

B7

TWA Flight Center Hotel, New York, USA

Beyer Blinder Belle &
Lubrano Ciavarra Architects

C1

La Fábrica, Sant Just Desvern, Spain

Ricardo Bofill

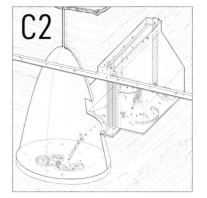

C2

Turda Salt Mine, Turda, Romania

Ecopolis

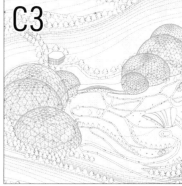

C3

Eden Project, Cornwall, United Kingdom

Nicholas Grimshaw

C4

Seaventures Dive Rig Resort, Sipadan, Malaysia

Morris Architects

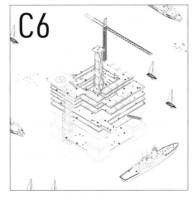

C5

InterContinental Shanghai Wonderland, Shanghai, China

Martin Jochman

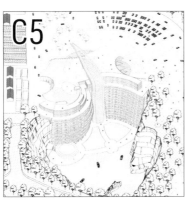

C6

Cromarty Firth, Scotland, United Kingdom (Proposal)

王俐雯｜Liwen Wang

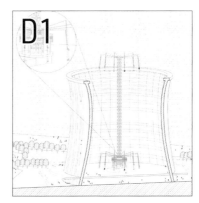

D1

Wunderland Nuclear Reactor Amusement Park, Kalkar, Germany

Hennie van der Most

D2

Tate Modern, London, United Kingdom

Herzog & de Meuron

D3

Kaohsiung Metropolitan Park,
Kaohsiung, Taiwan

皓宇工程顧問股份有限公司｜
Cosmos International Inc.
Planning & Design Consultants

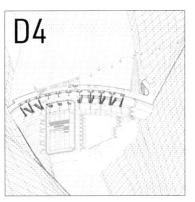

D4

Ronghua Dam, Taoyuan, Taiwan
(Proposal)

陳天健｜Tianjian Chen

W1

Watch Tower Sint Jansklooster,
Jansklooster, The Netherlands

Zecc Architecten

W2

Water Tower Pavilion (Renovation),
Shenyang, China

META-Project

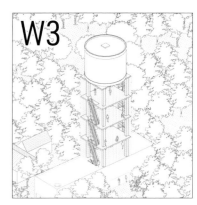

W3

Water Tower Housing-Brasschaat,
Belgium

Binst Architects

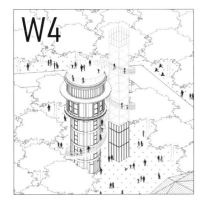

W4

BIORAMA-Projekt, Brandenburg,
Germany

Frank Meilchen

W5

House in the Clouds, Suffolk,
United Kingdom

Glencairne Stuart Ogilvie &
F. Forbes Glennie

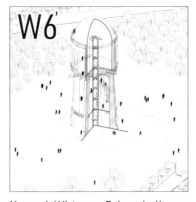

W6

Nagyerdei Víztorony, Debreczin, Hungary

architects Zoltán Győrffy &
Róbert Novák

S1

Frøsilos SILO apartment,
Copenhagen, Denmark

MVRDV

S2

Silo 468, Helsinki, Finland

Lighting Design Collective

S3

Kanaal, Wijnegem, Belgium

Stéphane Beel Architects

S4

Gasometer, Vienna, Austria

Jean Nouvel & Coop Himmelblau &
Manfred Wehdorn & Wilhelm Holzbauer

S5

King's Cross Gasholders,
London, United Kingdom

Wilkinson Eyre

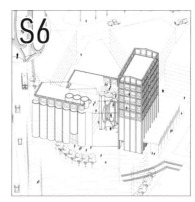

S6

Zeitz MOCAA, Cape Town, South Africa

Heatherwick Studio

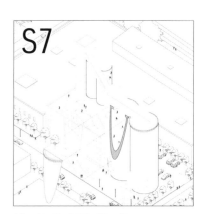

S7

Allez Up Rock Climbing Gym,
Montreal, Canada

Smith Vigeant Architects

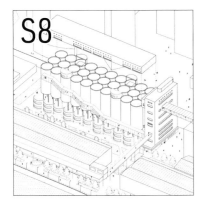

S8

80,000-tonne Silo Warehouse,
Shanghai, China

大舍（柳亦春／陳屹峰）｜
Atelier Deshaus

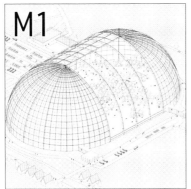

M1

Tropical Islands Dome Resort,
Krausnick, Germany

Oscar Niemeyer

M2

Fichte Bunker, Berlin, Germany

Paul Ingenbleek

M3

Renewable Wilhelmsburg Energy
Bunker, Hamburg, Germany

HHS Planer + Architekten AG

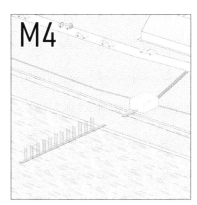

M4

Bunker 599, Zijderveld, The Netherlands

RAAAF + Atelier de Lyon

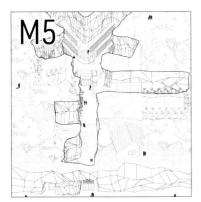

M5

Data Center Pionen White Mountain,
Stockholm, Sweden

Albert France-Lanord Architects

M6

Haus des Meeres Flak Tower aquarium,
Vienna, Austria

Pesendorfer & Machalek Architects

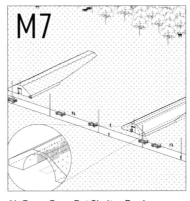

M7

Air Force Base Bat Shelter Bunker,
Maine, USA

Unknown

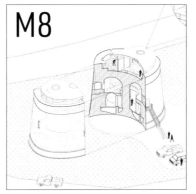

M8

Martello Tower Y, Suffolk coast,
United Kingdom

Piercy & Company

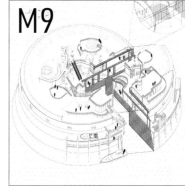

M9

Palmerston Sea Forts Hotel,
Solent, United Kingdom

Amazing Venues

M10

Fort Alexander, St Petersburg, Russia

Louis Barthelemy Carbonnier d'Arsit de
Gragnac

M11

The Oil Storage Classroom JianGong
Primary School, Hsinchu, Taiwan

九典聯合建築師事務所｜
Bio-Architecture Formosana

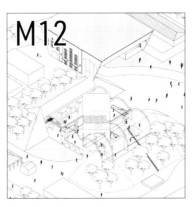

M12

Fongshan Wireless Communications
Station, Kaohsiung, Taiwan (Proposal)

林靖淯｜Jingyu Lin

M13

The Intrepid Sea, Air & Space Museum,
New York, USA

Perkins+Will

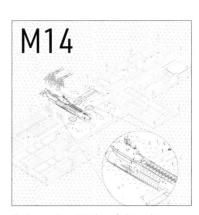

M14

Typhoon class Nuclear Submarine
(Proposal)

陳詠暄｜Yongxuan Chen

A Transportation

A0 Metropolitan Expressway Market, Tokyo, Japan (proposal)

A1 Viaduc des Arts, Paris, France

A2 High Line Park, New York City, USA

A3 House of Vans London, London, United Kingdom

A4 A8ernA, Zaanstad, the Netherlands

A5 2k540 Aki-Oka Artisan, Tokyo, Japan

A6 Refurbishment Viaduct Arches, Zurich, Switzerland

A7 The Huanhe S. Road Hardware Street, Taipei, Taiwan

A8 Viaduct Petrobras Jungle Park, São Paulo, Brazil (proposal)

A0 Tokyo Highway Market, Tokyo, Japan (proposal)
DESIGNER: Po Wei Lai

The Metropolitan Expressway was built in the preceding years of the 1964 Summer Olympics to help transport forthcoming visitors from all over the world from Narita International Airport to central Tokyo. Consequently, many historic bridges over the Edo River, including Nihonbashi (literally "the Japan Bridge," on the middle of which is the Kilometre Zero of Japan signifying the starting point of the national road network), are now in the shadow of the steel monster. The view Nihonbashi once offered is obscured, and the landscape in the surrounding area is also greatly transformed.

In 1991, Boston began the Big Dig that rerouted the chief elevated highway through the city into an underground structure to vacate the space above the ground and give the city back to the pedestrians; another famous megaproject that aimed to release urban space took place in Seoul in 2003, when the elevated highway built in 1968 over the river Cheonggyecheon was removed. The combination of Nihonbashi and Mount Fuji is not an uncommon theme in ukiyo-e prints from the Edo period. If similar projects like the Big Dig or the Cheonggyecheon can be carried out on the Metropolitan Expressway, it's not impossible that the bridge restores its splendor of bygone days. To comply with the wish of local citizens, the section of the Expressway over and near the bridge will have to be moved underground, beneath the Edo River. And it is plausible – the underground part can certainly make a detour to avoid the complex subway and pipeline systems already there. However, due to the enormous cost, the project hasn't been approved by the Tokyo Metropolitan Government, even though it's technically viable.

In this proposal, I'm not trying to argue whether or not such improvement and beautification projects are worthwhile. Rather, when I put forward the proposal as my take on the assignment to come up with some project for Tokyo for a course at Harvard during the school year 2006-07, I made an assumption that the Nihonbashi section of the Metropolitan Expressway had already went underground. That being said, the Expressway has served the city for more than 50 years, and it does not deserve to be wiped out entirely. Therefore, I propose to keep a section of the elevated highway and redesign it into a market. In my opinion, the part passing over the Yamanote Line is ideal, not least because it's far from Nihonbashi and thus won't compromise its landscape. As a matter of fact, since this section is a joint structure with the railway and removal will be difficult as well as have an adverse impact on the incoming and outgoing of trains, it is advised not to change it.

One of the challenges is to bring people up there. This section of the Expressway is 18 meters high, the equivalent of 6 stories, and is more or less parallel to the sidewalk on the ground level. All year round temporary maintenance platforms built with wooden planks can be seen at various points along the Expressway. These can well be remodeled into escalators and stairways that lead people to the market. Moreover, fixtures like lampposts, signals and road signs usually go to waste after a removal, but they are actually perfect to be reused as the lighting and signs in the market. Last but not least, the cars themselves can be renovated as shops selling all kinds of goods, returning to the Expressway with a new purpose. After all, highway infrastructure was originally designed for cars, so they should fare pretty well in an expressway market.

The open space above the market has lots of potential too. I add an observation deck level so that visitors can take in the scenic overlook. In the past, the infrastructure was exclusively for vehicles; now, people can go up to a few stories high, enjoying a different way to move on the Expressway while shopping in the market. There are still car-turned-shops on the market level, but the level above it and its urban view belongs to people and people only. One can even say that the whole structure looks like the Eiffel Tower lying flat.

A1 Viaduc des Arts, Paris, France
DESIGNER: Patrick Berger

The Promenade Plantée in Paris is probably the earliest example in the world that an obsolete elevated railway was turned into public space. Since it is a pioneer in old infrastructure remodeling, I decide it has to be the first case study of my collection of reborn infrastructure.

The Promenade Plantée is a 4.5 km elevated park located in the 12th arrondissement of Paris, with Opéra Bastille at the western end and Bois de Vincennes on the Boulevard Périphérique at the eastern end. It can be divided into five sections, including Viaduc des Arts, Viaduc Nouveau, Reuilly, Reuilly-Picpus, and Picpus-Boulevard Périphérique. During the remodeling process, more than one types of infrastructure were reborn and given new functions as part of the public space. The way the project was implemented made both the High Line in NYC and Seoullo 7017 in Seoul that appeared years later look all too unnatural.

The predecessor of Promenade Plantée is the Paris-Vincennes railway line built in 1858 to meet the transportation demand at the time of urban sprawl. Back then, the railway transported passengers in the daytime and freighted goods at night. In 1929 when the Paris Métro was under construction, the line began to decline (cause of death: emergence of alternatives / change in the industry). Sections gradually ceased operation between 1939 and 1969 until the entire line was abandoned.

For the future of a useless railway, the government at that time had two options: to tear it down or not to tear it down (to reuse it).

Option A: to tear it down

… so as to release space for new buildings. For the government, this option is more economic and convenient, and it can also heal the division between the northern and southern regions caused by the railway once and for all. Nevertheless, the viaduct was only 12 meters wide, too narrow to accommodate any new buildings. For this and other reasons, the government turned to option B.

Option B: not to tear it down

… and remodel it into an elevated park, a green belt with a view. Combined with the ample space under the arches, the whole structure can be used as a venue for lots of activities and events, and as an added bonus, the Avenue Daumesnil will gain popularity.

In this drawing is the first and the most famous section of Promenade Plantée: Viaduc des Arts, a 1.4 km viaduct made of 71 arches. In 1990, Patrick Berger's proposal "Le Viaduc des Arts" was chosen to convert the viaduct to a series of galleries. He gave the viaduct a thorough cleanup, walled off the space under the 64 arches with a glass wall on each side, then reno-vated the 64 vaults whose size varies between 150 and 450 square meters as galleries and cafés. The rehabilitation attracted the rich from the north who wouldn't have wanted to get across to the south. As a result, although the government kept the structure without tearing it down, the division was still healed.

A2 High Line Park, New York, USA
DESIGNER: Diller Scofidio + Renfro / James Corner

In 1934, the High Line elevated railway began to be used by freight trains to transport meat, dairy products and other produce directly from the port to factories without the disturbance of street traffic. Due to the construction of interstate highways during the 1950s, trucks gradually replaced trains to transport freight and the railway was face to face with its doom (cause of death). Eventually, the High Line was shut down in 1980. Factories along the line became desolate while weeds sprang up all over the tracks, making the area a criminal haunt. At the beginning, the government of New York City wanted to tear down this old viaduct and its surrounding facilities because they didn't exactly go hand in hand with the idea of urban planning and would obstruct the development of the Lower West Side of Manhattan. After more than twenty years of debate, during which the southernmost section of the railway, covering five blocks, was demolished under protest from the neighborhoods, a nonprofit organization called Friends of the High Line was founded in 1999 by two local residents. They advocated the preservation of the railway as New Yorkers' "secret garden" and held a design competition in 2003 to collect various opinions for the renovation. What they did not expect was that the canvassing of ideas that was meant to be of neighborhood scale turned out to be much more than that – they received 720 proposals from 36 countries. In the end, the joint proposal by NYC architects Diller Scofidio + Renfro and landscape architect James Corner won the competition.

High Line Park is 2.4 kilometers long and 9.1 meters above the ground, covering a total of 22 blocks from north to south. There are curvy concrete tracks, vegetation, fixed and movable seats, and lightings in the park. As for entrances to the park, they are scattered over the blocks in the form of stairways (some have an elevator for people with disabilities).

In 2006, the first phase of the construction began; in 2014, the whole line opened to the public. The entire project cost 153 million in US dollars, including designing and construction. The park is owned by the government of New York City and operated by the New York City Department of Parks and Recreation. In the 90s, the department established a public-private partnership (PPP) with Friends of the High Line, so the latter has been responsible for the operation and maintenance of the park. The organization raises its operating budget with private funds, 70 percent of which comes from donation.

High Line Park is a proof that grassroots organizations like Friends of the High Line can wield their influence on urban planning and raise private funds for public space, thereby enhancing the appearance of a city considerably. Today, this green belt built on top of an old elevated railway not only provides the citizens a multifunctional outdoor recreation ground, but also attracts numerous tourists. in turn, tourists bring development and create massive economic benefits with new jobs. It can be said that High Line Park is New York City's most important and influential open space project since the opening of Central Park.

A3 House of Vans London, London, United Kingdom
DESIGNER: Tim Greatrex

The Old Vic Tunnels, a ghostly industrial relic from the 19th century hidden beneath the Waterloo Station and evocative of Jack the Ripper, was an obsolete railway tunnel covering more than 2,500 square meters of space under the Waterloo Station in London. During WWII, the tunnels were used as a morgue and called the Necropolis Station. Later, the Old Vic producing house acquired the sections 228 to 232 from the British Railways and renovated them as an underground exhibition and performance center. It was open to the public in 2009 and had held many events for a few years since then, only to be closed down on the 15th of March, 2013, to many people's dismay.

In 2014, Vans won the bidding to operate sections 228 to 232 and refurbished them again into an underground skate park named House of Vans. Any forms of demolishing the original structure and the brick walls are prohibited throughout the entire space; while the walls remain intact, the floors were all renovated. The lighting and the signs on the walls highlight the history-rich bricks on which the history of the Vans brand is also displayed.

Now, the park operated by the street style brand, located in the tunnels beneath the Waterloo Station in central London, is a space with a distinct character – it is hard to find unless you know what you are looking for. Following the directions of the map, you will find the Leake Street first, originally a pedestrian tunnel under the tracks of the railway but gradually turned into a graffiti street where artists perform gigs, conduct street photography and basically just create. Then, in this street, you will find a small, narrow staircase that goes upward. Dubious you may be when you climb the stairs, but eventually it will lead you to the House of Vans, a free-entry skate park open to the public.

Vans intends to provide a cultural center for skateboarding, arts, movies and music. Taking advantage of the tunnels' layout, the whole space is divided into four main parts, each to fulfill one of the above functions. They are: an arts tunnel, an experimental gallery for artists where they can showcase their creations; a movie tunnel, including a cinema and a screening room; a music tunnel that can accommodate 850 people; and, finally, the skateboarding tunnels for skaters to display their skills. The general aim is to provide a venue that encourages creativity and to connect skateboarding with architecture – specifically, to demonstrate an association between a skateboarding ground and a tunnel structure.

There are three skateboarding tunnels for different purposes. The main one is made up of concrete "bowls" for professional skaters, the second one features a street setting for skaters with an intermediate level, and the third one is a mini ramp for beginners. All of the five individual tunnels in the House of Vans are covered with rubber flooring for unification. Before entering the skateboard area, visitors have to fill out an emergency contact form in case something happens. The first place visitors are going to arrive at is the mostly flat area for beginners and children. While heading to the professional area, they will pass a few holes on the walls through which they can have a preview of the bowls and get a glimpse of skaters showing their marvelous skills. Proceeding to the end of the first tunnel, visitors will arrive at the space composed of two big bowls. The highest point of the tunnel arch is nearly two and a half stories high, enough to engage in some serious skateboarding. By introducing these street activities, the architects managed to keep the characteristics of the old tunnels. This project not only explores the link between skateboarding and indoor spaces, but also combines a skate park with a derelict tunnel arch.

A4   A8ernA, Zaanstad, the Netherlands
DESIGNER: NL Architects

The Park of Highway A8 in the Netherlands (or A8ernA) designed by NL Architects is located in Koog aan de Zaan, a lovely little village by the river

Zaan near Amsterdam. In the early 1970s, Highway A8 was built in order to cross the river. Consequently, the fabric of the village was cruelly severed. The elevated highway is seven meters high and underneath it, there used to be a parking lot 40 meters wide and 400 meters long, forming a fault line that separated the church and the state. Indeed, there is a small church at one side of the highway and the former city hall at the other.

In 2003, the municipal council of Zaanstad decided to reconnect the two sides and invigorate the space under the highway by creating a recreation area and public facility to the local residents. Based on optimism, the government views the gigantic infrastructure as an opportunity instead of an obstacle. And it is an opportunity without doubt: due to the shape of the space and its proximity to the river, it can be seen as a large public corridor that satisfies all kinds of needs the residents may have. From a supermarket, a flower and fish shop and some entertaining facilities, to a park, a skate park, and a water sports marina, the space beneath the elevated highway not only unites both sides of the village, but also better connects them to the nearby river. Therefore, the space under a bridge has a lot of potential: it can even extend a church.

Two cross streets divide the corridor (the space under the bridge) into three sections.

The middle section is a covered square with a supermarket, a flower shop, a pet shop and a light fountain; to the east, across a street, there are a sculptural bus stop and a mini marina with a panoramic observation deck. The marina brings the water closer to the street and, in sunny days, fills the newly-constructed ceiling with lively light; the panoramic observation deck provides a special window for the residents.

The west section is a kid and teenage zone. There are a graffiti gallery, a skate park, a hip-hop dance stage, some tables for foosball and pin-pong, a 7-a-side soccer pitch, a basketball court and some loveseats.

In addition to the area right under the highway, there is also some public space near the city hall and the church, forming a series of public spaces along the highway. In front of the church, the authorities opened up the existing greenery square as a venue for outdoor fairs and celebratory occasions. At one side of the city hall, there is a new park with bowling lanes, a hill planted with birches, a barbecue area, a soccer cage, and shooting range under the ramp of the highway.

The local government invited the residents to collaborate with the developer and the end result is A8ernA. It is the cooperation between the local government and the community that is the key to its success in reusing a space neglected for more than thirty years.

A5   2k540 Aki-Oka Artisan, Tokyo, Japan
DESIGNER: JR East Urban Development Corporation

The project 2k540 Aki-Oka Artisan, born in December 2010, is one of the examples in Japan in recent years that have succeeded in reusing the space under old bridges. It is called 2k540 because the distance from Tokyo Station to the elevated bridge between Okachimachi Station and Akihabara Station is exactly 2 kilometers and 540 meters. 2k540 is an urban regeneration project based on regional development, carried out by JR East Urban Development Corporation, a subsidiary of East Japan Railway Company. In the past, the space under elevated railways was usually chilly, gloomy and even dirty, and it was the same for the one between Okachimachi and Akihabara.

Because the stations and tracks along railway lines are owned by railway companies (public and private alike), like JR East, Tokyo Metro Co., Ltd., Keio and Tokyu, it is not surprising that, to increase their revenues, there are department stores and shopping centers built into stations (such as Shibuya Station and Shinjuku Station) or under the elevated railway (like 2k540 and

the mall under the Tokyu Toyoko Line near Naka-Meguro Station).

Black and white are the prevailing tones in 2k540. White, clean shops are neatly arranged along the white pillars, while the black flooring with an asphalt texture extends the street to the space under the bridge. The ingenious design of embedded lighting on the floor to accentuate the whiteness of the pillars freshens up the once gloomy and dirty unused space and turns it into a bright and stylish place. With some tidy-up and design, all the wires and fixtures for the shops are displayed in a corner, as if they are some sort of art installation. Besides the artisan shops, the entire arcade feels like a gallery too: there are about fifty shops here, selling artworks made of various kinds of materials ranging from textile, paper, ceramic, leather and silver, hand-made craftworks in general and home roasting coffee, as well as bistros, connecting Okachimachi Station and Akihabara Station 886 meters apart with a leisurely walk.

In the Edo period, the area of Okachimachi was a small town home to samurais. The salaries of low-ranked samurais were often very meager, so they usually had to sell home-made products to local people in order to make ends meet. As time went by, particularly during the 1860s when Emperor Meiji started to promote industrialization and modernization, these samurais abandoned their original jobs and began to make and sell jewelry and other craftworks. Nowadays, a lot of senile artisans still live here and you can see many jewelry stores, leathercraft stores and the like on the street. However, most of the stores, like their owners, lost their vigor and opulence in the mists of time, so young people have stopped paying them visits. Therefore, the appearance of 2k540 means that both artisans of the traditional crafts and craftsmen of the younger generation now have somewhere to show their ardor. Every shop owner takes full advantage of the small studio and displays to the full the spirit of artisan unique to the Japanese culture.

A place for the innovation of traditional arts with an aim to facilitate regional development, 2k540 is not purely a shopping center. It values the participation of locals and thus holds various activities in the passages between shops and on the central square. The shop owners even work with the nearby market and organize local events to attract tourists and enliven the economy. This project is an absolute proof that urban leftovers can be reused to inherit the history and tradition of a place and incorporate community development, bringing back the prosperity of the region.

A6   Refurbishment Viaduct Arches, Zurich, Switzerland
DESIGNER: EM2N

A viaduct and its arches with a long history was turned into a trendy shopping center in Zurich.

All around the world, leftover space under bridges and elevated highways are being reused in the most avant-garde ways. Swiss design firm EM2N and landscape firm Zulauf Seippel Schweingruber have refurbished a series of 19th century viaduct arches in Zurich into the most exciting shopping street, introducing modern stores and restaurants into the space beneath the arches while maintaining the complete structure of the bridge at the same time.

The viaduct becomes a linear structure that connects; the infrastructure that was used as a railway is now a linear park which becomes part of the culture, work and recreation in this city. What used to be a barrier that separated was transformed into something brand-new that connects. The architects and landscape architects inserted in the leftover space – but not entirely, so that people can see the roof of the arches and natural light can enter the interior – structures made of black steel and with bubble-like skylights. There are a total of 36 such black structures under the viaduct, including galleries, shops and restaurants, while at the heart of the project lies the market hall with its bubbly skylights where 20 local farmers and food suppliers display and sell their products. All of these newly-built structures link and coexist with the old viaduct.

The buildings make use of contemporary and minimalist materials and designs so as to focus on the existing arches and exposed Cyclopean masonry. To keep the spotlight on the stonework, the architects set out some guidelines on the refurbishment: to limit the introduction of new elements to a certain degree in the purpose of accentuating the arches. When it comes to interior design, lessees can either choose from a series of default elements or design everything by themselves. As a result, the historic stone walls are kept intact and the bridge itself remains almost unaltered.

In this case, the leftover space under a historical building was transformed into a usable facility with cultural and commercial values. An architectural design competition was proposed in 2004, and the finished building was dedicated in 2010. Architects from EM2N claimed that they hoped to tackle two fundamental problems in this project: How can an infrastructure facility protected as a monument be reprogrammed and become part of the urban environment? How can a project be carried out with low budget nowadays when there are all kinds of regulations on energy, hygiene and fire control and when demands for comfort are getting greater and greater? In the 19th century, the elevated railway separated the downtown area of Zurich from the other neighborhoods; today, because of EM2N's design, the two parts are united by a cultural and commercial center.

A7 Huanhe S. Road Hardware Street, Taipei, Taiwan
DESIGNER: unknown

This friendly hardware street under the elevated bridge will soon be celebrating its hundredth anniversary.

Maybe it is because a river is close by and there are not many pedestrians, so goods and merchandise always occupy much of the pavement. However, this typical view of the riverside commercial district can actually be traced as far back as the year 1922, when Taiwan was still under Japanese rule. At that time, this area was right next to the Bangka Pier where cargos were transported here and there. Therefore, it is not surprising that it would become a commercial zone. According to the president of the autonomous organization, there are 135 stores in this area, with more than ten thousand pieces of merchandise. Things are not as orderly arranged as you may see in supermarkets, but customers love the feeling of treasure hunting and some even visit here just because of its reputation. Here, you can find auspicious red lanterns, disco-style neon lights, pinball machines, all kinds of unusual gadgets, electric saws, and even horns for ice cream trucks. The exact location of the hardware street is on Huanhe S. Road, a street where traffic is always busy. Back in 1923 during the Japanese rule, there were only two rows of bungalows, yellow dirt on the road, and wooden electric poles. Later, an elevated bridge was built over the street and shops selling hardware began to fill the space under it.

"The commercial district was officially founded on the 22nd of August in 1975," said Jiang Jinzhong.

After nearly 40 years since the autonomous organization was founded, there are now a total of 135 hardware stores. These stores all have a shabby look and they are not as spacious and bright as supermarkets, but their owners are not too worried about it because there are diehard fans. "They are cheaper and more professional. Anything we need, we can find it here," one of the regulars said.

A customer bought a blade and started assembling right on the spot, crouching there by the pavement. Presently, the apparatus was fixed. You don't have to throw away something because it is broken; you can buy the

parts and fix it yourself. This is the hardware street, with its long history, treasure trove, and the genuineness and friendliness emanated from its disorderly and casual way.

A8 Rio-Santos Highway Orangutan Jungle Park, Rio, Brazil (Proposal)
DESIGNER: Yaling Hou

Construction on the Rio-Santos Highway began in the 1960s and by 1976, the section of Viaduct Petrobras was to be linked with the existing section and became part of the Rio-Santos Highway. However, the plan was changed at the last minute due to a lack of funding. As a result, the existing section was connected with a coast route instead, while the newly-built but never used viaduct was given up. The asphalted viaduct is 300 meters long and 40 meters high above a jungle, and is characterized by retaining walls and gigantic concrete foundations.

The viaduct is situated in a vast swathe of natural environment. Because it has never been used before, nature has slowly crept up and clung to the structure. Conspicuous in such a surrounding, it begs us to rethink the values of man-made buildings and their ruins. A design proposal can turn the abandoned concrete into a guardian of the environment and a shelter for the animals.

The viaduct itself will be the focal point for animal conservation. There will be a gradual transition between different zones, and water bodies and greenery that form the boundaries of the whole park.

As the heart of the park, the viaduct must accommodate both human beings and animals. In this proposal, the designer makes use of the spatial properties of the bridge surface (horizontal) and the bridge pier (vertical) to divide the space into zones that are for humans only, for both humans and animals and for animals only.

Humans and animals can share the space through the inserting of new elements and their merging with the old ones. The man-made structure in a natural environment becomes a medium for a perfect human and animal symbiosis. Here, humans are no longer conquerors; they create a balance with nature by rethinking the meaning of construction and reversing the values of ruins.

B   Port facilities

B1   The Krane, Copenhagen, Denmark
B2   Maritime Museum of Denmark, Helsingør, Denmark
B3   Kraanspoor, Amsterdam, the Netherlands
B4   Faralda Crane Hotel, Amsterdam, the Netherlands
B5   Shipyard 1862, Shanghai, China
B6   Elbe Philharmonic Hall, Hamburg, Germany
B7   TWA Hotel at John F. Kennedy International Airport, New York, USA

B1 The Krane, Copenhagen, Denmark
DESIGNER: Mads Møller from Arcgency

Architectural studio Arcgency turned an industrial coal crane near the harbor of Copenhagen into a posh guest house – The Krane. It is a unique and luxury retreat where visitors can have an immersive, multisensory experience. Black is the main tone of the interior design, a reminiscence of its former identity as a coal crane.

In addition to the chief architect of Arcgency Mads Møller, there is another name that should be brought up as the driving force behind this project – the visionary developer Klaus Kastbjerg. Klaus Kastbjerg is the

current owner of The Krane who has also developed many other waterside projects in Copenhagen, including an apartment building that was once used as a silo. As a matter of fact, The Krane makes famous a nearby project Harbour House which is also owned by Klaus Kastbjerg – one of the reasons behind its rebirth.

Originally, the crane was not located where it is now. It was Jørn Utzon, the designer of the Sydney Opera House in Australia, who suggested that the crane be moved to a location with a more beautiful view and it was decided that it should be on the edge of Nordhavn, one of the last harbors that are under renovation in the capital of Denmark. The Krane is right beside the water and the access to it is through a drawbridge-like staircase.

The Krane is a multi-level structure, with the reception area on the ground floor, the conference room "Glass Box" on the first floor, the Spa and terrace on the second floor, and finally, the guest room with a lounge and terrace on the top floor. Each level and its facility can be rented separately. On the top floor is the 50-square-meter Krane Room with windows on all sides, so those who stay here can admire the sea, the sky, the harbor and a panoramic view of Copenhagen. Below the womb-like Krane Room is the spa area covered with slabs of grey stone from floor to ceiling, with huge glass curtain walls, so that guests can take in the grand view of the harbor and the sea while relaxing themselves in the two tubs.

Water is so important an element that it might constitute 80 percent of the experience staying at The Krane. With its rich history and beautiful view, this project succeeds in transforming a retired harbor crane into a modern space with soul.

B2 Maritime Museum of Denmark, Helsingør, Denmark
DESIGNER: BIG

Helsingør is a beautiful ancient harbor town not far from Copenhagen. In order to boost tourism, the local government launched the Kulturhavn Kronborg project which centered on the Kronborg Castle built in 1420 – the setting of Shakespeare's famous play Hamlet. In 2010, the former Elsinore shipyard nearby was renovated and reopened as Kulturværfte, or the Culture Yard; in October 2013, a 60-year-old dock situated between the castle and the yard was transformed into the Maritime Museum after 5 years of remodeling.

Out of respect for the Kronborg Castle, the architects made the museum almost wholly invisible on the ground. They turned an old concrete dry dock into a modern museum, adopting techniques never used in Danish architecture before and building everything below sea level. The 1.5-meter-thick concrete walls and 2.5-meter-thick concrete floors of the dock were cut open and rearranged, resulting in a structure precise as its steel bridges. A series of multi-level bridges cross the entire dock, not only connecting with the city, but also serving as passageways leading tourists to the different sections of the museum. The first bridge closes off the dock while functioning as a harbor promenade; the second bridge above the auditorium links with the adjacent Culture Yard and the Kronborg Castle; and the sloping and zig-zag third bridge takes the visitors to the main entrance. This last bridge unites the old and new, so when visitors are descending into the interior of the museum, they can overlook the grandeur of the space of the dock and how it incorporates the exhibition spaces, the auditorium, the lecture hall, the offices, and the café.

A two-story rectangular structure built along the walls of the old dock forms the underground galleries, displaying the maritime history of Denmark to the present day. The ship-shaped cavern is the heart of the exhibition. It is an outdoor area where visitors will have a chance to comprehend the sheer scale of ship-building and where the old dock full of stories is kept intact and renovated as a courtyard, bringing in light and air to the center of the submerged museum.

The museum is an independently operated substitute approved by the government and funded by eleven foundations. It has an underground area of 7,600 square meters and the construction cost USD 54 million. In this project, Denmark not only creates a world-renowned maritime museum and culture center, but also successfully advertises its glorious history and contemporary status as one of the world's leading maritime nations.

B3   Kraanspoor, Amsterdam, the Netherlands
DESIGNER: OTH Architecten

In the past, the port and industrial district at the west side of Amsterdam belonged to the Nederlandsche Dok en Scheepsbouw Maatschappij (NDSM). In this huge area, there is a concrete crane track designed and built in 1952 by JD Potsma to serve as a dock and a platform along which two cranes can move. However, during the 1970s, many docks fell into disuse because the shipbuilding industry was in decline (cause of death). Eventually, NDSM was shut down in 1978 and the permit for the demolition of the crane track was issued.

In 1997, architect Trude Hooykaas of OTH found that one can enjoy a commanding view on the old track and thus had a vision of a modern office building on top of it. As a result, the government decided not to tear down the track. Instead, they revised the overall plan to allow for the possibilities the "gold coast" may offer. This crane track is a relic of Amsterdam's shipping industry. It is composed of two horizontal beams and four double portal frames that support them from underneath. Based on this, architect Trude Hooykaas designed a three-story transparent office building that floats on top of the existing structure, with more than 10,000 square meters of unique working space where one can not only see the crane track of old from inside, but also absorb an uninterrupted view of the waters of Amsterdam.

The challenge is the limited load capacity of the original structure. To overcome it, the designer chose the lightweight steel as the construction material, which is easy to make the new structure transparent too. The glass-walled building has the same length and width of the track – 270 meters long, 13.8 meters wide – and is 10 meters tall, supported by slender steel columns three meters long. There is a movable outer layer of glass louvers, so its transparency is completely shown to outsiders. From the entrances, one can take the stairways and elevators that offer a panoramic view and be led to the upper floors.

Nowadays, the protected industrial heritage has a new purpose – it is not just an impressive building, but also a green one. Water from the IJ river is used for cooling in summer and heating in winter, there is a ventilation system through the opening on the floor, and it is also a breeding ground for wild birds. Old and new merge seamlessly in this project. As the remains of an industrial past and a modern architecture, Kraanspoor brings back the bygone days without losing any liveness.

B4   Crane Hotel Faralda NDSM, Amsterdam, The Netherlands
DESIGNER: IAA Architecten Amsterdam

The Faralda Crane Hotel is the first hotel in the world that uses a crane to be its structure. Containers were inserted into the frame of the 50-meter-high and 250-tonne crane as guest rooms. Considering the limited space of the crane frame and the difficulty in making much alteration, the hotel has only three rooms. Nonetheless, there is an attic in each room and the rooms are all decorated with different styles – luxurious, modern and

industrial. On the exterior, the original color of the steel was recovered while all the new elements were painted red, so there is a clear distinction between old and new.

The crane, numbered 13, was built in the 1950s by Hensen and belonged to the shipbuilding company Nederlandsche Dok en Scheepsbouw Maatschappij (NDSM). After the bankruptcy of NDSM in 1984 (cause of death), the crane had been idle. Twenty-five years later, the dilapidated crane number 13 faced a fate met by many other old cranes and was going to be blown up or torn down since it posed a threat to public safety due to lack of maintenance. Luckily, developer Edwin Kornmann Rudi saved it from destruction in time and gave it a new life. In August 2011, he obtained a permit granted by the government of Amsterdam and started making plans to reconstruct this steel giant. On the 7th of June, 2013, the National Restoration Fund signed a contract with Edwin Kornmann Rudi. Meanwhile, the government of Amsterdam approved his proposal about the renovation and the relocation of the crane.

The project is highly-acclaimed because it provides unique solutions to a lot of practical architectural and construction problems. For example, the frame of the tower that encases the three suites continues to spin around the pivot bearing, but in the tiny opening of merely a few centimeters, all pipe work, the drag link as well as the fire and safety measures work smoothly. On 22nd of July, 2013, the crane was disassembled and then transported to Franeker, to be restored on a 100-meter-long floating bridge. Then, on 22nd of October, 2013, it returned to Amsterdam fully restored. Finally, on the 4th of April, 2014, it officially became the Faralda Crane Hotel.

On the ground floor, there is a newly-built entrance next to the crane. A lift and a staircase leads to a platform at ten meters and the transition level to the upstairs suites. The ten-meter-high platform houses a studio and offices. From the transition level, guests can reach three suits via another lift. The spa is located at the top of the hotel. For those who love extreme sports, it is even possible to go bungee jumping at the end of the cantilever. From their rooms or spa at the top, guests can look out at the busiest streets of Amsterdam or its famous river view. It surely is an unforgettable experience for anyone.

B5   Shipyard 1862, Shanghai, China
DESIGNER: Kengo Kuma and Associates

Shipyard 1862, located in Lujiazui in the district Pudong of Shanghai, was a shipbuilding factory originally built in 1972 and is now an arts and commercial complex. The number 1862 points out a significant year in the history of China. It was in this year, when western countries forced the Chinese government to open up its ports and engage in commerce during the late Qing dynasty, that the British built Xiangsheng Shipyard in Lujiazui, Pudong, ushering in the glorious era of shipbuilding in the Shanghai Bund. Initially, the shipyard was used to manufacture ammunition; later, it was used to build ships. By acquiring several shipyards in the Puxi area and then taking over the German Ruirong Shipyard during World War I, the British expanded its shipbuilding industry in China rapidly. After the Pacific War broke out, the shipyards changed hands a few times, passing their ownership from the UK to Japan, then to the National Government of the Republic of China. Eventually, it became part of the Shanghai Shipyard Co., Ltd.

In 2005, after the entire Shanghai Shipyard was relocated to Chongming (cause of death), developers worked with the China State Shipbuilding Corporation and turned the site, covering 1.36 million square meters, into a finance and trade zone. This particular shipyard that is now Shipyard 1862, being closest to the river, was treated as the heart of the development and expected to reinvigorate the economy of Lujiazui through rebirth.

Teaming up with Arup, Kengo Kuma and Associates was commissioned to covert the shipyard and give it a poetic new life. Shipyard 1862 is an arts center featuring a theater. The industrial atmosphere is not lost in the 26,000-square-meter indoor mall in the five-story building (plus one level of basement) where you can find shops, restaurants, galleries and other secondary facilities, along with the theater with eight hundred seats and a panoramic view of Puxi. The whole project covers an area of 31,600 square meters.

The original structure is well-preserved because the designer adopted a strategy of renovating the old and the new separately. This also makes the original structure more like an exhibit on display. However, the introduction of necessary machinery and electrical appliances for the multiple purposes of the building was complex and difficult to solve, since it has to fulfill modern needs while trying not to offset the visual effect achieved by all the history. In addition to the design of the space as a whole, the architect reused some of the fixtures in the shipyard, such as the old steam pipes that were transformed into air-conditioning vents. The advice of theater experts was sought to ensure everything was in line with the special function and layout of a theater.

The façade is the most impressive element of the project. The existing beams and part of the brick walls are kept while lightweight and modern glass curtain walls were added to these nostalgic elements. On the outside of the walls, there is another layer of pixel-like brick system suspended by steel cables. Overall, the façade links the old and the new surfaces with an innovative yet antique design.

B6   Elbe Philharmonic Hall, Hamburg, Germany
DESIGNER: Herzog & de Meuron

Where is the most expensive concert hall in the world history? And the most expensive renovation project of industrial heritages too!

In 2001, an urban regeneration project, the biggest of its kind in Europe called HafenCity, was launched on an island of the Elbe in central Hamburg to transform the old port. Within HafenCity, there is Speicherstadt (literally "city of warehouses"), a UNESCO World Heritage Site that is maintained very well and still in use. One of the warehouses is the red-brick Kaispeicher A built between 1963 and 1966 by Werner Kallmorgen. For nearly fifty years, it was used for the storage of cocoa and tea. As such, the warehouse was designed to bear the weight of thousands of bags of cocoa. Now a new sight in the daily life of Hamburg citizens and a cultural center for world travelers, the Elbe Philharmonic Hall reborn from the former warehouse attracts attention to a location that most people never really took notice before.

The architecture firm Herzog & de Meuron proposed winning more space by "layering the new on top of the old." That is, keep the entire warehouse to be repurposed as a multistory car park and make the concert hall floating on top of it. The existing red-brick walls of the lower part and the steel structure of the upper part provide a sharp contrast. Because of the high bearing capacity of the original structure, it is able to cope with the vertical load of the new steel structure. The structural engineers added 1,761 reinforced concrete pillars beneath the water, each of which has a bearing capacity of more than 200 tonnes. The existing red-brick warehouse is 37 meters high, which used to be the average height of buildings at the port of Hamburg. On top of it, the architects designed a gigantic iridescent glass structure that is modern, lightweight, transparent, and looks like waves and icebergs at the same time, with the highest point at 70 meters. The exterior keeps changing its appearance according to the reflection of the sky, the waters and the city on the glass. The highest point of the entire building is 26 stories high, or 110 meters high, the total floor area is

120,000 square meters, and the total weight is 200,000 tonnes.

To solve the inconvenience brought by the concert hall's location on the periphery of the city, the government of Hamburg not only constructed a bridge that connects the old town and Sandtorhafen, but also extended the MRT U4 line closer to the area. Besides, there are many public transportation and ferry stops in front of the Elbe Philharmonic Hall. The Elbe Philharmonic Hall is not just a place for music lovers; it is a full-fledged accommodation and cultural complex too. In addition to the Grand Hall with 2,100 seats and the Recital Hall that can house an audience of 550, there are luxury apartments and a five-star hotel with a restaurant, a gym and a conference center. To make the concert hall a real tourist attraction for all, there is even an observation deck where you can enjoy a panorama of Hamburg. With all these different functions combined in one building, the Elbe Philharmonic Hall feels like a micro-city.

The Elbe Philharmonic Hall has become an iconic architecture not only of Hamburg, but also of the entire Germany. However, the most well-known aspect of the project is not its design, but the long construction period and the enormous cost and effort to accomplish it: the construction time was delayed for seven years, the budget increased tenfold, and there were a number of controversies and lawsuits to deal with while overcoming safety and technical issues that continually occurred. In 2007 when the warehouse was put out to bid, the government of Hamburg allocated a budget of 77 million euros. But after Herzog & de Meuron won the bid, the budget was increased to 114 million euros. In 2012, the final budget was set at 575 million euros, but in the end the entire construction cost 789 million euros! The construction time was also extended from three years to ten years. For many years, the construction site had been surrounded by huge cranes, so that during the last years of construction, some of the citizens even proposed to keep a few scaffolds because after all those years, they had already become part of the cityscape. On the 11th of January, 2017, the Elbe Philharmonic Hall was finally inaugurated and held its first concert.

B7 TWA Flight Center Hotel, New York, USA
DESIGNER: Beyer Blinder Belle & Lubrano Ciavarra Architects

The Trans World Flight Center located in John F. Kennedy International Airport, New York, USA was designed for Trans World Airlines by architect Eero Saarinen and launched in 1962. The thin shell of the concrete roof that seems to spread outward and upward like a pair of wings, the streamlined structure flowing from the interior to the exterior, the unusual tube-shaped red carpet corridors, the skylights inlaid in the slits of the roof, and the glass façade with a panoramic view where people can see the departure and arrival of the planes – all of these details form an inseparable unit that makes the flight center the symbol of the aviation industry in the mid-20th century.

The TWA Flight Center was recognized as a New York City Landmark in 1994 and listed in the National Register of Historic Places of the United States in 2005. However, due to Trans World Airlines' financial difficulties in the 1990s and the recent change in the operation patterns of the aviation industry, the flight center faced closure in 2001. In 2008, a new terminal, Terminal 5, was completed. To make space for it, the peripheral structure of the original complex was removed while a newly-built crescent-shaped structure partially encircles the site to be used as the actual terminal for JetBlue Airways. At that time, the main building of TWA Flight Center was still being renovated and its future fate was not determined yet.

Fifteen years after its closure, the governor of New York Andrew Cuomo, along with developer MCR/Morse and two architecture firms Beyer Blinder Belle and Lubrano Ciavarra Architects, launched an eagerly-anticipated renovation project, aiming to protect and restore the flight center and to

reflect the aesthetics of 1962 in its new design. This project incorporates a new hotel on both sides, the refurbishment of the main structure, the interior design by Charles Eames, Raymond Loewy and Warren Platner, and a museum that exhibits the role of New York City as the birthplace of the jet era, the history of Trans World Airlines, and the modern design movement.

The TWA Flight Center started its new life in 2019. After renovation, the old terminal now houses a lobby, a commercial area, restaurants and bars; on the plaza outside the lobby sits a 60-year old Lockheed Constellation airplane L-1649A that is turned into another bar. Passengers from all terminals can take the AirTrain and then walk through the tube corridors designed by Saarinen to reach the hotel and pay a visit to the classic TWA Flight Center more than half a century old.

C Mines

C1   La Fábrica, Sant Just Desvern, Spain
C2   Salina Turda, Turda, Romania
C3   Eden Project, Cornwall, United Kingdom
C4   Seaventures Dive Rig, Sipadan, Malaysia
C5   InterContinental Shanghai Wonderland, Shanghai, China
C6   Oil rig, Scotland, United Kingdom (proposal)

C1 La Fábrica, Sant Just Desvern, Spain
DESIGNER: Ricardo Bofill

This is probably the dream of all architects: to convert an old cement factory to a place for living and working, like what the Spanish architect Ricardo Bofill did.

In 1973, Ricardo Bofill and his team came across an obsolete cement factory in a town near Barcelona. He found that the factory was in quite a good condition. Most people might have found an abandoned factory a gigantic concrete waste not worth a second glance, but for Ricardo Bofill and his team, it was full of potential. The factory, deserted and partially in ruins, was a combination of surrealistic elements: powerful reinforced concrete structures crawling here and there, iron blocks hanging in midair, and vast space filled with magic.

The architect and his team decided to buy the factory and renovate it, turning it into his own residence and a studio for the team. The first phase took two years, starting from tearing down part of the old structures and leaving what remains now intact. It is as if the concrete was sculpted by an artist. Once the entire space was redefined, with the old excess removed and the new greenery taking root, the process of "adapting the old building to its new functions" officially began. The eight silos were left untouched; only part of the inner walls were demolished to obtain the open space for the studio. The enormous entrance of the factory remains intact too, to introduce natural light into the spacious interior. The silos were repurposed and became offices, a models lab, an archive, a library, a projection room, and a huge space named "The Cathedral," used for exhibitions, concerts and other cultural activities related to the architect profession. Landscaping elements are integrated with the new functions of the old building in accordance with the architect's will.

After 45 years' continual renovation, La Fábrica has been transformed from magic to reality. Its roof and surroundings are covered with dense vegetation, while the original forms and architectural styles of the different sections of the factory remain intact, with new functions. Overall, the project proves that an imaginary architect can find new use for any space, no matter how different it may be from the original one.

C2 Turda Salt Mine, Turda, Romania
DESIGNER: Ecopolis

This impressive underground space used to be a huge salt mine in Turda, Romania. Now, it has been turned into the largest underground amusement park and museum of salt mining history in the world. From the Middle Ages (it was first mentioned in 1075) to the year 1932, the mine had been producing table salt. The purity of the salt extracted in Salina Turda was as high as 80 percent, but because the equipment was not sophisticated enough here and the amount of production kept decreasing, it eventually lost the competition to another Transylvanian salt mine and stopped extracting salt in 1932 (cause of death).

In 1992, Salina Turda was reopened for tourism and therapy. The machinery from the Middle Ages is in good condition and worth seeing. In 2009, with the financial support from the United Nations to stimulate local tourism, Salina Turda began extensive renovations. After two years of construction and an investment of six million euros, Salina Turda was completely transformed. Now, there are a spa treatment room with natural aerosols, an amphitheater, a gym and a small panoramic Ferris wheel on which tourists can admire salt efflorescences and stalactites.

The lakes of Durgau on the surface of the mine is a popular destination that attracts a lot of tourists all year round. From there, the tourists take a lift down the shaft that once transported thousands of tonnes of salt and appreciate the sheer scale of the old mine in the process. Reaching the bottom of the shaft, visitors will find an underground lake and a space covering with sand-like salt. The facilities on the island on the lake look like they are inspired by the aesthetics of some deep sea creature, while man-made structures at the bottom of the shaft are mostly made of timber and brightened up by hanging tube lightings. Inside the mine, the temperature is maintained at twelve Celsius degrees and the humidity at 80%. With a unique micro-climate almost devoid of any allergens and bacteria, the mine becomes a great place for those suffering from allergic respiratory diseases.

There are actually three mines in Salina Turda: the lowest point of the whole mine is in the Terezia Mine (120 meters deep); the Anton Mine is 108 meters deep; and the Rudolf Mine is 42 meters deep. In each of these mines, you can see the layers of rock that look like solidified fluids formed over a very long time span. Every structure in the mine was returned to its original location after being renovated, so as to keep intact the underground landscape carved by industrious workers for hundreds of years. The layout is determined by the paths that have been preserved all these years.

Different sections of the amusement park are either converted from existing structures or designed with uniform elements. The amphitheater, sports ground, Ferris wheel, mini golf course and bowling alleys display various aspects of the mine; stalactites and salt efflorescences complete the inert equilibrium of the giant bell of the Terezia Mine; the underground lake at the bottom of the mine has a depth between 0.5 and 8 meters; in the center, there is an island formed from residue salt deposited here since 1880. You can even enjoy the fun cruising the lake at the lowest point of this huge conical bell on a rented boat.

From 2012 to 2014, a tunnel 50 meters long was built to connect the Terezia Mine and the Losif Mine. The deepest point of the mine is at 120 meters. The underground salt mine of Salina Turda is a good example that ingeniously repurposes an artificial underground structure and at the same time makes it authentic to the original. Artificial it may be, but can't we say that it is also rendered very natural?

C3 — Eden Project, Cornwall, United Kingdom
DESIGNER: Nicholas Grimshaw

The Eden Project, located in the county of Cornwall, United Kingdom, is the largest indoor botanic garden in the world and one of the most popular visitor attractions in Britain. Viewed by some to be the Eighth Wonder of the World, it is situated in a pit as big as 30 football pitches. More than just a tropical garden, the Eden Project links plants and humans, symbolizes our obsession with flowers and trees, and preserves relevant knowledge and resources. After more than 160 years of china clay extraction, the pit where Eden sits now was exhausted in the mid-1990s (cause of death) and left behind, obsolete and useless. Architect Nicholas Grimshaw, confronting with the ever-changing landscape of the pit, came up with the idea of bubbles. Bubbles are pliable and can adapt to any surface; when two or more bubbles link together, the seams are always straight and at right angles to the surface. Therefore, installing bubble-like biome structures on uneven surfaces like the bottom of the pit is the perfect way to build a greenhouse.

Based on this idea, the Eden Project is made up of multiple giant bubbles and designed to emulate natural biomes. The domes of the greenhouses are composed of hexagons and pentagons supported by hundreds of steel frames. To make them lightweight, each of the window within the hexagonal or pentagonal space is filled with the material ethylene tetrafluoroethylene (ETFE). Once inflated, it forms a 2-meter deep transparent cushion. The ETFE windows are lighter than their glass counterparts, but they are strong enough to support the weight of a car. Besides, they can transmit UV rays from the sun and are easy to clean, with a longevity of more than 25 years.

The three main domes belong to the Outdoor Gardens, the Rainforest Biome, and the Mediterranean Biome, covering a total of 13 hectares. The largest of the three is the Rainforest Biome, where visitors can have a taste of the four major rainforest environments in the world: tropical islands, Southeast Asia, West Africa and South America. There is also a Canopy Walkway that offers a treetop view across the entire biome; visitors can take a walk and take in the grandeur of a tropical rainforest. The first part of the Eden Project, the visitor center, opened to the public in May, 2000; the whole site didn't open until the 17th of March, 2001.

In addition to contribute over £1 billion to the economy of Cornwall where it is located, the Eden Project is also a pioneer in green energy – its power comes from local wind turbines and geothermal plants. It is named the Eden "Project" rather than the "Garden" of Eden because like many landscape projects, it continues to grow and change, with every new phase and new objective symbolizing the ongoing regeneration of the land.

C4 — Seaventures Dive Rig Resort, Sipadan, Malaysia
DESIGNER: Morris Architects

Called The Rig back then, Seaventures Dive Rig used to be an oil platform moved from place to place to extract petroleum and natural gas. It was constructed in Panama, owned by Petronas Oil Corporation of Malaysia, and then sold in Singapore after retirement in 1985.

The retired rig showed great potential of being converted to a hotel, and since there was no need to chop down a single tree or need for earthworks in the middle of the ocean, the rig can be renovated without destroying marine life. The challenge was to turn an industrial oil platform into a comfortable hotel. In the end, there are air-conditioned guestrooms furnished with toilet and hot shower, and a central lounge where the guests can enjoy fresh air and occasional live performances (because it's a very spacious area!). Initially a hotel that focused on fishing, it transformed into a diving

resort. The Suzette Harris' family transported the rig from Labuan to one of the top dive destinations in the world eight hundred kilometers away. The new location is off Mabul Island near Sipadan, an island and world-renowned divers' heaven where you cannot visit unless you've got a permit issued by Sabah Parks.

The rig has stayed where it is now since 1997 and the structure has become a shelter and haven for the fish. Although not a luxurious type of resort per se, this industrial hotel and its twenty-five rooms does provide all the facilities you will need for an accommodation. Meals, transfers, dives and equipment rental are all included in the package. Without doubt, the Suzette Harris' family has expanded a small diving hotel into a popular and famous holiday resort.

The reuse of a retired oil platform is unprecedented. However, the hotel itself and the journey to it are not entirely environmentally-friendly. Apart from the fact that the power of the rig comes from a foul diesel generator, it is very energy-consuming to get there: first, you need to take a flight to the capital of Malaysia, Kuala Lumpur. Then, you have to take another flight to Tawau and arrive at the eastern coast of Sabah on the island of Borneo. Once there, a one-hour trip takes you to the small port city Semporna. Finally, you can get on a ship and reach the rig an hour later. One day, maybe another hotel reborn from an oil platform can try to install a solar array or wind turbines.

Having said that, it is still pretty cool to discover an extreme recycling project like this.

C5 — Inter Continental Shanghai Wonderland, Shanghai, China
DESIGNER: Martin Jochman

Shanghai is a place with no mountains and rocks. However, the pit of Tianmashan used to be a mountain, with Sheshan to the northeast. The mountain was blown up to excavate its rocks and by the end of 1950s, the whole mountain was gone. In 1959, the people's commune in the area dug another quarry in Xiaohengshan. In the wake of urban development and economic growth, demand for stones increased. The quarry became bigger and bigger while the pit became deeper and deeper. In the end, a huge pit more than eighty meters was formed. In 2005, Shimao Group bought the biggest of the seven pits in the Sheshan area. They chose not to erase the history but to acknowledge that the destruction of the original landscape is a part of civilization. They decided to take advantage of the features of a pit and be creative, and the result is a five-star hotel that blends perfectly with its surroundings.

The project began in 2006 after the proposal from the British architecture firm Atkins won the international design competition. Unlike traditional hotels, InterContinental Shanghai Wonderland was built below ground level. This brought a lot of technical difficulties, such as preventive measures regarding fire, flood and earthquakes, to name but a few. After seven years of scientific research, the construction finally started in 2013.

The entire building is about seventy meters high with nineteen stories of which three are above the pit and sixteen are in it (two of them are underwater). There are guest rooms, conference rooms and restaurants. Each guest room was built according to the shape of the cliff with mainly steel plates and glass where the guests can take in a panoramic view of the cave. The designer also created an artificial lake in the abandoned pit and, through a pumping system, makes sure that the water level never rises by more than five centimeters to guarantee safety. As for the prevention of fire, since fire engines cannot reach the bottom of the pit, there is an access point on every balcony of the hotel.

There are 337 guest rooms and one presidential suite. For standard

rooms that are located in fourteen stories above the water level, each has a balcony connecting to the corridor to form a "sky garden." From their rooms, guests can see on the northwest face of the cliff the nearly 100-meter waterfall, one of the most impressive features in this project. Those with a spirit of adventure can even enjoy rock-climbing. Suites and the presidential suite are situated in the lowest two stories beneath the water level with a marvelous underwater view that makes you feel like you are in the sea.

Martin, the designer of InterContinental Shanghai Wonderland, believed that it was one of the most remarkable examples of turning trash into treasure to repurpose a deserted industrial relic that was also a scar on the surface of the earth and turn it into a brand-new architecture. During construction, the engineers had to tackle many issues. For example, to solve one of the problems, they developed two major systems more than 77 meters long to pump concrete. The construction took ten years to complete and mobilized more than five thousand architects, engineers, designers and workers. The team filed a total of 39 patents and made hundreds of technical breakthroughs. InterContinental Shanghai Wonderland reuses an old quarry and thus becomes another successful enterprise that aims at solving the problem of industrial relics in China. It is also the lowest hotel in the world.

C6 — Oil rig, Scotland, United Kingdom (proposal)
DESIGNER: Liwen Wang

Oil platforms are large structures on the sea used to extract petroleum and natural gas and providing temporarily storage before crude oil is transported to refineries on land and further manufactured into petrochemical products.

On oil platforms, there are extraction facilities and accommodation space for the workers. There are different types of platforms for different needs, such as mobile rigs and fixed platforms anchored onto the seabed. Platforms can be interconnected to form a group.

In this proposal, the platform that is to be renovated is of the semi-submersible type. It has a drilling unit with columns and pontoons that can be submerged to a certain depth by flooding the tanks. Although the drilling part is submerged, the whole structure is still floating. When the lower hull is filled with water, it provides stability for operations on the sea.

This proposal is inspired by the abandoned oil rigs off the Scottish coast. In recent years, due to environmentalism and the ever-decreasing oil price, more and more oil rigs are falling into disuse. They cost a huge amount of money to construct, but to demolish them is also very expensive. Therefore, after retirement, they are mostly left there on the sea. When reusing this kind of infrastructure, we should make good use of its existing equipment and fixtures.

Since oil platforms can be moved from place to place, each of them can be designed to fulfill a different function (such as housing, doing research and meeting daily needs) and then connected with one another via bridges renovated from existing structures to form a city on the sea, creating a new place for humans to live outside the land. Furthermore, these platforms can utilize bioenergy, solar energy and tidal energy that are available on the sea, thus achieving self-sufficiency and autonomy.

First of all, the designer carried out a thorough analysis of the function and equipment of the platforms. Essentially a huge piece of machinery, the platforms show great potential for renovation and reusing. After conducting a careful research, the designer rearranged and redesigned those objects, equipment and spaces that can be kept and given new purposes, so that not only can they become useful again, but the cost of discarding them can also be lowered considerably.

By reusing and finding new purposes for existing facilities, the designer solved the problem of either having to demolish the oil platforms at a high cost or leaving them on the sea and pollute the environment. The rigs can be turned into self-sufficient units for research and housing. If the proposal can be put into practice for all oil platforms that are going into retirement in the future, it is likely that a new lifestyle will be created.

D Power Supply

D1   Wunderland Kalkar, Kalkar, Germany
D2   Tate Modern, London, United Kingdom
D3   Kaohsiung Metropolitan Park, Kaohsiung, Taiwan
D4   Ronghua Dam, Taoyuan, Taiwan (proposal)

W Water Towers

W1   Watch/Watertower, Sint Jansklooster, the Netherlands
W2   Water Tower Pavilion, Shenyang, China
W3   Water Tower House, Brasschaat, Belgium
W4   Biorama Projekt, Brandenburg, Germany
W5   House in the Clouds, Suffolk, United Kingdom
W6   Nagyerdei Víztorony, Debrecen, Hungary

D1   Wunderland Kalkar, Kalkar, Germany
DESIGNER: Hennie van der Most

This amusement park is sitting on the site of a former nuclear power plant in Germany that had cost 5.3 billion US dollars to build. Now, it attracts several hundred thousand visitors every year.

Wunderland Kalkar is located in North Rhine-Westphalia, Germany, north of Düsseldorf. Its predecessor was SNR-300, the first fast breeder nuclear reactor in Germany that began construction in 1972. It was designed to be fueled by plutonium and cooled by sodium and can output 327 megawatts. However, compared to the other more common types of reactors, the risk of nuclear disasters was also higher.

After the Chernobyl disaster in 1986, the locals of Kalkar became even more concerned about the existence of the nuclear reactor. Protests against it kept delaying the construction to the extent that it became a political nightmare (cause of death). When the plant was finally completed, it had cost 5.3 billion US dollars, but due to the enormous operating cost and the ongoing protests from the locals, the authorities eventually decided not to take it into operation. As a result, the nuclear power plant became the most sophisticated and expensive waste there ever was in the world. The money spent is enough to build twenty thousand houses, the concrete used enough to build a highway from Amsterdam to Maastricht (the complex is as big as 80 football pitches), and the wire installed long enough to loop around the globe twice when link together.

In the end, the plant was sold to a Dutch investor, Hennie van der Most, at a rumored price of US$3 million. He decided to renovate it as an amusement park and named it Wunderland Kalkar. In this 136-acre park, there are 40 different rides, including a log flume similar to the famed Splash Mountain in Disneyland. The architect utilized the spatial arrangement of the plant to create a vertical swing inside the cooling tower and a 40-meter-high climbing wall on its exterior. Many facilities that were meant for the plant are incorporated in the park and its attractions. There are also four restaurants, eight bars and six hotels (one of them has 450 rooms) in the amusement park. Visitors do not have to worry about radioactive con-

tamination at all because the plant never went online.

The nuclear power plant was reborn and has a new life now. Every year, it receives about 600 thousand visitors from all over the world. After the Fukushima Daiichi nuclear disaster in 2011, the German government decided to gradually close down all the nuclear power plants in the country and make Germany a nuclear-free nation by 2022.

D2   Tate Modern, London, United Kingdom
DESIGNER: Herzog & de Meuron

Tate Modern is successful in three aspects: the reuse of desolate infrastructure, the development of the waterside area, and the exhibition and maintenance of modern art.

It is hard to believe that, in the last 20 years, one of the driving forces behind London is an old power station, or what they call Tate Modern nowadays. The former power station, located on the bank of the River Thames, was designed and built by architect Giles Gilbert Scott near the end of WWII, but it was shut down in 1981. Due to the economic development, the station faced a fate to be torn down. However, through many people's efforts, it was kept untouched and then repurposed as a modern art gallery.

In 1995, Jacques Herzog and Pierre de Meuron from the Swiss architecture firm Herzog & de Meuron beat all the other well-known competitors in the international design competition. They were acutely aware of the potential of the power station and proposed to make minimal changes on its exterior, which was in agreement with the vision of the museum: when designing the art gallery, make use of the original spaces and structures as much as possible. For example, they stripped off all the electricity generators and created a grand space called the Turbine Hall. It is used as an exhibition space for large works and a vast hall for visitors to hang around.

Herzog and de Meuron kept most of the station's original layout, structure and the names of different rooms. The main building can be divided into three parts: the Boiler House, the Turbine Hall and the Switch House. On the basement floor of the Switch House is the Tanks while on the top of the huge chimney on the north side there is the "Swiss Light," a glass corridor sponsored by the Swiss government that runs the full length of the building.

In 2004, the extension project of Tate Modern began. The plan was to build a sort of revolving pyramid on the south side to accommodate more visitors. It cost 260 million pounds to build the ten-story extension that added 20 thousand square meters of exhibition space for the museum. In 2012, three large oil tanks from the old power station were temporarily open to the public; they form another type of exhibition space unique to the art gallery. In 2016, the Tanks reopened. Now, they not only provide extra exhibition space but are also used to hold classes and DIY workshops.

More than an icon in London, Tate Modern has also played a crucial role in reinvigorating the waterside area of the River Thames. With the Millennium Bridge linking Tate Modern on the south and St. Paul's Cathedral on the north of the river, the area has become very popular with tourists, attracting numerous of them each year. It can be said that Tate Modern is one of the most successful art galleries with the most visitors in the world. It transformed London and has an impact on its urban design, art, culture and social life.

D3   Kaohsiung Metropolitan Park, Kaohsiung, Taiwan
DESIGNER: Cosmos International Inc.
Planning & Design Consultants

Most infrastructure won't be infrastructure once reborn. This project is

one of the few cases when an infrastructure facility was given another infrastructural purpose. And it should totally be considered the pride of Taiwan, being a landfill that was turned into a park and biogas power plant.

The whole park measures an area of 95 hectares. A landfill used to cover 48 hectares of the current park, but it was closed down in 1999 because incineration had become the more popular method of disposing waste in favor of environmentalism (cause of death). It was the first landfill that had been reused in Taiwan. In 1989 when authorities started to think about building a metropolitan park, besides the main goal to create a place for leisure activities, they also wanted to improve the landscape and the environment of the region. The two phases of the construction were complete in 1996 and 2009 respectively; since then, the park has become a new site for local people to do exercise or go cycling. Of the two phases, the aim of the second one was to reuse the old landfill. To that end, a slurry wall and an impervious bed were constructed to serve as insulation from the waste and the earth right above it, and also to prevent any collapse or permeation. Wastewater is directed to the sewage treatment zone so as not to pollute groundwater. The original waste and the earth that covers it have formed two large hills with gentle slopes; the highest point was at an elevation of 43.5 meters. On the hills, various species were planted to beautify the park with the concept of urban forest and ecology in mind. The kinds of trees chosen are of indigenous species that grow well in the climate of southern Taiwan. They were brought up from seedlings and are either nice to look at or can provide nectar and food for animals. In addition to the plants, there are hiking trails on the hills as well as bikeways and paths in the rest of the park. Several detention basins were installed at the feet of the hills to collect rainwater flowing down the slopes. They also create an aquatic ecosystem and beautiful landscape.

In the past, the biggest problem caused by the landfill was the gas produced from the decomposition of waste. This not only posed a real risk of a fire and explosion, but might also pollute the air through the emission of methane, a greenhouse gas. Therefore, you can see methane collection facilities in the park: with a layer of impervious material and another layer of earth to prevent methane from leaking (and obtain better results), methane is collected in a well linked with a station and then transported to the biogas power plant through underground HDPE pipes. To attain the goal of protecting the environment and generating renewable energy, methane thus collected is burnt before it is used to generate electricity. The biogas power plant here is estimated to have more methane than any other landfills in Taiwan. Its anti-contamination period is as long as twenty years and the amount of methane stored in the landfill can provide electricity for twenty years. As the amount gradually decreases every year, monthly treatments have fallen from two million cubic meters to about a million. Being the largest in Asia, the biogas power plant in Kaohsiung is the first of its kind in Taiwan and among top ten around the world. When it can no longer generate electricity in the future, the extraction pipes and collection facilities will be preserved in situ as the evidence of this part of the history.

D4   Ronghua Dam, Taoyuan, Taiwan (proposal)
DESIGNER: Tianjian Chen

Water is the origin of civilization. As early as 4000 BC, people were building dams. Modern, large dams were mostly built in the first half of the twentieth century, so many of them had lived their lifespan in the 1980s. But, to demolish such huge structures not only requires a colossal amount of money, but also entails a certain degree of challenge and risk. Therefore, most of the retired dams were just left where they had been. At the beginning of the 21st century when environmentalism was gaining traction, people in Europe and the US started to demand that the retired dams be torn down. As a result, some of them were demolished under pressure.

However, other than being dismantled, is there a way retired dams can be repurposed and given new lives? It is a question that should be answered. In this proposal, the designer tries to provide another option for retired dams by redesigning a dam that has lived its lifespan in Taoyuan, Taiwan.

Having been online since 1983, about 97 percent of the reservoir of Ronghua Dam has been filled by sediment as of 2012. Because its location makes it difficult to clear the silt, the dam can no longer arrest the sand or control floods; now, it can only be used for power generation. To renovate the dam without losing its capability of generating power or blocking water during the wet season, the designer proposes to cast the ten floodgates of Ronghua Dam in different roles, based on a return period of 200 years. Of the ten floodgates, four are used for regular flooding, two for auxiliary flooding, two for emergency flooding, while the other two floodgates should fall into disuse entirely.

Ronghua Dam is located in a remote area in the mountains where you can find several mountaineering training institutes and camping sites. Combined with the fact that the dam itself is a curvy wall 160 meters long and 82 meters high, it is perfect for outdoor activities that take place in a vertical dimension. Therefore, the aim of the proposal is to add some structures on the dam wall and turn Ronghua Dam into a mountaineering training center where the outdoorsy type can go bungee jumping, rock-climbing, springboard diving, scuba diving, indoor parachuting, and get some training for mountain climbing.

Ronghua Dam is situated in a gorge, so there is a 100-meter drop between the main road, Provincial Highway 7, and the top of the dam. Therefore, one of the challenges is to transport visitors to where the new training center will be. Once there, they also need to be able to reach the different facilities all over the 82-meter-high dam wall. Finding inspiration from the gondola system common in ski resorts, the designer proposes to expand the original control center by the highway into a gondola lift station from which visitors can be transported to the tourist center at the top of the dam. They can then use another transport system similar to a Ferris wheel to reach all the facilities located at different height of the wall.

Dams are great infrastructure that enables civilizations to expand. By adding minor structures to a dam, people can engage various kinds of activities and witness the sheer scale of it. Most important of all, after accomplishing its original mission, the dam can keep playing a role in human civilization.

W1
Watch tower, Sint Jansklooster, the Netherlands
DESIGNER: Zecc Architecten

In a nature reserve De Wieden, Overijssel, the Netherlands, an old water tower was renovated by Zecc Architecten in 2014. The architect built a series of winding staircases in the tower, turning it into a 45-meter high watchtower.

The exterior of the water tower, now called the Viewpoint Sint Jansklooster, was largely kept as it had been, except for the four huge windows installed on the top floor. As for the interior, three sets of stairs were built to connect the different sections of the tower. To overlook the beautiful De Wieden from the observation deck on the top floor, visitors have to ascend a series of stairs. To begin with, there is a closed staircase which leads to the first floor 4 meters above their heads; next, stairs made of oriented strand boards (made by layering and compressing wood strands) provide a stark contrast with the warmth of its wooden texture and the rigidity of the original concrete walls. Intertwined with the original steel stairs that climb up

the structure against the walls within the cylindrical space, this newly built set of stairs rises towards the underside of the old concrete tank at 28 meters. While the steel stairs make its way from the edge to the center of the bottom of the tank, the new stairs shot straight through the middle of it.

Lastly, visitors have to go up a new set of steel stairs to reach the observation deck at the top. The designer removed part of the tank structure and replaced it with transparent floor boards to add some extra space. People can walk through it, above the heart of the tank, and experience the dizziness looking down the space. In addition to the four existing windows, the architect inserted four more larger ones; jointly, the eight windows create a 360-degree view. During the climb, visitors can experience all kinds of surprises and fun, then be presented the beauty of De Wieden as a wonderful ending.

The water tower is a national monument and viewed by the local residents as a public good. In this project, the architect refrained from altering the appearance of the tower. Through ingenious remodeling of the interior, the local residents' doubt and concern about the renovation is eliminated while their bond with and their sense of belonging for the neighborhood is deepened.

W2
Water Tower Pavilion, Shenyang, China
DESIGNER: META-Project

The Water Tower Pavilion is located in Tiexi District, Shenyang. Its predecessor was the People's Liberation Army No.1102 Factory, an arsenal built in 1959 during the Great Leap Forward period. As one of the most important heavy industry regions in China, Tiexi District has numerous industrial relics from that time. Water towers, seen all over the place, have hence become a unique symbol of its industrial history. In 2010, Vanke bought several factories there for its Blue Mountains project. Meanwhile, a water tower remained untouched and was then handed over to Wang Shuo's design team META-Project to be renovated. The team kept the exterior of the water tower out of respect for its role in history and refurbished only the interior. It was turned into the Water Tower Pavilion, a public space and exhibition gallery for the community.

Taking both history and practicality into careful consideration, META-Project tried to insert new elements into the old industrial building in a sophisticated and delicate way. They designed two "shells" for the tower, with the outer one representing the past and the inner one the present, or what the architects want. The two shells are combined with colorful windows.

The team tried their best not to alter the completeness of the water tower and only conducted essential structural reinforcements as well as partially treated the original window frames. That being said, a complicated device composed of two funnel-like structures linked end to end was indeed added to the tower. While the smaller one at the top of the tower functions as a skylight, the bigger one creates stretched interior space with multiple light boxes that look like camera lens. These boxes indirectly introduce natural light from the surroundings into the tower and seem to have "grown" out of every possible opening on the tower. At the bottom, between the entrance and the elevated observation deck, there is a series of steps made of recycled red bricks. This is the mini theater where people can hold activities of smaller scale or just gather together. When you look up from here, the tower becomes almost like a sensory organ with which you sense the outside world. Light enters the central funnel in the tower through the funnel at the top and through those windows each in a different shape and color, so that there is a continuous yet subtle change of light during the day. On the other hand, if you look at the surrounding area from the observation deck protruding from the tower, the device becomes a viewfinder

for you to observe the outside world.

The Water Tower Pavilion provides a new public space for the residents in the Blue Mountains project. It can be seen as an urban art installation that is full of spirit.

W3
Water Tower House, Brasschaat, Belgium
DESIGNER: Crepain Binst Architects

The water tower in a manor in Brasschaat, Belgium was originally constructed to provide water for the house and its outbuildings, but a new water supply system composed of four newly-built water towers replaced the completely outdated old tower in 1937 (cause of death).

In 1950, the government bought the manor and the rest of the property, including the water tower. The structure and the layout is simple enough: four large concrete pillars support a cylindrical tank that is itself four meters tall but at a height of 23 meters; the dimension of the rectangular platform within the pillars is 4m x 4m; there are a filter pool and a reservoir under the ground; the property, covering an area of 680 square meters, is surrounded by woods; a brook at the foot of the water tower separates it from the woods.

Eventually, the government must decide whether to tear the water tower down or to sell it. Fortunately, although the tower was stripped bare, a person with a good taste somehow came upon it and was able to appreciate this seemingly despicable object. In fact, he liked it so much that he viewed it as his house of dreams. Working with a local architect Jo Crepain, he turned the tower into his own residence.

The two created a simple design of a six-story house with a winter garden right beneath the old concrete tank. The four stories in the middle of the tower sitting on the six-meter-high living space are encased by semi-transparent glass curtain walls (U-shaped glass), just like how you would use a semi-transparent sausage casing to enclose the pork filling. A steep stairway leads to all the floors. Views from the house overlook the adjacent woods of Brasschaat.

The building is seen as the lighthouse of the town because the entire tower seems to shine brightly in the evening with its interior lightings turned on. While the glass curtain walls introduce sufficient natural light, their semi-transparency maintain a certain degree of privacy. Coexistence of old and new industries is realized here, with the combination of the old concrete structure and the new glass walls. A new space was created; meanwhile, the original quality of the water tower is enhanced.

Due to a tight budget, Crepain and his client had to complete the project with a limited amount of money. In retrospect, the architect said in 2008, "We were lucky we didn't have much budget. Now we've got this simple, clean and lovely house." Sometimes, to accomplish a simple thing is more difficult than to do something complicated.

Theirs is a work that adheres to the philosophy of Mies van der Rohe – Less is More.

W4
Biorama Projekt, Brandenburg, Germany
DESIGNER: Frank Meilchen

In 2002, 48-year-old product designer Richard Hurding visited the Schorfheide-Chorin Biosphere Reserve under the UNESCO Man and Biosphere Reserve Programme in the State of Brandenburg, northeast of Germany. Here, Hurding found an obsolete water tower which later became his home surrounded by trees on a hill. At that time, he and his 47-year-old wife Sarah Phillips, an industrial designer, were looking for a new place to live after working in several different cities around the world.

What they had been looking for was something devoid of the enterprise culture, friendly to the environment, and good for the society. Moreover, they already had some experience in creating living space out of industrial buildings before when they lived in London. This mustard yellow water tower is a historical landmark from the former East Germany. A row of windows at the top with a vast view of the forest and the idea of living in a tower were too tempting to resist for Sarah Phillips who saw the potential of reusing it.

To make their dream come true, the couple filed an application to the local authority in the district of Barnim for the renovation of the water tower. To make their application more persuasive, they proposed that part of the water tower be turned into a tourist attraction. On the top of the 21-meter-high tower, they could build an observation deck where visitors could overlook the reserve and observe its forests, lakes, precious birds and amphibians.

In 2003, their application was granted. The couple paid 75 thousand euros for the water tower and its surrounding area, a land covering about three acres. However, since it is a designated historical building, the government decided not to sell but to let it to the couple for 500 euros a year for a period of 99 years.

In the same year, they commissioned Frank Meilchen, an architect based in Berlin, to help them carry out the renovation. For Meilchen, the biggest challenge was to remove the gigantic concrete water tank in the tower. The worker cut the tank open with a diamond blade while a crane drew out the fragments piece by piece from the window. It took them a whole month just to finish this task. Then, Meilchen built a tower inside the original one in order to bear the weight of the future floors of the home. Yet another tower was built next to the water tower. It has red glass windows and an elevator to take the visitors up to the observation deck. The entire project cost about 600 thousand euros and took them eight months to complete. The UN donated 270 thousand euros for the construction of the elevator tower to make the observation deck accessible to wheelchair users.

In 2006, Richard and Sarah moved in. Now, the water tower is a 1,500-sqaure-feet apartment with six stories. On the first floor is Sarah's study; on the second floor is Richard's study; on the third floor are their bedroom and bathroom; on the fourth floor are the second entrance to their home, a toilet for guests and a storage room; on the fifth floor is the kitchen and dining area; on the sixth floor, the top floor, is the living room with a 360-degree view.

The couple named the tower Biorama, a perfect combination of the words "biosphere" and "panorama."

W5 House in the Clouds, Suffolk, United Kingdom
DESIGNER: Glencairne Stuart Ogilvie & F. Forbes Glennie

An old water tower in England was transformed into a unique holiday accommodation in the clouds.

The predecessor of the House in the Clouds is a water tower located in Thorpeness, Suffolk, England. It was constructed in 1923 by Braithwaite Engineering Company of London with a capacity of fifty thousand gallons to store the water pumped by the Thorpeness Windmill. Since then, it was guarded by William Knights until 1938. In 1943, the tower was hit by a bomb in an air raid during WWII. Although engineers repaired the tank with the tower's own steel material, its capacity was reduced to thirty thousand gallons.

In 1963, modern water supply system was introduced to Thorpeness, so the tower was now only used for water storage. In 1977, the Ogilvie family became the owner of the house. In 1979, even the storage function of the tower was deemed redundant, so the main tank was demolished (cause of death) and the tower building was completely converted to a regular

house. Meanwhile, the interior of the tower also underwent renovation in order to release more living space.

When the renovation first began, the local residents of Thorpeness were worried that the view would be spoiled. To enhance the appearance of the water tower, Glencairne Stuart Ogilvie and architect F. Forbes Glennie brilliantly disguised the giant as a house, so that when seen from afar, the cottage on the horizon seemed to be on top of a 21-meter-high tree.

In 1987, Ogilvie redecorated the interior again to be a private home and holiday rent, this time for his friend and writer of children's books, Malcolm Mason. The House in the Clouds provides spacious accommodation for holiday goers with five bedrooms, three bathrooms, a living room, a dining room and a marvelous games room on the top floor. There are 68 stairs from bottom to top in this 21-meter-high building and many newly-added windows that offer good lighting, air circulation and beautiful views.

The holiday village in Thorpeness founded by Glencairne Stuart Ogilvie in 1920 is nothing like the other resorts elsewhere, because what he wanted to create was an ideal place for people who wished to experience the real English life. Since 1995, the House in the Clouds has been a Grade II Listed Building, sitting on a one-acre private property. Overlooking the Thorpeness Golf Club and the Suffolk coast, it is one of the most famous holiday accommodation in England, with the most stunning view of Suffolk.

W6 Nagyerdei Víztorony, Debrecen, Hungary
DESIGNER: Zoltán Győrffy & Róbert Novák

Nagyerdei Víztorony, initially built in 1913, is now a new tourist attraction in Debrecen, Hungary.

In the early 20th century, water towers played a unique role in engineering, thanks to the construction of railroad networks, the development of factories, and the expansion of cities. But nowadays, most of them have fallen into disuse. One of the rare exceptions is Nagyerdei Víztorony, a water tower owned by the University of Debrecen and still in operation. The renovated tower is one of the most important items in the regeneration project of the Debrecen Great Forest. Two architects, Zoltán Győrffy and Róbert Novák, took nine months to refurbish the working water tower. They created spaces for a restaurant, a bar and café, a shop, a gallery and an observation deck, all within the concrete frame of the tower. People can even climb up the building via the central pillar slash rock-climbing wall.

The rebirth of Nagyerdei Víztorony did not happen just once. In 1914, the next year of its completion, a café was designed on the first floor. In 1933, a cafeteria appeared. In 1969, the building was handed over to the University of Debrecen. The first floor was turned into a housing unit, but later, owning to the water tower preservation movement, it no longer functioned as one. Prior to the year 1994, there were no major alterations to the building except for some routine maintenance work. Then in 1994, as part of the water tower preservation movement, the local government and university submitted a UN joint programme and carried out a comprehensive renovation. The goal was to preserve the existing structure and its architectural features while at the same time create an attraction. Therefore, the housing unit was demolished and a restaurant appeared again, along with other public spaces and a rock-climbing wall.

The arches on the first floor were filled up with glass, so both the indoor and outdoor areas of the restaurant seem bright and light. There is a bar in the restaurant that forms a part of the original central pillar; the rest of the pillar is a rock-climbing wall. The gallery is located under ground level with a reception desk, the exhibition space and a ring of skylights that provide sufficient natural light for the underground space. To produce more outdoor space, the architects dug downward and created an artifi-

cial mound surrounding the water tower. People can enter from outside and into a semi-circular court protected by the original retaining wall and a sturdy gate. The entire tower is 42 meters high and the observation deck, at 34 meters, has a dome that covers the tank in the middle. The engineers from the past built this amazing terrace around the tank, so now visitors can have a panoramic view of the forest and the city.

The rebirth of Nagyerdei Víztorony gave birth to a multi-purpose, young space for the community and an alluring tourist site. Visitors can admire the charm of Nagyerdei Víztorony in the daytime and its artistic lightings at night, as well as attend various festivals and events, such as those held by the university's movie club, exhibitions and concerts. Every month there are numerous people coming to pay a visit.

S Storage

S1 Frøsilo, Copenhagen, Denmark
S2 Oil Silo 468, Helsinki, Finland
S3 Kanaal, Wijnegem, Belgium
S4 Gasometers, Vienna, Austria
S5 Gasholders London, King's Cross, UK
S6 Zeitz Museum of Contemporary Art Africa, Cape Town, South Africa
S7 Allez-Up, Montreal, Canada
S8 The 80,000-tonne Silo Warehouse, Shanghai, China

S1 Frøsilo, Copenhagen, Denmark
DESIGNER: MVRDV

Ancient ports all around Europe are being transformed into affluent residential areas. The exceptional view, waterside playground, closeness to the city center and their original uniqueness are some of the reasons why old ports are undergoing fast development.

The Gemini Residence Frøsilo, converted from two grain silos and completed in 2005, is located in Islands Brygge, a waterfront area in Copenhagen. Originally constructed in the 1960s, the twin silos fell into disuse in the 1990s due to the decline of soybean processing industry (cause of death). More than a decade later, the architecture firm MVRDV took on the task of renovating the silos and turned them into apartment buildings.

The structural constraint of the cylindrical silos posed a challenge for the designer. It is quite troublesome in terms of construction to open up huge holes in the circular concrete load-bearing walls while at the same time trying to keep the interior intact.

To solve this problem, therefore, the housing units of the apartments are designed to "suspend" on the outside of the silos, rather than installed inside them. In other words, the entirety of the housing units actually huddles against the original concrete silos, like an accessory. This way, the residents of every unit can also take in an optimal view of the city. There are eight stories in the building and there is a private balcony for every housing unit. Meanwhile, the vacant interior of the silos is used as public space and for vertical movement, with two spacious atriums in the center.

On the ground floor, there are no housing units; instead, part of the concrete silos is exposed. Frøsilo is different from other silo refurbishment projects commonly seen in the area that keep most of the original appearance because MVRDV designed it in such a way as to integrate a futuristic residence building into the exposed concrete structure. Unlike those silo conversions that took the natural approach and filled the inner space, this project chose to build around the structure and thus made the whole design more flexible.

S2
Oil Silo 468, Helsinki, Finland
DESIGNER: Lighting Design Collective

In winter, nights in Finland can be as long as eighteen hours. During the months when daylight is so precious and rare, public space with sufficient lightings becomes almost like a sanctuary in the city to help the citizens while away the long wintertime.

Oil Silo 468 is a permanent public space designed by Lighting Design Collective (LDC). It was originally used as an oil silo that faced toward the city center at the seaside. Later, the company re-painted its dingy exterior and drilled 2012 small holes to honor the designation of Helsinki in 2012 as the World Design Capital. Together with its delicate lighting system, this tower-like installation marks the beginning of the regeneration of the blighted areas in Helsinki.

The lighting design was inspired by the sea breeze and the light reflected on the sea so familiar to the local people. Behind 450 of the 2012 holes are steel mirrors that flutter with the wind, so that in sunshine, they look like the glistening waves. For another 1280 holes, there are white LED bulbs at a color temperature of 2700K. LDC developed a special lighting system making use of swarm intelligence and nature simulating algorithms, so it will achieve a lighting effect according to the weather data at the moment (such as wind speed, wind direction and temperature) that are updated every five minutes. This produces numerous unpredictable combinations and people can literally "see" the change in the weather of Helsinki. At night, these flowing and natural patterns that neve repeat can be seen even from several kilometers away.

During the day, the light on the interior walls penetrated through the holes moves its position over time. At midnight, the white LED bulbs turn red for an hour, illuminating the interior with cozy light. The red light and the dark red interior of the silo drop a hint that it used to be a sealed container that stored petroleum. With the floor that was newly paved and the wires, pipes and emergency lights that were put in order, the space becomes a gathering place for people to hang out and chill. At 2:30 a.m. every day when the last ferry leaves the shore, Silo 468 is turned off, waiting for the coming of the next twilight to brighten and warm up the hearts of the citizens.

Through a design that incorporate both natural and artificial light, Silo 468 brings light and hope to an area with 11,000 residents. This unique commune space and public art is a pioneer in the renewal of the blighted areas in a city.

S3
Kanaal, Wijnegem, Belgium
DESIGNER: Stéphane Beel Architects

Kanaal is a residential and commercial project located in Wijnegem just outside the city of Antwerp. A former abandoned but still valuable 19th-century industrial site along the Albert Canal was turned into a multi-purpose complex. Most of the complex was renovated as apartments while the rest of it was turned into workshops, museum space, offices and a subterranean parking lot.

Initially, the complex was part of a jenever distillery founded in 1869. Back then, the distillery was one of the largest in Flanders, exporting its products as far as Australia. In 1919, the government suppressed the sale of hard liquor in Belgium, which practically shut down the distillery (cause of death). After WWII, Heineken purchased the complex and turned it into a brewery. Eight silos were added to its premises.

Stéphane Beel Architects was in charge of the renovation of the grain storage silos of the old brewery, as part of a bigger development project.

Since the spring of 2017 when the renovation was completed, the complex has been home to an arts and antique foundation, a high-end shopping center and a hundred artistic apartments.

It is an enhancement of the grey silos: without losing their original features, they now function as apartments that guarantee a great view and sufficient natural light. Two of the grey silos, 31 and 28 meters high respectively, were replaced each by a slender and transparent newly-built rectangular structure. The other six silos were preserved, with new rectangular windows on their façades. Together, the eight buildings form two L-shaped housing units, each composed of three old silos and one rectangular structure. The bottom sections of the existing silos are kept, while an elevated platform serves as the exhibition space and an entry to the housing units. The buildings are linked together with glass bridges to connect the four individual spaces that form a complete housing unit.

The new function as a place to live makes it possible for the old silo complex to continue its existence; on the other hand, the old silos provide an opportunity for the architects to explore a new form of modern houses. The layout of a traditional house almost always combines different sections in one big space. Nevertheless, in the case of the reused brewery, each of the old silos is an individual section of a housing unit. Therefore, you have to connect at least four silos/sections to fulfill the basic functions of a house. For modern houses, this is not the most economic layout, but viewed from an experimental angle, you will find that there is good ventilation and light in each of the separate sections. Most important of all, Kanaal is a successful example of giving a new purpose to a deserted space. It restores the brewery to its former glory.

S4
Gasometers, Vienna, Austria
DESIGNER: Jean Nouvel, Coop Himmelblau, Manfred Wehdorn & Wilhelm Holzbauer

A gasometer should not be confused with a gas meter. While a gas meter is a round meter that measures the volume of gas, a gasometer is actually a large container structure in which gas is stored. Referring to a gas holder as a gasometer is a practice that continues to this day.

The predecessor of the Gasometers in Vienna were four old gasometers that were constructed between 1896–99 and belonged to the gas works Gaswerk Simmering located in the 11th district, Simmering, of Vienna. The facilities of the gas works were built by a British company called Inter Continental Gas Association (ICGA) to supply Vienna with town gas. These gasometers were the largest of their kind in Europe at the time, each with a capacity of 90 thousand cubic meters. Although the gas works provided more than a hundred million cubic meters of gas per year, eventually they could not meet the demand from the ever-growing population of Vienna. After several upgrades, the gasometers still could not keep up with modern technologies. In 1969, the city began to replace town gas with natural gas. On top of that, new technologies for the construction of gasometers made it possible to store the gas underground or in high-pressure storage spheres under higher pressures and in smaller volumes (cause of death). In the end, the four gasometers were retired in 1984.

A few years before their retirement, the gasometers were designated as protected historical buildings. The government of Vienna refurbished and restored the historic site before inviting the public to put forward any opinions for the reuse of it in 1995. The gasometers are four cylindrical telescopic gas containers, each with a capacity of about 90 thousand cubic meters, a height of 70 meters and a diameter of 60 meters. During the renovation, the interior of the gasometers was stripped bare; only the brick walls and the roofs were kept intact. Nowadays, these structures have found new residential and commercial use.

In 2001, four internationally acclaimed architects were each commissioned to renovate and design a gasometer, numbered A, B, C and D. Jean Nouvel was responsible for Gasometer A, Coop Himmelblau Gasometer B, Manfred Wehdorn Gasometer C and Wilhelm Holzbauer Gasometer D. Every gasometer is divided into three leverls: the apartments (at the top), the offices (in the middle) and the shopping mall (on the first floor). The shopping mall levels of the four gasometers are interconnected by skywalks. Indoor facilities include a concert hall which can accommodate two to three thousand people, a cinema, a student dormitory and the municipal archive of Vienna. There are about eight hundred apartments, two thirds of which within the historic brick walls, with 1,600 tenants and about seventy student apartments with 250 students.

With their uniqueness, the Gasometers can be deemed as a city within a city. It can certainly be categorized as a community, whether a physical community of households with residents or an active virtual community on the Internet. It is a phenomenon which journalists have written about and scholars in the fields of psychology, architecture and urban planning have published papers on.

S5
King's Cross Gasholders, London, UK
DESIGNER: WilkinsonEyre

Gasholder Park is part of an urban regeneration and upscale residence project centering on four retired gas holders located at King's Cross. Although it is rather small compared to other parks, its unique industrial character makes up for that. It is the remaining cast iron frame of Gasholder 8, the biggest of the four standing by the canal, with a capacity of 1.1 million cubic feet. In the past, this 82-foot-high cylindrical giant was also a symbol of King's Cross, dominating its horizon. Now, after its rebirth, there is a circular lawn within it that looks like an art installation with dynamic lights. It is part of the urban landscape and public space for recreation in the neighborhood.

These large gas holders were constructed in the 1850s to supply most regions in London with town gas for heating, cooking and lamps both at home and on the street. They had been working until 2000, when they were finally retired. In an early phase of the regeneration of King's Cross, the complicated cast iron frames of Gasholders 8, 10, 11 and 12 were dismantled and transported to Yorkshire where they underwent restoration for two years before returning back to King's Cross in 2013. They were relocated to a new home by the canal and Gasholders 10, 11 and 12 became a triplet housing unit named Gasholders London.

The triplet belongs to the industrial heritage of King's Cross. The architecture firm WilkinsonEyre incorporated its features into a deluxe housing complex that is filled with modern designs. Its various public spaces and facilities include a gym and spa, a studio for rent, a conference center and a banquet room. A total of 145 apartments in different styles are distributed in the three concrete cylinders with eight, nine and twelve stories respectively. The use of glass and aluminum is a way of showing respect to these Grade II-listed cast iron pillars from the Victorian Era. As for the indoor space, a combination of industrial, artisan and luxurious elements aims to match the residential and commercial interior design commonly found in the modern-day central London. A series of retailing stores is located on the ground level, so that the public get a chance to see the inside of the complex. It can be said that Gasholders London has achieved an optimum balance between industry and luxury, old and new, private and public, interior and exterior.

The large cast iron frames of Gasholders London are the products of water-sealed gas holder technology. Their angular and distinct form is gradually disappearing from London's horizon. However, to preserve these beautiful industrial relics requires a large sum of money. Although it is quite understandable that it is the continued existence of historic buildings that people love so much about their reborn selves, it is a matter for debate whether it will be a reasonable investment if similar projects of the same scale are carried out elsewhere.

S6
Zeitz Museum of Contemporary Art Africa, Cape Town, South Africa
DESIGNER: Heatherwick Studio

Zeitz Museum of Contemporary Art Africa (Zeitz MOCAA), the biggest art museum in South Africa, was created by carving out the inside of a historic grain silo complex. Described by English designer Thomas Heatherwick as the oldest building in the world, its successor Zeitz MOCAA is the most important exhibition space for African art in the whole world. Located in a grain silo complex built at the seaside of Cape Town during the 1920s, the museum covers 9,500 square meters of custom-made interior space that spans ten stories. The complex had been a monument to the industrial history of Cape Town until it fell into disuse in 1990.

Now, the transformation by Heatherwick Studio has given the old structure a new life. By carving out huge portions of the original tubes, a complicated network of 80 galleries was born. In addition to the 6,000 square meters occupied by the galleries, there are a rooftop sculpture garden, a storage area, a bookshop, a restaurant, a bar, a hotel and reading rooms on the premises. There are also different centers for Costume Institute, Photography, Curatorial Excellence, the Moving Image, Performative Practice and Art Education. The concrete structure seems to be of a uniform configuration, but in fact it is composed of two parts: a tower structure that is clearly divided into multiple levels and a large beehive-like silo made up of 42 tubes. The biggest challenge is to turn these concrete tubes that were densely put together into an exhibition space for art while at the same time preserving the existing industrial style of the complex.

Most silo renovation projects avoided taking vigorous action on the concrete silos or other cylindrical structures. However, Heatherwick did the exact opposite. He boldly dug out an atrium that looks like a cathedral with a high vault from the inside of the structure, making it the heart of the museum and granting entries to the different exhibition levels built around the atrium. First of all, the designer used reinforced concrete to thicken and strengthen the concrete tubes that were only 17 centimeters thick, so that the tubes now have a thickness of 42 centimeters. Then, he carved through the tubes and created a 4,600-cubic-meter atrium with curvy, geometric patterns. The tubes were polished where it had been cut, providing a contrast to the original rough concrete. On top of each tube there is laminated glass with a diameter of six meters to bring natural light into the atrium. The underground tunnels are used to create artworks that need to be done in special spaces.

On the exterior of the museum, there are bulged windows made up of multi-faceted glass panels which reflect the Table Mountain, Robben Island, bustling crowds and clouds in the sky. The panels, installed within the existing concrete frames, not only introduce light into the atrium, but also produce some sort of kaleidoscope visual effect. By digging out and opening up the space meant for the atrium, the identical tubes have differences now: while some are loosely packed, others seem to be denser; while some are more open, others are more restrained; while some are more fluid, others are more like solid. It is as if the visitors are making their way through, lingering in and exploring the inside of a huge sculpture, and by doing so, waking up the grain silo that has been sleeping for a long time and witnessing its rebirth.

S7
Allez-Up, Montreal, Canada
DESIGNER: Smith Vigeant Architectes

The rock-climbing gym Allez-Up is at the heart of the reinvigoration project of the Southwest Borough in Montreal, Canada. Its predecessor was the silos of the old Redpath sugar refinery situated by the Lachine Canal. Now, it has been transformed into a unique indoor rock-climbing facility and immensely increased the recreation and tourism values of the canal. It is an exceptional idea to turn abandoned silos into a rock-climbing gym. Allez-Up has exploited the enormous potential of the Montreal industrial heritage to the full.

Opened in 1825, the Lachine Canal provided a shortcut to cross Saint Lawrence River and thus ushered in the industrial era of the region. By the 1860s, the area had grown into a prosperous industrial zone with a greater variety of industries than anywhere else in the country. In 1952, the Redpath sugar refinery built four silos here to be used as the storage space. However, at that time, the canal was already in decline because it could not be widened anymore and thus bigger vessels could not navigate on it. In 1959, the Saint Lawrence Seaway was opened, practically ending the golden age of the Lachine Canal. In 1970, the canal was officially closed. A lot of factories were relocated and the neighborhood eventually fell into decline too (cause of death). Later on, the derelict buildings in the industrial zone were being renovated and reused one by one, but due to the special shape of silos, they were completely neglected for forty years.

Allez-Up was originally on a site that had been renovated from another industrial structure near its current location, but the proprietor of the gym wanted to house more facilities and expand the premises threefold, so he decided that the Redpath silos and the surrounding area were perfect for such a project. The architect connected the existing cylindrical spaces with a rectangular volume, thus linking the two pairs of silos together. The entire renovation displays a very unique way of carrying out a project like this. Inspired by the dynamics of rock-climbing walls, the architect attempted to make use of the vertical space to the maximum for the climbers. Originally, the inner walls of the silos were covered with two layers of cedar wood planks to prevent the stored sugar from getting dampened. During the refurbishment, some of them were removed to be reused and made into furniture and interior decorations.

The rock-climbing walls in the main building resemble sugar cliffs, so visitors are often reminded of the fact that the silos were once part of the Redpath sugar refinery. The pure-white, angular walls provide many different routes for beginners and experienced climbers alike. The colorful holds across the face of the walls add to the dynamic charm of this extraordinary interior space. The main façade of the building was cut diagonally into several openings to be replaced by large pieces of glass and to provide the interior with natural light in all weathers. Most important of all, practical functions were carefully incorporated into the silos themselves: the reception desk is located in the west side of one of the silos, while there are rock-climbing routes all over the inner and outer walls of the concrete structures to take full advantage of the vertical space. The silos absolutely become part of the rock-climbing experience at Allez-Up and as an industrial heritage, they continue to remind the visitors of what they used to be.

S8
The 80,000-tonne Silo Warehouse, Shanghai, China
DESIGNER: Atelier Deshaus

As the organizer of Shanghai Urban Space Art Season (SUSAS) 2017, Shanghai Design and Promotion Center for Urban Public Space, after careful deliberation and taking in account of all factors (a preference for com-

paratively undeveloped regions, the compatibility of architecture and environment, the feasibility of the project and so on), decided to put an 80,000-tonne silo complex located in Minsheng Wharf in the Pudong New Area at the heart of the event, making both the complex and the warehouse 257 close by the main exhibition venue of SUSAS. The aim was to revitalize the region with a boost from the art season and turn the buildings at the venue into a new cultural hub with the regeneration of the wharf.

In the region of Minsheng Wharf, there are fifteen existing buildings. Among them, the 40,000- and the 80,000-tonne silo complexes are grain elevators while warehouse 257 is the original bagging workshop. The predecessor of Minsheng Wharf was built in 1908 by Swire Group, who was commissioned by the biggest British shipping company at the time, Blue Funnel Line, who engaged in overseas trade between Europe and Asia. In 1924, it was the most advanced wharf in the Far East. It was renamed several times during its history and finally, in 1956, it was changed into Minsheng Wharf and the name has stuck ever since. In 1975, the 40,000-tonne silo complex was constructed. Later, due to the economic development in Shanghai, the demands for storage increased. Between 1991 and 1995, the 80,000-tonne silo complex was built. However, a revolutionary change in the way of storing grains (cause of death) caused the grain elevators to be shut down ten years later in 2005.

After more than a decade of disuse, the silos suffered serious damage. Therefore, the objective in the first phase of renovation was to ensure the safety of visitors, strengthen the structure, implement preventive measures in the event of a fire and evacuation, and create a good atmosphere for visitors. The main entrance to the 80,000-tonne complex and the warehouse 257 is covered with white, semi-transparent panels, making it both modern and distinct. Also, the translucent and light material makes a strong contrast to the solid and heavy concrete structure. The unique configuration of a silo and its magnitude bring much shock to first-time visitors. Also, the funnels, the tips of which are at a height of ten meters, draw the visitors, making them both curious about what is inside and touched by all the artworks around. The complex itself is 40 meters high and the exhibition space covers an area of more than three thousand squares meters. Since moving vertically forms a major part of the experience, despite the stairway and elevator installed in four of the thirty silos for emergency only, the architects from Atelier Deshaus also built an escalator that could take you from the first to the third floor through the heart of the building, another escalator suspended on the exterior walls that could lead you to floors four to seven, and spiral ramps and stairs built along the interior walls. These arrangements not only make vertical movement possible, but also create a special experience for visitors who go here and there in the silos. Particularly, when they arrive the platform on the third floor, they will be drawn to the view through the transparent glass and come to the escalator on the outside of the northern façade. From here, the escalator will carry them to the sixth floor at 32 meters high in three stages. During the process, visitors can read the inner walls of the silos at close range and find some grain fragments on one side, and take in the beautiful view of Huangpu River through the glass panels with patterns designed by artist Yi Ding on the other side.

All of the alterations and additions are done for the purpose of establishing a better relationship between the silo complex, the river and the visitors, so that the industrial heritage can be reconnected to the city. Thanks to the development of riverside areas and the transformation brought about by the event, other industrial relics along Huangpu River are getting more attention and more opportunities to be reborn.

M Military structures

M1 Tropical Islands Resort, Krausnick, Germany
M2 The Fichte-Bunker, Berlin, Germany
M3 Energy Bunker, Hamburg, Germany
M4 Bunker 599, Culemborg, the Netherlands
M5 Pionen White Mountains, Stockholm, Sweden
M6 Haus des Meeres, Vienna, Austria
M7 Bat Hibernacula and Sanctuary, Maine, USA
M8 Martello Tower Y, Suffolk coast, United Kingdom
M9 Spitbank Fort, Solent, United Kingdom
M10 Fort Alexander, St Petersburg, Russia (in ruins)
M11 The Silo Classroom, Jian Gong Primary School, Hsinchu, Taiwan
M12 Fengshan Communication Center, Kaohsiung, Taiwan (proposal)
M13 Intrepid Sea, Air & Space Museum, New York, USA
M14 Typhoon-class Submarine (proposal)

M1 Tropical Islands Resort, Krausnick, Germany
DESIGNER: Oscar Niemeyer

The Tropical Islands Resort is a tropical theme park located in the former Brand-Briesen Airfield in Germany, 50 kilometers from the southern boundary of Berlin. The whole park is built in a former airship hangar called the Aerium, which is also the largest free-standing hangar in the world. The hangar is 360 meters long, 220 meters wide and 106 meters tall, with a bowl-like steel dome and a 180-meter cutting table to manufacture the shells of the airships. Overall, the whole structure is big enough to fit the Eiffel Tower lying down. The construction of the hangar finished in November 2000 and cost 78 million euros. Initially, it belonged to CargoLifter and the company managed to build the prototype of airship CL75. However, the CL160 which the hangar was to house was never constructed since CargoLifter went bankrupt in 2002 (cause of death: lack of funding). In June 2003, Malaysian company Tanjong bought the hangar and 500 hectares of land with 17.5 million euros and renovated the hangar as a holiday retreat named the Tropical Islands Resort. The official opening date is 19th of December, 2004.

Inside the hangar, the temperature is controlled at 26 Celsius degrees and the humidity at around 64%. There are beaches and rainforests, along with some 50,000 plants out of 600 different species, some of which very rare. Moreover, there are waters that look to be of a coral island, a lagoon, fountains, a canal, whirlpools and many slides, swimming pools, saunas and spas – in fact, this is the biggest tropical sauna facility in Europe. What's more, there are bars, restaurants, a camping site, a hotel, a playground for kids and a wild golf course. The hangar is open all year round.

The Tropical Islands Resort can host 8,200 visitors a day at the most. It is the largest indoor rainforest and water park in the world.

M2 The Fichte-Bunker, Berlin, Germany
DESIGNER: Paul Ingenbleek

The Fichte-Bunker in Berlin, Germany is a very good example of an infrastructure facility that has been reborn multiple times to meet the needs of different historical periods. Initially, it was an engineer, Johann Wilhelm Schwedler, who built this structure as a gasometer in the Kreuzberg district of Berlin in 1874. Modeled after a church with a height of 21 meters and a diameter of 56 meters, it is the last remaining brick gasometer in the city. In the 1920s, electric street lighting replaced the original lamps in Berlin, which rendered the well-designed gas holder useless (cause of death).

Eventually, it was shut down in 1922. In 1940, an engineer and senior Nazi Fritz Todt commissioned Siemens-Bauunion to turn the gasometer into an air raid shelter with six levels (the first rebirth). The walls were reinforced with three-meter-thick concrete. At first, it was designed to shelter six thousand people, but during a raid on the third of February, 1945, approximately thirty thousand people were crammed into 750 individual rooms varied from five to seven square meters. Contemporary artists called it a "luxury bunker." Despite heavy bombardment, the bunker survived the war.

After the war, the Fichte-Bunker was used as a shelter for the displaced, especially the refugees fleeing East Berlin to the West. Later, it was also used as a youth detention center and a nursing home. Finally, it became a shelter for the homeless (the second rebirth) and the rent was 2.5 DM per night. Because the conditions in the shelter were horrible, it was called the "bunker of the hopeless."

In 1963, the facility was closed down due to its sanitation and safety issues. Between then and the reunification of Berlin, the Senate used it as a storehouse to stockpile emergency provisions (the third rebirth) until the end of the Cold War. After 1990, the Fichte-Bunker once again fell into disuse.

In 2006, SpeicherWerk Wohnbau GmbH purchased the bunker and began its renovation. Although the locals objected to the idea on the basis of regulations concerning historical buildings, permission was granted to the project. Engineer Michael Ernst and architect Paul Ingenbleek worked their magic and turned the air raid shelter into Circlehouse in 2010 (the fourth rebirth). Today, there are thirteen two-story condominiums and private rooftop gardens at the top of the structure. Residents can enter the building via an external tower with an elevator and stairs. Beneath the apartments is an exhibition hall open to the public where you can see artifacts related to the construction technologies at the time and the grim fate faced by all the refugees and the displaced during the war. After its final rebirth, the dark past shows itself again as a warning to humanity.

The change from gasometer and shelter to storehouse and luxury apartment building means that the structure has always played a central role in the evolution of Berlin.

M3 Energy Bunker, Hamburg, Germany
DESIGNER: HHS Planer + Architekten AG

The Energy Bunker, a veteran from the war and a pivot of regional energy source, is one of the rare cases when an infrastructure facility became another infrastructure facility after rebirth.

The predecessor of the Energy Bunker, located in the Reiherstieg district of Wilhelmsburg in Hamburg, Germany, was a large air raid bunker constructed by the German army in 1943 to protect local residents from bombings and to function as a fortified building to aid soldiers at the front line. By 1947, the inner structure was almost destroyed by the British army. Since then, it was deserted for sixty years.

In 2006, the government decided to renovate it. In 2010, preliminary investigations were conducted. In 2011, construction that focused on improving safety measures and repurposing the space started. Funded by the European Regional Development Fund and Hamburg Energie, an energy company based in Hamburg, and carried out by IBA Hamburg, an urban development company, the renovation project cost 27 million euros. The aim was to transform the bunker into a regional renewable energy center.

Although the interior was severely damaged, the three-meter-thick walls and the four-meter-thick roof, luckily, were both quite intact and solid. To make the entire structure safe for future use, concrete was sprayed on the weathered façade to stabilize it. Thermal insulation layer was also added. The indoor staircases destroyed during bombardment were dis-

mantled and replaced with new elevators and stairs. Solar panels of a huge amount were installed on the exterior.

The heart of the project is a reservoir with a capacity of two million liters to be used as a large heat storage facility. The heat comes from thermal energy produced from a nearby bioenergy plant, the solar panels on the roof and the south side of the building, a wood combustion system and the waste heat from local industrial facilities. By gathering thermal energy from different sources, the Energy Bunker can then redistribute heat to the neighborhood. In the future, it is going to provide heat for most of Reiherstieg and to directly supply green power to the public electrical grid.

After more than sixty years of disuse and seven years of development and construction, the Energy Bunker now becomes the first ever energy storage and redistribution facility in the world that was renovated from a great war monument. This public building and local landmark, reborn to commemorate the past and to serve an environmentally-friendly future, not only supplies the region with clean, renewable energy, but also displays how it produces and stores heat with local sources. There are a tourist center, permanent exhibition space, and a café with an observation deck at 30 meters high. Nowadays, it is a popular tourist attraction in Hamburg.

M4 Bunker 599, Culemborg, the Netherlands
DESIGNER: RAAAF & Atelier Lyon

Bunker 599 is part of the 20-year plan that aims to turn the Dutch Water Line into a national park, starting from 2000. Completed in 2010, Bunker 599 is an example of interpreting landscape and history through tangible objects. It displays a WWII bunker, bringing the catastrophic tactic of the Dutch military defense at that time into play. By cutting it in half and exposing its interior, the bunker is transformed into a sculpted work and poignant ruins.

This project reveals the secret of the New Dutch Water Line. The water line was used as a military defense between 1815 and 1940, protecting cities including Muiden, Utrecht, Vreeswijk and Gorinchem with flood zones. It was an ambitious construction that utilized the Netherlands' geography while learning both from the success of the Eighty Years' War of Independence (1568–1648) and the bloody failure of Napoleon in Russia (1812). Incidentally, the New Dutch Water Line was also a helpless mistake in the Dutch history: in 1940, when the Dutch army was fighting on ice, German airborne troops bombarded Rotterdam and installed their paratroopers behind the front line. As a result, the defensive New Dutch Water Line was never really put into practice.

The original bunker was built in 1940 and could accommodate 13 soldiers during bombing. In 2010, Dutch studios RAAAF and Atelier Lyon worked together to disclose the cramped dark space of the bunker's interior to the public. What they did is this: with a diamond wire saw, they sliced open the heart of the structure and thus dissected one of the seven hundred seemingly indestructible bunkers along the New Dutch Water Line. It took them forty days to cut through the solid concrete. Then, a crane pulled the two halves apart and a narrow gap appeared. Visitors can walk through the small yet weighty military structure via the newly-paved wooden passage which takes them to the flooded area and the sidewalk of the adjacent nature reserve. Now, as part of the New Dutch Water Line, Bunker 599 has been elevated from a municipal monument to a national historic site. The architects also designed a set of stairs to connect a nearby street and a path that goes through the bunker and leads to the wooden boardwalk above the water. The boardwalk and its supportive wooden stakes remind the visitors of the fact that the surrounding waters did not come from the removal of mud, but from flooding plains during wartime.

This aggressive intervention opens up a new poetic possibility for the cultural heritage policy in the Netherlands. Meanwhile, time provides a new perspective for people to read their surroundings.

M5
Date Center Pionen White Mountains, Stockholm, Sweden
DESIGNER: Albert France-Lanord Architects

Bahnhof AB, the biggest Internet service provider in Sweden, launched an ultra-secure data center in 2008. A 40-centimeter-thick gate separates it from the outside world. As a matter of fact, it was built in a former anti-atomic bunker thirty meters below granite rocks underneath the Vita Berg Park in central Stockholm. During the Cold War, it was used as a military anti-atomic bunker and command center. Now, its original code name during wartime, Pionen White Mountains, is retained. Pionen White Mountains is the biggest of the five data centers Bahnhof AB has in Sweden; if you want, you can also store your server there, just like what WikiLeaks is doing.

In addition to server rooms, there are greenhouses, waterfalls, German submarine engines and simulated daylight in this underground data center. The designer is Albert France-Lanord Architects. Between 2007 and 2008, the architects did a total redesign of the 1,200-square-meter space and turned it into the current data center. The biggest challenge was to work with something with no square angles: the rock itself. To create extra space for Bahnhof AB to install their backup generators and servers in the cave, they spent more than two years blowing up over four thousand cubic meters of the hard rock.

Pionen White Mountains has fifteen senior technical staff. The German submarine engines are used as backup generators: these are two diesel engines that can produce 1.5 megawatt of power. Since they were designed for submarines, the employees installed the original submarine sirens on the engines just for fun. Cooling depends on Baltimore Aircoil fans which can produce a cooling effect of 1.5 megawatt, enough for hundreds of rack-mountable servers. Also, in the data center, they have three access points for CRC Internet backbone: the network of optical fiber and copper cables uses three different paths to enter the cave, thus making Pionen White Mountains a place with best connection in northern Europe. Furthermore, to create a comfortable working environment, there are simulated daylight, greenhouses, waterfalls and a fish tank with a capacity of 2,600 liters.

Jon Karlung, the CEO of Bahnhof AB, explains that designers of data centers more often than not forget to care about those who work with something like servers. However, in his opinion, this bizarre space should be able to inspire a designer. "Since we got hold of this unique nuclear bunker in central Stockholm deep below the rock, we just couldn't build it like a traditional – more boring – hosting center," he said. "We wanted to make something different. The place itself needed something far out in design and science fiction was the natural source of inspiration in this case."

Because of the unusual design, this Internet service provider becomes famous and attracts more attention than before. Moreover, since clients often need to come here for business, they naturally share what they see and feel with other people.

M6
Haus des Meeres, Vienna, Austria
DESIGNER: Pesendorfer & Machalek Architects

The Haus des Meeres is another infrastructure facility that has been repurposed for several times in different phases of its history (this often happens to war relics). The predecessor of the aquarium was one of the six concrete flak towers that formed a defense system during WWII. Originally designed to defend against low-level missiles but due to the fact that the Allies never used this kind of weapon on it, the system didn't go online. Despite its defensive nature, the flak towers were also used to shelter local residents and as a hospital. Nowadays, all of the six towers are protected as historical monuments and partly belonged to the nation. People proposed different ways to reuse them, including as server rooms, cafes and libraries.

After the war, the tower located in Esterhazy Park was temporarily used as a hotel with 38 rooms (the first rebirth), then as a fire station (the second rebirth). In the 1960s, the underground part of the building was used as a youth hostel named Stadtherberge Esterhazypark. For the levels aboveground, there are the fire station occupying half of the habitable stories, the aquarium gradually taking over the others, and both debris and waste filling the rest of the tower. As a result, the tower became a place where occupants make use of the building in different ways, all trying to further expand their own territories. In the end, the aquarium got hold of six of the stories. The firefighters withdrew from the building and the aquarium staff started to clear the war relics in the basement (the third rebirth).

In 1991, a lightweight wraparound box was added to the structure with a slogan "Smashed to Pieces in the Still of the Night" (in English and German). It was designed by Lawrence Weiner as a memorial against war and Fascism. Initially, the local government wasn't sure if they should treat the box as a transient sign or an art heritage, but in 2005, the trend leant towards the latter and the government increased its budget for the maintenance of the tower.

The old elevator had been used since 1944, until it stopped working in 1997. It was replaced during an extensive renovation, so the new elevator can reach the new levels 7, 8 and 9. In 2000 and 2007, the glass wings which house reptiles and tropical birds were built. By 2010, ten stories and a rooftop observation deck have been in place. On the tenth story, a flak control vault was restored and renovated as an exhibition room to display WWII relics. It is open on weekends only and requires reservation in advance.

Further expansion and ownership of the flak tower has always been controversial. The building has always belonged to the government of Vienna who is responsible for the maintenance cost but wants nothing more than getting rid of this financial burden. In 2008, the authorities proposed selling the tower to an outside investor and approved the project of building a private hotel on top of the tower. Local residents strongly opposed this plan. Plus, the Haus des Meeres wished to retain its right to expand the aquarium upward in the future. To win the trust of local residents, a less radical approach must be sought (the fourth rebirth). The Haus des Meeres proposed an extension project of six million euros to add a million-liter fish tank and an outdoor restaurant to the aquarium in 2011. The government expressed their willingness to sell the flak tower to the Haus des Meeres as long as they kept Weiner's artwork and picked up the maintenance cost from the taxpayers.

Currently, the Haus des Meeres, covering 4,000 square meters, is home to more than ten thousand marine beings. There is also a five-story-zoo managed by Aqua Terra Zoo. In this tall tower, there are a multi-story greenhouse with a boardwalk and a rope bridge, mini water features with tropical plants, fish and turtles, various kinds of tropical birds and monkeys roaming freely among the visitors, and a shark tank. In 2007, the aquarium installed a shark tank with a capacity of 300 thousand liters. Because of the global economic crisis, art museums in Vienna have seen a sharp decline in their patrons. However, instead of following suit, the ticket sale of the Haus des Meeres increases in time of adversity. Every year it attracts more and more visitors who come to admire the beautiful urban view from the top of the tower.

M7
Bat Hibernacula and Sanctuary, Maine, USA
DESIGNER: unknown

In northern Maine, Cold War ruins become bat hibernacula and sanctuary.

The United States Fish and Wildlife Service (USFWS) took over 43 Cold War bunkers located in Loring Air Force Base, Maine. Before its closure in 1994, the base was used to deliver and store atomic weapons; now, it becomes part of the Aroostook National Wildlife Refuge. For many years, USFWS had tried to find a new use for these long-abandoned Cold War ruins covered all over with weed. In 2012, they decided to turn two of the bunkers into artificial caves for sick bats.

The white-nose syndrome (WNS) is one of the deadliest wildlife diseases that have erupted in recent years. So far, it has killed more than 6.7 million bats in North America, with an agricultural loss of 53 billion US dollars. In 2012, USFWS decided to remodel the bunkers into bat shelters, providing contamination-free artificial hibernacula for healthy bats and taking care of the sick ones.

These bunkers were designed as storehouses for weapons, so secrecy, temperature and humidity were all taken into account in the beginning. With some straightforward renovation and installation of appliances, these artificial caves became very much like their wild counterparts.

After carefully monitoring the temperature and humidity in the 43 bunkers, USFWS decided to renovate two of them first, creating a suitable winter habitat for 30 little brown bats from Vermont and New York. Heating pads were installed on the roof of the bunkers and a small pond was dug out in the cave to maintain the temperature at 37 to 39 Fahrenheit degrees and the humidity at over 90%. Logs, wires, plastic grids and roosts were also put in place before everything was ready to go.

In December 2012, male bats were transported to their winter homes. The scientists monitored their activities with thermal imaging cameras in the caves and found that the bats liked to gather on the vertical wall at the back of the caves. Only one bat used the logs. Occasionally when they were awake, they drank from the pond.

In late March, the bats were released to their original habitats. In the end, nine bats survived. In an article written for USFWS, Steve Agius, the assistant manager of the Aroostook National Wildlife Refuge and the Moosehorn National Wildlife Refuge, said that the experiment proved that bats were indeed more likely to survive the white-nose syndrome in artificial hibernacula. This finding has encouraged scientists to do more research on the field.

M8
Martello Tower Y, Suffolk coast, United Kingdom
DESIGNER: Piercy&Company

Converting a Napoleonic defensive tower and Scheduled Monument in an Area of Outstanding Natural Beauty (AONB) built in 1808 into a 21st-century private residence was no easy task. The said Martello tower, with three-meter-thick solid brick walls, is located near the village of Bawdsey, watching over the sea and the beach. Its renovation was carried out by architect Stuart Piercy from Piercy&Company and designer Duncan Jackson from Billings Jackson Design and deemed by English Heritage as a fine example of the rebirth of important historical buildings.

The renovated house can be accessed from a metal staircase which leads to the porch of the second floor. From here, the two-story high interior which looks like a giant umbrella sculpture is laid out before you. There are a toilet and a cloakroom in this floor, as well as the reception space with a wood burning stove. A stairway leads to the bedrooms and bathrooms downstairs, while two existing, symmetrical stairways at each side mysteriously shoot through the umbrella to the top floor where the kitchen, dining and sitting areas are situated. Here on the top floor, you will feel like you are on the deck of a ship. The roof/ceiling is not flat and the space offers a 360-degree view of the surrounding sea and countryside.

On the first floor (or ground floor in the UK), there are three bedrooms, two bathrooms and two studies. This is where the water storage area used to be. The strategy was to make a clear distinction between old and new, introducing modern interventions on the original fabric as lightly as possible and making the coarse bricks the centerpiece. The newly-added curving roof expands and reshapes the interior, a design keen to maintain the original monument and its setting. The tower's fabric is composed of 750 thousand bricks. As it was revealed, this set the standard for the design. The new roof is a lightweight 3D structure made of steel and laminated plywood and supported by five pairs of Macalloy bars. A detailed 3D model was used to make 2D cutting patterns for off-site manufacture. The frameless curved glass just below the roof distinguishes the old and the new while providing a panoramic view. To make the roof a minimum visual impact on the heritage, it was covered with a single membrane with three skylights. The whole system is very basic because everything had to be stored and installed from the gun platform. In order to bring light into the basement, six light shafts with a diameter of 450 millimeters were drilled through the thick brick walls from inside the windows to one of the bedrooms and both the bathrooms. For the other two bedrooms, there are light shafts with a diameter of 60 millimeters, providing a Camera Obscura view of the countryside to the southwest and northwest. Moreover, several holes with a diameter of 200 millimeters were also drilled from the top of the parapet to accommodate the passive ventilation system leading to the basement and the first floor. These brick ducts supply water, electricity and gas for the kitchen, and heating for the roof. There are also supply and exhaust ducts for the heat recovery ventilation system in the basement and the first floor.

By turning the Martello tower into a private home, the principle of preserving historical buildings is pushed beyond its boundary. The building at risk is not only preserved, but also given a new life.

M9   Spitbank Fort, Solent, United Kingdom
DESIGNER: Amazing Venues

Spitbank Fort, or simply Spit Fort, is a Victorian sea fort built as a result of the Royal Commission in 1859. Located in the strait of Solent (near Portsmouth) that separates the mainland from the Isle of Wight, it has a diameter of fifty meters and was designed to defend the port of Portsmouth. The fort is smaller than Horse Sand Fort and No Man's Land Fort, so that when ships successfully passed the two main forts, it could form another line of defense for them.

In 1898, the role of Spitbank Fort was changed to defend against light craft. In 1962, the authorities proclaimed that it was no longer needed (cause of death) and the Ministry of Defense eventually disposed of it in 1982. In 2009, it was put on sale at a price of 800 thousand pounds. Before the auction, it was bought for more than a million pounds. Now, Spitbank Fort, along with the other two forts, is operated by Amazing Venues under a luxury hotel brand and as a museum. The three forts have a total of fifty rooms. For Spitbank Fort, there are nine suites, as well as a dance hall and a restaurant for private banquet or wedding event.

Upon arrival, the guests will experience everything the original fort has to offer. The shuttle boat is hoisted up by a lift and then goes deep in via the only entrance. Eventually, the passengers are taken to the isolated world of Spitbank Fort which is divided into three levels: the basement magazine, the spacious battery/deck and terrace, and the rooftop light-

house and the front battery. The special layout of a fort and the fact that its original structure cannot by altered demand a smart strategy to design an uninterrupted and sensible floor plan. Another challenge the designer needed to address is the limited amount of windows that open outwards and the exposure of inner structure. In the end, eight of the canon rooms became guestrooms while seven were turned into a cellar, a ping-pong room and other facilities. However, there remains another problem: where should the wires and pipes go in this 19th century sea fort? Not wanting to make the historical building ugly with all these wires and pipes, the designer tried to find as much room for them as possible within the newly-added floors and spaces, while at the same time reutilizing and giving a new purpose to the existing vents.

Inside Spitbank Fort, it is clear that functionality overrules everything; its exposed structure is a proof of the robust construction at that time. Rows upon rows of red brick arches radiate from the center of the fort, creating a lovely dining area; the central passageway leading to the battery/deck guides the guests around the brick corridors and take them to their rooms; the new, more modern design of the roof offers a place of peaceful relaxation and meditation as well as wonderful views. Keeping the original character of the fort, the designer renovated the roof and now there are not only a new dining area on the rooftop but also a spa where the original concrete walls surrounding the two front batteries are.

Spitbank Fort has been a Scheduled Monument since 1967. However, giving it a new life is obviously a much more active and attractive way of preservation than just maintaining it as an important historic site. The designer knew how to make use of the original architectural features and the different spaces without losing the historic value of the fort. With a powerful and practical aesthetics at its core, this commercial project has successfully turned these sea forts into unique landmarks.

M10   Fort Alexander, St Petersburg, Russia (in ruins)
DESIGNER: Louis Barthelemy Carbonnier d'Arsit de Gragnac

Fort Alexander, or Plague Fort, in Russia used to be an institute of medical research for plague disease.

When St Petersburg was established in 1703, the waterways in the Gulf of Finland became strategically important to Russia. To defend the city from the Baltic Sea, Russia gradually strengthened its fortification and eventually deployed more than forty forts off the north and south shores of the gulf in the following two centuries, forming an impenetrable defense line against enemy from the sea. One of the forts, Fort Alexander, was built on an artificial island between 1838 and 1845. Its formidable appearance alone was enough to intimidate and deter any fleets to invade St Petersburg. The blueprint of Fort Alexander was drawn up by Louis Barthelemy Carbonnier d'Arsit de Gragnac, but after his death, Jean Antoine Maurice (aka Moris Gugovich Destrem), a Russian military engineer of French origin, took over and altered the design. During the construction, another Russian military engineer Mikhail von der Veide was the supervisor. Emperor Nikolay I officially dedicated Fort Alexander on the 27th of July, 1845 and named the fort after his brother, Emperor Alexander I.

This oval-shaped fort measures ninety by sixty meters with three levels and a court in the center. 55,355 twelve-meter-long piles were driven into the sea bed for enforcement. The total floor area is over five thousand square meters, enough to house a thousand soldiers. There are 103 canon ports with additional space on the roof to accommodate 34 large guns to provide the fort with military advantage. During the Crimean War, Fort Alexander played a key role in preventing the British and French fleets from invading the Russian naval base at Kronstadt. It played a similar role twice

afterwards: during the conflict with the British Empire in 1863 and the Russo-Turkish War between 1877 and 1878. However, by the end of the 19th century, its defensive role also came to an end as a result of the development of rifled artillery (cause of death). Since then, Fort Alexander was mainly used to store ammunitions.

In 1894, Alexandre Yersin discovered the pathogen of plague (Yersinia pestis). In the wake of the discovery, the Russian government set up a special Commission on the Prevention of Plague Disease and a suitable site was needed to facilitate the research. Since Fort Alexander was no longer used as a military base and its location was isolated, Russian scientists chose to come here and study various kinds of diseases caused by deadly bacteria, such as cholera, tetanus, typhus, scarlet fever and Streptococcus infections, as well as the plague itself and the preparation of serum and vaccine. In 1897, the Imperial Institute of Experimental Medicine commissioned the fort as the new research laboratory. Duke Alexander Petrovich of Oldenburg donated a lot of money to renovate the building for its new purpose.

Overall, the research in bacterial infections in Fort Alexander was quite successful. However, the laboratory was closed down when the communist regime took over in 1917. The fort was transferred under the care of Russian Navy while the research assets to institutes in Moscow and St Petersburg. In 1983, the fort was abandoned. From late 1990s to early 2000s, it was a popular venue for rave parties. Since 2005, it had been managed by the presidential conference center, or Constantine Palace, in Strelna. In 2007, the administration department of Fort Alexander expressed their willingness to find an investor for a renovation proposal and estimated the budget to be at 43 million US dollars. Nowadays, people called it Plague Fort. It has become a favorite haunt of city explorers and photographers. In winter, when the sea is frozen, tourists can walk to the fort; in summer, they can visit it on boat tours.

M11   The Silo Classroom, Jian Gong Primary School, Hsinchu, Taiwan
DESIGNER: Bio-Architecture Formosana

In this project, an oil silo from when Taiwan was under Japanese rule was renovated as an exhibition space and learning center on campus.

The "Big Silo," situated near the "Little Lovers" hill at the west side of Jian Gong Primary School's campus, is one of the three remaining oil silos left of the Imperial Japanese Navy's Sixth Fuel Factory in Hsinchu and thus one of the few WWII arms industry relics left in Taiwan. During the extension project of the school, Bio-Architecture Formosana was commissioned to turn the abandoned silo that had been used to pile recyclables into a multi-functional classroom.

The Imperial Japanese Navy's fuel factories are institutions that specialized in manufacturing, processing, researching and studying fuels and lubricants of use to the navy. Their forerunner were the temporary coal briquette-manufacturing plants built during the Russo-Japanese War. In the tenth year of Taisho (1921), these plants were converted to fuel factories for the navy and in charge of fuel-related affairs. In the sixteenth year of Showa (1941), the fuel factories were reorganized into six factories. Four of them were located in Japan, while the fifth was in Pyongyang, Korea, and the sixth in Zuoying District, Kaohsiung, Taiwan. In fact, the oil refinery built in Zuoying was not the only facility belonging to the Imperial Japanese Navy's Sixth Fuel Factory. There were also the isooctane plant (isooctane is an important ingredient in the fuel of aircraft) that covered more than 400 hectares but had never really been built as well as another piece of land covering more than 500 hectares that would become the cornerstone of the scientific and technological development of the area, both in Hsinchu.

Prior to the establishment of Jian Gong Primary School, the site of the current campus was part of the premises of the Imperial Japanese Navy's Sixth Fuel Factory in Hsinchu. Soon after the oil silo was built and before it went online, the US bombed Japan. The Japanese army withdrew from Taiwan shortly, leaving the silo behind (cause of death). Now, it has become the Silo Classroom of Jian Gong Primary School.

Although the appearance of the silo (bare concrete walls without any paintwork) cannot be said to be much appealing, this circular structure, with a diameter of sixteen meters, a height of six meters and a thickness of sixty centimeters to render the walls explosion-proof, doubtlessly has stories to tell – and a high adaptability.

To refurbish the silo, the design team first of all demolished the corrugated iron roofing. Then, they built a new roof that can introduce some fresh air and natural light into the reborn interior and at the same time keep it dry and warm. Furthermore, steel pillars were erected to support the frame of the roof and the exhibition space on the mezzanine. The original load-bearing walls are preserved and the existing windows were reopened to let in more fresh air. The central space is used as a mini amphitheater, while the peripheral space forms a mezzanine that is divided into an upper part for exhibition and a lower part for playing. After a thorough wash-up, the original concrete exterior reappeared and could be seen as made up of some porous material. The team planted climbing plants around the walls and built a rainwater harvesting system on the roof to water the plants.

There is an axis deviation of 12.5 degrees toward the south in regard to the new buildings at the west side of the campus in order to highlight the silo. By doing so, the team from Bio-Architecture Formosana successfully brought the silo from the background to the focal point.

M12   Fengshan Communication Center, Kaohsiung, Taiwan (proposal)
DESIGNER: Jingyu Lin

The Former Japanese Navy Fengshan Communication Center was built between 1917 and 1919 as a tactical move to jam the communication of the enemies and to monitor the whereabouts of the British and American fleets and crafts in the Far East. Being an important communication center at the southernmost part of Japanese colonies, the radio station in Fengshan, together with two other radio stations Funabashi and Hario (both in today's Japan), formed the backbone of the communication network of the Imperial Japanese Navy. Fengshan Communication Center is the most well-preserved of the three.

The main building of the radio station complex is the great bunker, or the First Station. Other auxiliary facilities include the dormitory, the office building, a small bunker and a cross-shaped signal station which was added later. The antenna network used to detect the direction and distance of fleets was shaped like an umbrella, thus the complex and its surrounding area formed a unique concentric configuration: the radius of the outermost circle measures four hundred meters, there are a total of 54 foundation blocks in the outermost and middle circles to support the secondary radio tower, and the diameter of the innermost circle (that is, the complex itself) measures three hundred meters. The Former Japanese Navy Fengshan Communication Center has undergone four phases in its history in four different roles: Fengshan Radio Station of the Imperial Japanese Navy, a navy boarding house, the Mingde Disciplinary Camp, and the Kaohsiung Military Dependent's Village Cultural Association. Due to its special status as a military facility, there was almost no hiatus between each rebirth. When the National Government of the Republic of China took over Taiwan in 1949 after WWII, this place was also used as an intelligence station where servicemen accused for political and thought crimes were interrogated and im-

prisoned. To conceal its true nature, it was called a boarding house. In 1962, this place was once again reused as the Navy Discipline Centre. In 1976, it became the Mingde Disciplinary Camp where persistent disobedient servicemen were taken into custody. Meanwhile, the navy was still using the cross-shaped signal station. In 2001, as a result of the military reform, the complex fell into disuse. In 2010, it was made a national heritage site.

In this proposal, the issue of urban sprawl and its demand for more land is addressed. Specifically, what the designer is trying to is to open up the forbidding military complex to the public; to obscure the boundaries between the complex, the city and its citizens; to put emphasis on the role the complex has played in the military history of Taiwan; and to introduce modern facilities into the existing building. In short, the aim of this proposal is to encourage people to return to the scene where historically significant events had happened and by doing so, reshaping the value of the place.

The unique concentric configuration is at the core of the project. First of all, part of the walls that have caused an urban disruption will be removed. Then, the premises will be divided into a north-south axis and an east-west axis. Along the north-south axis, the old military dependents' village, the former Japanese military facilities and a museum will be connected to offer a complete cultural, historical and educational experience to the visitors; along the east-west axis, the existing foundation blocks and green spaces will link with the Fengshan Communal Market to create a social and leisure circle for locals and tourists alike.

The First Station of the complex of is the heart of the project. It will be repurposed into a cinema and museum where people can re-examine the history of the communication center through the three steps of reading, reflecting and recognizing. The cinema, a modern building for entertainment, will be installed inside the First Station. This black box that is open to the public but at the same time seems to close in on the audience will fragmentize the original spatial arrangement of the interior and thereby invite the visitors to read and reflect on its history. The box will be built with reflective metal, forming a structure in a structure. Historical events and modern activities will have happened in the same place; the past and the present will be overlapping. Part of the façade will be removed to create an entrance and an opening along the north-south axis. Inside the bunker, there will be an event hall, an artifact gallery, an image gallery, an archives room and a reflection room, reading and reflecting on the history while recognizing a new self.

M13   Intrepid Sea, Air & Space Museum, New York, USA
DESIGNER: Perkins and Will

At the intersection of 46th Street and 12th Avenue in Manhattan, New York City, an aircraft carrier carrying various aircraft on its deck is permanently anchored at Pier 86. This is the Intrepid Sea, Air & Space Museum where you admire not only the aircraft carrier, but also a submarine, a Concorde jet, a space shuttle and typical military aircraft.

USS Intrepid (CV-11) is the fourth US warship to bear the name and one of the Essex-class aircraft carriers. Built in 1941, it was active during WWII and played a role in many a campaign on the Pacific, including the largest naval battle of WWII, the Battle of Leyte Gulf. Intrepid survived several suicide attacks from kamikazes and torpedo strikes during the war, continued to serve as the principal recovery ship for NASA programs like Projects Gemini and Mercury in the Space Race during the 1960s, participated in the Vietnam War, and finally retired in 1974 when the Cold War was at the climax. The warship was awarded the Presidential Unit Citation twice and earned five battle stars for its performance in WWII and the Vietnam War.

After its retirement, the US Navy decided to sell and dismantle the aircraft carrier, but the plan was strongly opposed by the general public. In 1978, Zachary Fisher, a New York real estate businessman and philanthropist, established the Intrepid Museum Foundation and started to raise funds to purchase the ship and repurpose it as a museum. Through his efforts, the Navy eventually handed Intrepid over to the organization in 1981. In 1982, Intrepid was struck from the Naval Vessel Register and renovated as a museum ship. The Intrepid Sea, Air & Space Museum was opened in New York in the same year and has since become an important landmark and tourist attraction in Manhattan. Later, Fisher bought more retired vessels and aircraft from the Navy to enrich the collection.

In 1986, Intrepid became a National Historic Landmark. The aircraft carrier forms part of the educational center that covers eighteen thousand square feet of the entire Pier 86, a public pier operated by Hudson River Park Trust. In addition to Intrepid, there are in the Intrepid Sea, Air & Space Museum: USS Growler, the only diesel-propelled cruise missile submarine that is open to the public; 28 reconstructed aircraft, including the reconnaissance aircraft, Lockheed A-12 Blackbird, and the then fastest transatlantic passenger airliner, the Concorde. In April, 2011, NASA announced that the space shuttle Enterprise would be relocated to the museum. On the 27th of April, 2012, Enterprise arrived New York; on the 6th of June in the same year, it was moved to Intrepid.

Today, the Intrepid Sea, Air & Space Museum receives more than a million visitors a year. Events and activities of the museum differ in theme every day and always attract large crowds. The retired space shuttle Enterprise is located in the central pavilion. Visitors can not only observe it from underneath but also step onto a platform to snap a selfie with the shuttle in the background at a perfect angle. In this exhibition area, there is also a "space capsule" zone where recordings, films, photographs and other artifacts are displayed, so that visitors can get a good grasp of everything about a space capsule and the role spacecraft play in space trips. The objective of the Intrepid Sea, Air & Space Museum is to increase people's understanding and knowledge of history and science with its collection, exhibition and activities.

M14   Typhoon-class Submarine (proposal)
DESIGNER: Yongxuan Chen

The largest nuclear-powered submarines humans have ever built, the Typhoons, are huge vessels 175 meters long that, when submerged, can reach a speed at 27 knots, travel at a depth of 500 meters, stay underwater for 180 days, with a displacement of 48,000 tonnes. The construction of the six Typhoon-class submarines was completed in 1989. They are the direct result of the doctrine of mutual assured destruction and the conflict between the United States and the Soviet Union during the Cold War. With twenty ballistic missile tubes, these submarines are powerful enough to destroy the hemisphere they are in. Since the dissolution of the Soviet Union in the end of the Cold War, three Typhoon-class submarines have been dismantled and two have been retired by the end of 2013; only one is still in service after a modern renovation. Currently, the UN is looking for a way to adaptively reuse the three retired submarines.

In this proposal, the designer puts forward a solution that hopefully can settle an international issue that has long existed once and for all:

To fight for the land between the Mediterranean Sea and the Jordan River, the Jews and the Arabs have been quarreling with each other for almost a century. Both claim that the land is theirs because it's where their race and religion originated. In particular, the Gaza Strip to the west of Israel suffers the most in this seemingly endless conflict. In this densely-populated region that doesn't really belong to any government, flames of war are everywhere, land is scarce, and resources are meagre. Most families can only barely make ends meet with the aid from the UN. During the attack from the Israeli army in July and August of 2014, more than two hundred schools were destroyed and at least 1,500 children were orphaned. The air raid damaged the only power station in the region, which disrupted fresh-water supply.

To end the misery of the Gaza Strip, the designer proposed to repurpose the retired Typhoon-class submarines and turn the lethal weapons designed to destroy the planet into a supplier of vital resources and a guardian of children.

First, the unique two-core nuclear reactors of the Typhoons are able to provide sufficient electricity for a coastal city or even a small nation. Since the submarines will be offshore, earthquakes and tsunamis won't be a problem. Even when a meltdown takes place, sea water will be right at hand and can cool the system immediately. Therefore, it is quite a feasible scheme to deploy the submarines now belonging to the UN offshore and use the reactors as power stations and desalination facilities to bring electricity and water back to the Gaza Strip.

Next, an artificial island will be built around the submarines and used as an orphanage, providing education and shelter for the orphaned. In short, the island of hope and peace will be a floating town where kids can grow up safely. Sadly, many children in this region were handicapped or orphaned due to the incessant wars, so a place like this is just what they need. The UN will be in charge of the island, giving humanitarian aid of all kinds to the inhabitants, including medical care, provisions, shelter, education, vocational training and counselling.

After renovation, the Typhoon-class submarines will be transformed from destructive weapons into life-saving facilities. In the long term, allocating resources to and taking care of the orphans devastated by the ethnic conflict will pay off. Once they grow up, the educated orphans can help rebuild the infrastructure ruined during the war, solve the problem of low employment and deal with the economic crisis. In any way, this is doubtless one of the best solutions to bring about lasting peace to the society in the future.

製作團隊：（建模與製圖團隊）

王俐雯
王致崴
史可蘋
何相儀
李政儒
李　齊
周曉晴
林欣慧
林秉翰
林靖淯
林馨熒
侯雅齡
高亮慈
莊翔程
張淳茜
張瀚元
陳天健
陳怡萍
陳婉宇
陳詠暄
曾睿宏
黃皓篪
黃　薔
蔡承宇
蔡昀庭
賴伯威
薛尹端
謝仁恩

（依姓名筆劃順序排列）

圖片來源 Image Credits

A0　p.34 賴伯威
A1　p.36 薛尹端
A2　p.38 曾睿宏
A3　p.40 薛尹端
A4　p.42 Luuk Kramer
A5　p.44（左）Forgemind ArchiMedia/Flickr；
　　（右上）othree/Flickr；（右下）akn48/Flickr
A6　p.46 Denis Linine/iStock
A7　p.48 蔡承宇
A8　p.50 侯雅齡
B1　p.54 吳宜倫
B2　p.56 謝家融
B3　p.58（左）Fred Romero/Flickr；（右）廖昱嘉
B4　p.60（左）Fred Romero/Flickr；（右）吳宜倫
B5　p.62 陳怡萍
B6　p.64 陳潔
B7　p.66（左）何相儀；（右）謝雁如
C1　p.70 Forgemind ArchiMedia/Flickr
C2　p.72（左）Ungureanu Adrian Danut/Wiki, CC BY-SA 4.0；
　　（右上）Gabriel Tocu/Wiki, CC BY-SA 4.0；
　　（右下）Strainu/Wiki, CC BY-SA 3.0 ro
C3　p.74（左）Andy L/Flickr；（右上、下）陳麗雯
C4　p.76 Ville Palonen/Alamy Stock Photo
C5　p.78 李世安
C6　p.80 王俐雯
D1　p.84 dpa picture alliance archive/
　　Alamy Stock Photo
D2　p.86（左下）vgallova/flickr；（上、右）薛尹端
D3　p.88（左）薛尹端；（右）Liaon98/Wiki, CC BY-SA 3.0 tw
D4　p.90 陳天健
W1　p.92 Stijn Poelstra (Zecc Architects)
W2　p.94 META 王碩
W3　p.96 Binst Architects
W4　p.98（左）Thomas Zimmermann (THWZ) /CC BY-SA 3.0 DE
　　(https://creativecommons.org/licenses/by-sa/3.0/de/deed.en)；
　　（右）Daniela Kloth/kloth-grafikdesign.de
W5　p.100（上）Andrew Dunn, 1 November 2005. {{cc-by-sa-2.0}}；
　　（下）whitemay/iStock
W6　p.102 Nagy Balázs

S1　p.106（左上）Jonas Smith/Flickr；（右下、下）吳宜倫
S2　p.108 Tuomas Uusheimo
S3　p.110 Michael Jacobs/Alamy Stock Photo
S4　p.112 陳婉宇
S5　p.114 薛尹端
S6　p.116 Esther Westerveld/flickr
S7　p.118 Allez Up and Alexa Fay
S8　p.120 大舍柳亦春
M1　p.124（上）Technouwe/Wiki, CC BY-SA 3.0
　　（左下）Stefan Kühn/Wiki, CC BY-SA 3.0
　　（右下）Tropical Islands Resort/Wiki, CC BY-SA 3.0
M2　p.126 葉俐琪
M3　p.128 葉俐琪
M4　p.130 Frank Van Laanen/Flickr
M5　p.132 Jonathan Nackstrand/Getty Image
M6　p.134 陳婉宇
M7　p.136 USFWS/Steve Agius/CC BY-SA 2.0
M8　p.138 geogphotos/Alamy Stock Photo
M9　p.140 Amanda Retreats/Wiki, CC BY-SA 3.0
M10　p.142 Rozakov/iStock
M11　p.144 九典聯合建築師事務所
M12　p.146 林靖淯
M13　p.148 曾睿宏
M14　p.150 陳詠暄

重生之路　基礎設施的死與生，全球經典案例圖解

2020年11月初版　　　　　　　　　定價：新臺幣1500元

著　　　者　賴伯威
英文譯者　羅亞琪
叢書主編　李佳姍
校　　　對　馬文穎
整體設計　黃暐鵬

出　版　者　聯經出版事業股份有限公司
地　　　址　新北市汐止區大同路一段369號1樓
台北聯經書房　台北市新生南路三段94號
電　　　話　（02）23620308
台中分公司　台中市北區崇德路一段198號
暨門市電話　（04）22312023
台中電子信箱　e-mail：linking2@ms42.hinet.net
郵政劃撥帳戶　第0100559-3號
郵　撥　電　話　（02）23620308
印　刷　者　文聯彩色製版印刷有限公司
總　經　銷　聯合發行股份有限公司
發　行　所　新北市新店區寶橋路235巷6弄6號2樓
電　　　話　(02)29178022

副總編輯　陳逸華
總　編　輯　涂豐恩
總　經　理　陳芝宇
社　　　長　羅國俊
發　行　人　林載爵

行政院新聞局出版事業登記證局版臺業字第0130號

本書如有缺頁，破損，倒裝請寄回台北聯經書房更換。　ISBN　978-957-08-5524-1（精裝）
聯經網址：www.linkingbooks.com.tw
電子信箱：linking@udngroup.com

國家圖書館出版品預行編目資料

重生之路：基礎設施的死與生，全球經典案例圖解／
賴伯威，WillipodiA 都市研究團隊著.
初版. 新北市. 聯經. 2020年11月
200面. 30×28公分
ISBN　978-957-08-5524-1（精裝）
1.基礎工程　2.公共建築
440.1　　　　　　　　　　　　　　109004862